*St Antony's Series*

General Editor: **Jan Zielonka** (2004– ), Fellow of St Antony's College, Oxford and **Othon Anastasakis**, Research Fellow of St Anthony's College, Oxford and Director of South East European Studies at Oxford.

*Recent titles include*:

Cathy Gormley-Heenan
POLITICAL LEADERSHIP AND THE NORTHERN IRELAND PEACE PROCESS
Role, Capacity and Effect

**St Antony's Series**
**Series Standing Order ISBN 978–0–333–71109–5 (hardcover)**
**Series Standing Order ISBN 978–0–333–80341–7 (paperback)**
(*outside North America only*)

You can receive future titles in this series as they are published by placing a standing order. Please contact your bookseller or, in case of difficulty, write to us at the address below with your name and address, the title of the series and the ISBN quoted above.

Customer Servics Department, Macmillan Distribution Ltd., Houndmills, Basingstoke, Hampshire RG21 6Xs, England

# The Politics of Emerging Strategic Technologies

Implications for Geopolitics, Human Enhancement and Human Destiny

Nayef R.F. Al-Rodhan
*Senior Member, St Antony's College, Oxford University, UK, and
Senior Scholar in Geostrategy, and Director of the
Geopolitics of Globalisation and Transnational Security Programme,
Geneva Centre for Security Policy, Geneva, Switzerland*

*In Association with palgrave macmillan*

First published 2011 by
PALGRAVE MACMILLAN

Palgrave Macmillan in the UK is an imprint of Macmillan Publishers Limited,
registered in England, company number 785998, of Houndmills, Basingstoke,
Hampshire RG21 6XS.

Palgrave Macmillan in the US is a division of St Martin's Press LLC,
175 Fifth Avenue, New York, NY 10010.

Palgrave Macmillan is the global academic imprint of the above companies
and has companies and representatives throughout the world.

Palgrave® and Macmillan® are registered trademarks in the United States,
the United Kingdom, Europe and other countries.

ISBN: 978–0–230–29084–6 hardback

This book is printed on paper suitable for recycling and made from fully
managed and sustained forest sources. Logging, pulping and manufacturing
processes are expected to conform to the environmental regulations of the
country of origin.

A catalogue record for this book is available from the British Library.

Library of Congress Cataloging-in-Publication Data

Al-Rodhan, Nayef R. F.
    The politics of emerging strategic technologies : implications for
geopolitics, human enhancement, and human destiny /
Nayef R.F. Al-Rodhan.
      p. cm.
    Includes index.
    ISBN 978–0–230–29084–6 (hardback)
      1. Geopolitics. 2. Technology – Political aspects. 3. Technological
innovations – Political aspects. I. Title.

JC319.A494 2011
303.48′3—dc22                                    2011004880

10  9  8  7  6  5  4  3  2  1
20  19  18  17  16  15  14  13  12  11

Transferred to Digital Printing in 2014

# Contents

# Acknowledgements

The author would like to thank the following people for their help: Amber Stone-Galilee, Liz Blackmore, Andrew Marsh, Bethany Reichenmiller, Lyubov Nazaruk, Christina Lycke, Beatrice Fihn and Julia Knittel.

He also would like to thank his colleagues at St Antony's College, Oxford University, and the Geneva Center for Security Policy, Geneva, Switzerland for their help and support.

The author and publishers wish to thank the following for permission to reproduce copyright material: LIT Publishers, for Figures 1–3, from Nayef R.F. Al-Rodhan (2009) *Sustainable History and the Dignity of Man: A Philosophy of History and Civilisational Triumph* (Berlin: LIT) and for Figure 4 from Nayef R.F. Al-Rodhan (2007) *The Five Dimensions of Global Security: Proposal for a Multi-sum Security Principle* (Berlin: LIT).

Every effort has been made to trace all copyright-holders, but if any have been inadvertently overlooked the publishers would be pleased to make the necessary arrangements at the first opportunity.

The views expressed in this book are entirely those of the author and do not necessarily reflect those of St Antony's College, Oxford University, or the Geneva Center for Security Policy.

# General Introduction

Technology touches every element of our lives. Sometimes, this happens in obvious ways – one need only to observe the pervasiveness of cell phones, the Internet or even more mundane commodities such as medicine, electricity and television – but technology also inhabits our lives in less obvious ways. For example, artificial intelligence technology plays a role every time we transfer money to a bank account, and nanotechnology and related particles are present in products as basic as sunscreen and cosmetics. Just as important are the – now hypothetical but soon-to-be real – applications of emerging and revolutionary technologies. Developments in these fields promise to dramatically affect our lives and the world around us. Scientists are currently working across disciplines on concepts like nano-particles that eliminate carbon dioxide ($CO_2$) air pollution, customized and highly-targeted drug delivery systems, computers with smarter-than-human intelligence, as well as more fantastical things such as uploading one's brain function, memories and personality on to a computer.

Although many of these emerging strategic technologies (ESTs) are still in development, their capacities for altering our geopolitical landscape and our human existence are enormous. As a broad concept, ESTs can be defined as technologies the basic science and principles of which are understood and which have some existing applications but the full potential of which has yet to be tapped. Failure to fully take advantage of ESTs could be linked to a lack of resources or investment, a lack of market interest or the need for additional research and innovation before the technology is ready to go mainstream or to be applied to its fullest potential.[1]

Across the board, ESTs have the potential to dramatically influence our daily lives, from applications in business and health care to politics

1

and strategic planning to facilitating and improving our ordinary routines.[2] In some instances, emerging technologies may be at odds with prevailing processes or products. In such cases, ESTs may become disruptive and bring about a fundamental shift in the way we as humans approach an issue.[3]

Rarely have these emerging strategic technologies been examined together in a comprehensive fashion with an eye not just to their day-to-day applications but also to a consideration of how these technologies are altering and will continue to alter multiple facets of our geopolitical landscape and our human destiny – often dramatically. This book seeks to fill that gap by examining ESTs with scientific sophistication, in a context of international regulatory frameworks and with an eye to the potential challenges and opportunities that the technologies present in everything from health care to climate change to the enhancement of the human body and mind.

For the purposes of this book, the technologies that fall under the umbrella of ESTs are diverse and include the categories of information and communications technology (ICT), energy and climate change, health care, biotechnology, genomics, nanotechnology, materials science and artificial intelligence. Each type of technology or group of technologies was chosen because of its status as a new, cutting edge field and because of the likelihood that it will affect many facets of international security, human nature and the broader global community.

After introducing nine of the most significant emerging fields and sectors of strategic technology of the twenty-first century, this book will go deeper into uncharted territory, analysing the impact of emerging strategic technologies on human nature and the overall destiny of the human race. Such issues are weighty and important from philosophical and practical perspectives, but the way we use technology to alter our bodies and the essence of humanity will also reshape the global political structure. Policymakers must act now to construct legal and ethical frameworks that ensure we use technology to benefit the human race – not to inadvertently curtail its dignity or even eliminate it.

Certain ESTs – nanotechnology, biotechnology, information technology, artificial intelligence and cognitive science in particular – have the potential to be used to modify, alter or enhance the human body and mind. Given this potential, the next natural question is whether such changes also affect the essence of human nature and human dignity, and if so, how can the human race control and regulate these technologies so that they are used for good? The stakes of these new ESTs are high, and it is not an exaggeration to say that humanity's entire destiny lies in the balance of how we decide to handle this technological revolution.

## Knowledge and its influence on technological innovation and development

Why begin an analysis of emerging strategic technologies, geopolitics and human destiny with a section devoted to knowledge? Knowledge and technology are deeply intertwined, so it is a natural starting point for this book to take a deeper look at society's constructs of knowledge and how we can understand what knowledge really is and how it is acquired. Knowledge is a term used frequently and often bandied about, and its definition and meaning are seemingly taken for granted. Knowledge is understood by most societies to be good and useful, and yet we rarely take the time to probe its origins beyond a general idea that it comes from experience, active learning and advice from others. However, a deeper look at knowledge and its origins is in order here, because the political nature of knowledge, combined with how it is physically acquired, are fundamental, underlying elements in technological innovation and development and in determining why technology develops in the way it does. Knowledge also influences how we use technology and the types of problems and challenges we hope, either consciously or implicitly, that technology will solve.

In general, we ask questions and frame challenges based on what we think we know. Whether it is through neurologically-based physical knowledge, dogma, cultural norms, religious values, or any number of factors, what we know and what we think we know play a significant role in determining what we *will* know or the knowledge we will pursue next.

What is knowledge? From a governance perspective, 'knowledge is the ultimate public good'.[4] This is because of its non-exclusive and open nature. Knowledge can become a private good through the definition of property rights or other legal means but, more importantly, some knowledge is a *global* public good; every country in the world can potentially benefit from the scientific and technological knowledge produced by other countries.[5] Once created, knowledge can be shared in many forms and across time and space.[6] In addition, since knowledge can take on different personalities (e.g., theoretical or applied, public or commercial, codified or tacit), it is an inherently dynamic and complex good.[7]

As a basis for technological innovation - in areas as diverse as cognitive science, nanotechnology and biotechnology, information technologies, energy and environmental sciences, genomics and proteomics – knowledge ultimately becomes a crucial part of furthering the social, economic and technological development of all nations. Moreover, according to the International Task Force on Global Public Goods, knowledge is the fundamental building block for other global public goods such as peace and security, the sustainable management of natural commons, financial stability and the control of communicable diseases.[8] Thus, the significance of knowledge in the context of both technology and society cannot be understated.

## The acquisition of knowledge: the theory of *'neuro-rational physicalism (NRP)'*

It is tempting to speak of knowledge as something universal and, moreover, as something universally good, but this is not necessarily the case. Because of factors such as different life experiences, cultural backgrounds and societal norms, as well as the unique neurochemical processes that underlie the thought processes of each individual, knowledge and what we think we know are, in many ways, personal and not at all universal. If knowledge is not concrete and we only presume to know things for certain, it follows that we should examine where knowledge comes from and then to define what we can say we know for certain versus what we think we know that is not necessarily true or provable.

Where does knowledge come from? In my theory of *'neuro-rational physicalism (NRP)'*,[9] I explain that, contrary to many philosophies of the origins of knowledge, knowledge is neither purely based on empiricism nor entirely based on rationalism. Rather, knowledge comes from a combination of employing both sense experience and reason. Importantly, both these foundations of knowledge are *subject to interpretation*.[10] How we interpret our sense experience and how we frame the questions that generate our accepted knowledge depend on many things, including prior assumptions as well as cultural, spatial and temporal settings.[11] In other words, we use what we think we already know and our understanding of the world based on our own personal experiences as the foundation for our pursuits of new knowledge. Sense data can make great contributions to knowledge, but it is important to keep in mind that such data have a high probability of being incomplete, thus making them subject to error.[12] What we know, we only know with a reasonable amount of certainty, and this is why interpretation of knowledge or perceived knowledge is such a critical part of the knowledge equation.[13]

I also maintain that the acquisition, analysis and retention of knowledge has 'a physical neuro-biological foundation, including thoughts, memories, perceptions and emotions' and that therefore 'mental states and thought processes are physical'.[14] They are rooted in chemical reactions and processes in the brain, all of which are physical. Metaphysics may tell us what a state of affairs may hold, but it does not necessarily tell us whether that state of affairs actually exists.

Non-physical knowledge does exist and is an integral part of our daily lives and experiences. However, because it cannot be physically verified, this type of knowledge is best described as *'possible truths subject to proof'*, a concept that I proposed previously.[15] In other words, this refers to the ideas that we believe to be logically true even if we do not currently have the scientific methodologies or physical resources to prove them.[16] It is in this area that reason becomes a central part of knowledge, as rationalism recognizes that all knowledge is at least partially based on concepts we take for granted as being true.[17] The key point to keep in mind is that 'physicalism' is the defining component of what we know for certain. In short, 'all the universe and its energies are physical' and thus real, verifiable knowledge has a fundamentally physical foundation.[18]

It is worth stressing that although reason cannot be entirely certain as a form of knowledge, and although it is subject to interpretation, reason is nonetheless a central component of human dignity and of living a dignified life. Indeed, 'a life governed by reason is likely to be more dignified than one shaped by dogma and unbridled emotions'.[19] A graphical presentation of 'NRP' can be found in Figure 1.

## 'NRP' and technological development

To draw a full circle between 'NRP' and technology, it is necessary to ask how the 'NRP' understanding of knowledge applies specifically to technological development. First, it is worth noting that even in science, knowledge is not acquired through pure observation.[20] Scientists first ask questions and, importantly, these questions are based on things that are already established or accepted.[21] For example, gravity is a known force, so scientists will seek to understand *how* gravity affects something; not *whether* gravity affects it. This is where the importance of interpretation and society comes into play, as different experiences or observations may lead to different frameworks for scientific discovery.

Much of what we often consider knowledge is actually a point of view held without sufficient grounds: in a word, dogma.[22] Whether this dogma comes from family values or religious backgrounds, dogma can undermine our rationality and, by extension, our dignity. Often, our dogmatic beliefs are based on the societal norms and customs we

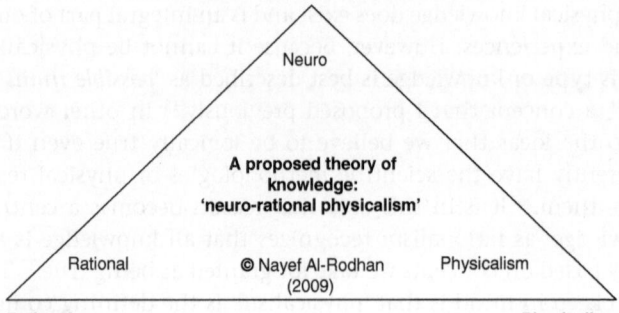

**Neuro**
All human knowledge has a physical neuro-biological foundation, including thoughts, memories, perceptions and emotions. Thus, mental states and thought processes are physical.

Neuro

A proposed theory of knowledge: 'neuro-rational physicalism'

Rational          © Nayef Al-Rodhan          Physicalism
                   (2009)

**Rational**

Knowledge derived from sense-data is not certain. Interpretation is the prism through which we order sense-data. Pure empiricism is therefore an inadequate means of acquiring knowledge. Knowledge is also inferred from what is accepted as established knowledge, with new knowledge being based on the best explanation. Knowledge about things beyond our immediate environment may be acquired through deduction, if the initial premises are believed to be correct. The notion of innate knowledge (including moral knowledge) is rejected. Thus, knowledge is based on both sense-data and reason, and has a high probability of being subject to error or incomplete understanding. While facts exist, all knowledge is to some extent interpreted and ultimately indeterminate, being perhaps also temporally, spatially and

**Physicalism**

All the universe and its energies are physical, although some matter and energy may be unobservable with our current technologies

*Figure 1*   A proposed theory of knowledge: *'neuro-rational physicalism' '(NRP)'*

*Source*: N.R.F. Al-Rodhan (2009) *Sustainable History and the Dignity of Man: A Philosophy of History and Civilisational Triumph* (Berlin: LIT), p. 131. Reproduced with permission from LIT.

observe in day-to-day life. This is not to say that such norms are necessarily bad in and of themselves – although a life governed by reason is certainly preferable – it is just to highlight the possible fallacies of these norms and the role they play in our approaches to knowledge in general and to scientific discovery in particular.

We will ask questions and frame technological challenges based on what we think we know. This is true whether that knowledge is neurochemically based physical knowledge, *'possible truths subject to proof'*, or merely dogma rooted in cultural norms, belief systems, and so on.

For example, whether a society decides to pursue anti-aging or life extension technologies will depend in part on how that society perceives old age and death. In China and in many Latin American cultures, the elderly are valued for their wisdom and experience, and the impetus to look younger or to reduce wrinkles is minimal (although this is changing rapidly). The opposite is true in many parts of Europe and North America. Youth is valued almost above all else, and the industries and technologies created to reduce signs of aging, such as wrinkles, age spots and loss of skin elasticity, are worth billions of dollars. In this roughly constructed framework, it could be argued that scientists in Europe and North America will be more likely than their Chinese or Latin American counterparts to seek the knowledge and to pursue technological innovation that might minimize or reverse the signs of aging. The reasons for this discrepancy can be attributed to different cultural frameworks, yet it has real implications for technological innovation and development.

In short, we can only seek to use technology to control, manage or improve the issues, challenges, and opportunities we know about or what we believe we know from society and culture. This theme is explored in greater depth below.

## The theoretical foundation: approaching the study of science and technology through the lens of *'neuro-rational physicalism (NRP)'*

Science and Technology Studies (STS) is a relatively new, interdisciplinary field that looks at how social, political and cultural forces affect scientific and technological research and development.[23] It is possible to trace its roots to the early 1970s.[24] STS strives to understand science and technology not only from a technical and practical standpoint, but also from a social point of view. According to this theoretical approach, one should 'use the tools of the social sciences and humanities to study, understand and analyse science, technology and the work of engineers and scientists past and present'[25]. The STS school argues that knowledge is socially distributed. STS emphasises that both technology and science 'are social institutions that depend on social factors'[26]. This realization raises questions such as: Who or what is controlling science? Who is responsible for driving technological change? Why do we trust some experts and not others? What are the values and ethical dilemmas related to science and technology? How are science and technology related to questions of social justice?[27] How does our basis of knowledge fit into this framework?

In my opinion, knowledge is rooted in *'neuro-rational physicalism (NRP)'*, and, as a part of that, all knowledge is subject to interpretation. Given this fact, our societal, temporal, spatial and cultural contexts will define the questions we ask in pursuit of new sciences and technologies as well as how we respond to and apply scientific innovations. This analysis is compatible with the STS school of thought because how we respond to and understand metaphysical truths and *'possible truths subject to proof'* will be largely influenced by the realities of our daily lives and societies. Society will play a great role in defining which technologies we pursue and what problems we seek to solve through technology from a practical perspective, and *'NRP'* will define how we approach the development of these new, socially motivated sciences and technologies in a technical sense. In other words, we can only develop new technologies using the knowledge we already have, so *'NRP'* will guide how society's demands for technology actually develop and manifest themselves. Thus, the direction of strategic technological developments and innovation in the twenty-first century will, on the one hand, be driven by society and society's changing needs in response to changing geostrategic threats while, on the other hand, the basis for technological developments will also be limited to what we know for certain through *'NRP'*.

The theoretical framework of STS in conjunction with *'NRP'* is suitable for this analysis because this book looks deeply at the ways in which science and technology are shaping society and how we plan for our global future. It also looks carefully at how society's needs and desires are pushing certain technological developments at the cost of others. For example, one of the ongoing themes throughout this book is the technological gaps between rich countries and poor ones. An extension of this trend is the fact that scientists often focus their energy on the more lucrative demands of people in developed countries.

Market forces of supply and demand are just one of the ways in which society shapes technology and vice versa. We see similar patterns with regard to protecting global public goods such as the environment. As climate change and environmental protection receive more and more attention from policymakers and activists, we see science and technology follow a similar course. The more society begins to care about $CO_2$ emissions and environmental degradation, the more research we are seeing into things like hydrogen fuel cells, environmental monitoring biosensors and other similarly focused technological innovations.

Our human nature is also a driving force in the developmental path of new technological innovations and developments, as is our means of acquiring knowledge. *'NRP'* stresses that the way we develop knowledge affects technological developments in much the same way that our societal and cultural frameworks do. I believe that scientific knowledge is premised on the best available explanation of a given phenomenon, and that knowledge is therefore somewhat approximate, and it is difficult, if not impossible, to be dogmatic about what we know for certain.[28] This puts a certain ethical obligation on knowledge and how we apply it to technological development. By not looking at knowledge through an ethical lens, we run the risk of undermining human dignity.[29] As I elaborate in Part II of this book, humans are inherently driven by emotional self-interest, and we are inevitably drawn to technologies that will correct perceived flaws or improve our natural capacities. This facet of our nature, combined with the way we acquire knowledge, is unquestionably pushing science and technology in certain directions. Cosmetic surgery, prosthetic limbs and fertility treatments are just some of the more common examples of technologies designed specifically to improve or enhance our human selves. By the same token, the availability of a broad range of new enhancement technologies is influencing society, our relationships with each other and how we plan our futures, both from an individual perspective and from a political and societal one. New technologies offer some unprecedented opportunities for the betterment of the human condition, but they could also have a strongly negative impact on the future of humanity. This fact must be kept in mind as progress is made with further technological and scientific advances, and is explored in this study.

All things considered, it is hard to deny the intimate and almost symbiotic relationship between society, science and technology, and for this reason STS, approached through the lens of *'NRP'*, is an appropriate and useful framework for this book and its message.

## An overview of this book

Part I examines nine specific emerging technologies and how they are likely to influence international relations, global governance and geopolitics in the near and distant future. Each chapter is divided along roughly the same lines. The technology and its potential are introduced, and the broader industry is reviewed. I then evaluate the specific and relevant technological innovations in each field with an eye to scientific depth, accuracy and technological relevance. Once the nature of the emerging strategic technology has been outlined, each chapter spends some time looking at the relevant international regulatory structures governing each technology – or, in several cases, the lack thereof. Finally, and most importantly, each chapter analyses how the respective EST is shaping geostrategy and global politics.

Part II looks precisely at these issues of technological achievement and their application to human enhancement. It is an area often neglected by policymakers, as many of the implications of technology on human nature are seemingly abstract and exist only in the more distant future. Thus, regulating these technologies and their applications seems to be a low priority. Add in, the fact that many of the issues surrounding these technologies can quickly become political, ethical and religious minefields, and policymakers will feel even less inclined to take the political risk by addressing them in-depth. However, Part II of this book strenuously argues that it is precisely because of the uncertain nature of human enhancement, as well as technology's incredible power to fundamentally change our essence as human beings, that it is imperative for us to establish a legal and ethical regulatory framework for dealing with these issues *now*. Any delay and we run an increasingly large risk that these technologies may get out of our control. In such a situation, it is not an overstatement to say that this could eventually threaten the future of the human race.

Part II starts with some of the key definitions used in the discourse on human enhancement, including the characteristics of trans- and post-humans, the difference between enhancement and eugenics, and my theories of human nature, human dignity, and *'sustainable history'*. Once these terms have been defined and explained, I spend some time setting out the arguments for and against enhancement, drawing on the writings of some of the world's leading bioethicists and philosophers. I then set out my assessment of human enhancement. In short, I believe that wide scale human enhancement is not a question of if but of when, and I outline in detail my proposal of *'inevitable transhumanism'*.

I then move on to describe some specific enhancement technologies used on individuals, and to take a more macro look at human enhancement. Using my previously proposed *'multi-sum security principle'*,[30] I analyse the ways in which human enhancement could threaten or at least alter the context of global security. Looking at topics such as human, environmental, national, transcultural and transnational security, the recommendations and practical examples in this section are again highly relevant to interested policymakers. Finally, this book analyses the current regulatory frameworks for dealing with human enhancement and makes some policy recommendations for improving international regulatory structures. The urgency with which such regulations must be adopted is a key theme in this final section.

Although often described as though they exist in separate policy silos, science and society are deeply interconnected and intertwined. This fact will only become more evident in the future, as emerging strategic technologies increasingly dominate and define major geopolitical challenges. For policymakers, technology offers endless opportunities but also many possible pitfalls. This book will help provide a framework and a structure for contextualizing these technologies and for helping policymakers navigate these rapidly advancing developments. From regulatory issues to the five dimensions of global security, it is only when we adopt a comprehensive, interdisciplinary perspective on the role that technologies play in our lives that we will be able to maximize the potential of technologies to better our societies and our global future.

# Part I

# Emerging Strategic Technologies: Geopolitical Implications

# 1
# Introduction

Technology has contributed to human health, development, community growth and economic stability since the dawn of history. For our primitive ancestors, the ability to create fire was a version of technological innovation that dramatically improved the quality of life, providing a heat source, increased protection from animal attacks and new ways to prepare food. To say that the power to make fire revolutionized life for our ancient predecessors would be a dramatic understatement.

Over the millennia, technology has become considerably more advanced, and its far-reaching impact on human lives has advanced just as considerably. History-changing innovations have ranged from the stunningly simple to the phenomenally complex. For example, at the time of its creation, the wheel was a revolutionary concept that greatly facilitated people's capacity to transport goods over long distances with less energy. Today, the wheel is such an integral part of our daily lives that we quite literally cannot conceive of our lives without it. In fact, virtually every mechanized system in existence today is dependent on the wheel in some form or another.[1] Its role in modern society is so deeply embedded that it is hard to fathom how different our lives would be without this simple technology, and yet, despite its seeming simplicity, at the time of its creation, the wheel was at least as momentous a change to daily life as the Internet has been in contemporary history.

The next generation of technology inevitably builds on previous generations, and each new discovery potentially paves the way for something even bigger and better. We have obviously come a long way since the discovery of fire and the invention of the wheel. Our contemporaries are able to manipulate matter at the subatomic level to create explosions that are so powerful and enormous in their effect that they can literally disable satellites in orbit around the world.[2] With Internet access at the click of a mouse, we have practically unlimited access to information at our fingertips, and someone in Paris, France, can communicate with someone in Paris, Texas, in real time and at virtually no cost. These examples are just the tip of the iceberg of already tested or established technological innovations that are influencing our daily lives.

Richard Lipsey, Kenneth Carlaw and Clifford Bekar define technological change in terms of categories of 'general purpose technologies' or GPTs. According to them, a GPT is 'a single generic technology, recognizable as such over its whole lifetime, that initially has much scope for improvement and eventually comes to be widely used, to have many uses, and to have many spillover effects'.[3] Previously developed GPTs include the domestication of plants and animals, printing, the steam engine and the computer.[4] Certain ESTs also have the potential to become GPTs, in particular nanotechnology and biotechnology.[5] In general, GPTs are rare in human history, occurring roughly two to three times per millennium over the past 10,000 years, but the pace of change is accelerating.[6]

Looking at how technology has influenced and shaped human life so dramatically and so consistently over the course of human history, it is natural to wonder what is next for our world and technology. What might the next big innovation system be and how will it affect us as human beings and as citizens of the broader world community? Part I of this book seeks to answer these questions. Over the past 50 years, we have seen our ways of interacting with the world change completely as a result of the information technology revolution. Now, we are on the cusp of a period of convergence in which technologies that were previously considered and studied separately (e.g., biotechnology, information technology and cognitive science) are being approached in an increasingly interdisciplinary fashion. The potential for these different technologies to come together and lead to new discoveries in the fields of health, energy and the environment are enormous. United, these technologies have tremendous potential.

Part I examines some of the most promising emerging strategic technologies of the twenty-first century and analyses ongoing developments in these fields. Importantly, it assesses how each technology could affect our geopolitical landscape both as a stand-alone entity and in conjunction with other emerging strategic technologies.

The technologies covered in this section were chosen in part because they are already having a major impact on our daily lives, but also because of their future potential to alter our individual existences as well as the fabric of the global society in which we live – from broad categories of technologies that influence the way we communicate, how we use energy, our environmental preservation efforts and the management of human health and quality of life to more specific technological fields such as biotechnology, genomics, nanotechnology, materials science and cognitive science. Part I provides policymakers and other interested parties with an understanding of how technology already influences geopolitics and how it may fundamentally shape issues of geopolitics and geostrategy in the near and distant future. It explores the implications of existing and potential technologies in order to provide a more sophisticated understanding of the interplay between technology and some of our biggest global threats. It assesses the less obvious ways in which high-profile technologies will influence our globe for better or worse.

Rooted in science, technology is often evaluated using a systematic, causal methodology. A new drug cures a disease, and therefore the impact of that new drug is assessed strictly in relation to its ability to perform its assigned task. Part I acknowledges the significance of such direct impacts of technology, but it also demonstrates the indirect impacts of technology. Not only can a new drug potentially eradicate a deadly disease, for example, but it may also enable a mother to survive an illness, thereby avoiding the tragedy of orphaned children. Growing up in a more stable home, these children are more likely to develop normally, learn family values, and so on. The new drug technology would admittedly have the primary effect of saving a mother's life, but its secondary and tertiary effects could mean healthier, happier children and a greater capability of both the mother and the children to contribute to their society's cultural and economic development. When the effects of such a technology are multiplied beyond one single mother to mothers around the world, it is clear that even technology designed to improve a single individual's health can have dramatic geopolitical implications.

Each chapter in Part I describes a different technology; the context of the technology's industry, with information relating to sales and major players in the field; the most relevant technological developments in the field, both existing and potential; the regulatory structures that exist for each technology as well as the regulatory shortcomings; and, finally and perhaps most importantly, the geopolitical and geostrategic implications of each of the selected technological fields.

# 2
# Information and Communications Technology (ICT)

## General overview

In the first ever issue of *PC Magazine* published in January 1982, Bill Gates predicted that computers and related technologies would change the way people worked. Twenty-five years later in the same publication, Gates reflected that this technology had indeed changed the way we worked but, more importantly, it was changing how we lived.[1]

Information and communications technology (ICT), more specifically digital ICT, is a central part of this change. ICT includes some of the most important technological innovations of the twentieth century.[2] A broad term, ICT and related services span the entire range of the production, consumption and distribution of information in all media, ranging from the Internet and satellites to radio and television.[3] They also include all technologies that collect, distribute, produce, consume and store information.[4] In recent decades no technology has had a global impact on the same level as ICT. Improvements in ICT have transformed the nature of production, communication, and dissemination of information and have expanded their reach. The increasing availability of information and the growth of communications networks have contributed greatly to the process of globalization.[5] Overall, rapidly evolving ICT makes it possible for us to live in a more globalized and interconnected world, and it is directly influencing everything from business to health to global discussions on politics, culture and religion.

## The global ICT industry

ICT is an increasingly prevalent facet of modern life, with a strong and growing presence in the daily routines of businesses, governments and private households around the world. In June 2008, nearly 1.5 billion people worldwide were accessing the Internet, an increase of over 300 per cent since 2000.[6] Overall, the global ICT marketplace was estimated to be worth approximately USD 3.7 trillion in 2008, and by 2011 it is expected to have exceeded four trillion USD, even in spite of the global economic slowdown.[7] The United States, Japan and China are the world's largest spending countries on ICT.[8] In total, countries outside the Organization for Economic Cooperation and Development (OECD) make up over 20 per cent of world investment in ICT, and they are responsible for about 50 per cent of all the ICT products manufactured.[9]

While the United States is currently the world's largest user of the Internet and related services, it is likely that this will change over the next decade. Internet usage in China, India and parts of Africa is expected to grow especially quickly, as are these countries' international Internet revenue streams.[10] Some of the major technologies and innovations driving ICT are outlined below.

## Relevant technologies

### The Internet

First it was the telegraph, then the telephone, the radio and the computer. All these communications technologies laid the groundwork for what is now regarded as one of the most significant technological revolutions of modern times: the Internet.[11] Essentially a widespread information infrastructure, the Internet first made its public debut at the 1972 International Computer Communication Conference. Designed to allow networked computers to communicate transparently across numerous linked packet networks,[12] the Internet is a 'network of networks', capable of delivering information and data to any part of the global network, often in just a matter of seconds.[13] The development of the World Wide Web in 1992 made accessing information over the Internet dramatically easier.[14] Defined by its creator Tim Berners-Lee, as 'an abstract (imaginary) space of information',[15] the World Wide Web was one of the major catalysts that made the Internet more user-friendly and a central part of daily life.

The other major component that helped transform the Internet to mainstream usage was the development of user-friendly web browsers such as Netscape Navigator, Internet Explorer and Firefox. Without the simplicity of design of these browsers, the Internet would never have taken off. Indeed, as J.R. Okin notes, the most remarkable thing about the Internet is how its engineering makes it simple, usable and able to be customized to a variety of needs.[16] Browsers in particular keep the technical components of the Internet 'hidden from view', making it accessible and user-friendly for those without a highly technical background.[17] Without the advent of these browsers, it is unlikely that the Internet would have become the widespread commercial force it is today.

The Internet constitutes a major shift in the communications realm, largely because of its interactivity and accessibility.[18] Combining computers and telephony, the Information Revolution – as it has been called – makes it possible to create, store, exchange and use information at any time and from anywhere.[19] Moreover, the Internet and related technologies represent 'the sum of all the private and public investment, activities, decisions, inventions and creativity of a billion users, over 23,000 autonomous systems, and countless creators and innovators'.[20] In short, the Internet is the fastest growing communication tool in human history. Since its inception, it has had a profound impact on the global political, economic and social structure. It has changed the way we communicate, the way we educate and the way we access and exchange information.[21]

## Broadband

Broadband refers to telecommunications that benefit from a wide band of frequencies with which to transmit information. High-speed Internet is one of the most prevalent forms of broadband, and it allows for faster transmission of data, higher quality services such as streaming video and interactive services, and constant Internet access that is not dependent on a phone line.[22] As broadband becomes more widely available, it is leading to greater economic opportunities, better distribution of Internet services to populations regardless of income or geographical location, and more reliable Internet access.[23] Broadband is largely responsible for the deeper integration of the Internet and related services into our daily lives, as well as for the growing convergence between technology and lifestyle. Broadband is also interesting as an example of technology that was initially designed for a restricted group of users – the military and scientists – that was later extended to the general public. This is a theme that is often seen in cases of emerging strategic technologies.

## Web 2.0

Web 2.0 is a term that was coined in the early 2000s for what was perceived to be a new era of the Internet. It was introduced by the Vice-President of O'Reilly Media Inc. Dale Dougherty, as he and other

Internet executives sought to assess the future potential of the Internet in the wake of the 2000 bursting of the 'dot-com bubble'.[24] Web 2.0 is associated with collaboration among users, and a high degree of participation, networking and creativity. Tim O'Reilly defines Web 2.0 as:

> The network as platform, spanning all connected devices; Web 2.0 applications are those that make the most of the intrinsic advantages of that platform: delivering software as a continually updated service that gets better the more people use it, consuming and remixing data from multiple sources, including individual users, while providing their own data and services in a form that allows remixing by others, creating network effects through an 'architecture of participation,' and going beyond the page metaphor of Web 1.0 to deliver rich user experiences.[25]

Sometimes called the 'living' or 'active' Web,[26] Web 2.0 is focused on collaborating, sharing, socializing and connecting. In many ways, Web 2.0 makes the Internet more personal as it offers a space where people have a chance to express themselves and to share their experiences with people around the world. The highly diverse applications and services include social networking, photograph- and video-sharing sites, weblogs (blogs), wikis, podcasts, tagging and social book marking. The use of Facebook, Twitter, YouTube, eBay and so on, is expanding exponentially. In 2009, Facebook directed 13 per cent of traffic to portals such as Yahoo or MSN, while Google and eBay accounted for about 7 per cent.[27]

While the term 'Web 2.0' has been dismissed in some circles as meaningless, it is still one of the best, most concise ways to summarize the changing role the Internet is playing in people's social lives. Web 2.0 is not an all-encompassing summary of the Internet's significance to modern lives, but in terms of social networking and helping people connect with each other at the individual and community levels, the term is quite useful. A more participatory, social Internet has the potential to promote better mutual understanding and positive empowerment, as well as to develop respect and forgiveness among and within cultures. At its best, Web 2.0 could promote the adoption of a set of shared values and, as a result, the creation of a more secure world.

## Blogs

Blogs are a key component of Web 2.0, and they are so important that they merit going into with some additional detail. In a previous work,

I designated blogs as the 'fifth estate' after the other modern estates that influence policy: the executive, legislative, and judicial branches of the government and the media as a whole.[28] Blogs are regularly updated online resources reporting about topics and events of interest to the writer. Other key components of a blog include the use of the Really Simple Syndication (RSS)-feed file format, which permits readers to see new content automatically; reverse chronological organization; links to other interesting blogs; and the option for readers to comment on individual blog postings, thus making blogs a very interactive form of media.[29] Blogs have a number of features that make them important for issues of international security. For example, their content often lacks any editorial oversight, which contributes to openness and freedom of speech but also undermines quality control and makes it easier to spread false information.[30] Blogs tend to be difficult to censor and are often reflective of the opinion of the 'masses' as opposed to the elites of a society.[31] As is outlined below, blogs have demonstrated that they are capable of effecting social change and contributing to peaceful dissidence. However, blogs, the Internet, the growing ease of communication and the ability of rogue groups to connect with one another are also increasing risks of terrorism, and contributing to organized crime and manipulation by extremist political groups or cults.

## Mobile Internet

Mobile computing refers to the 'ability to use technology that is not physically connected to a static network'.[32] Wi-Fi, a wireless technology using radio waves to offer Internet access, is one of the most popular and important technologies driving this mobile computing revolution.[33]

Mobile Internet technology has become more widespread and accessible through innovations like the Blackberry and the iPhone, and mobile Internet is expected to grow into 'a thriving, low-cost network of billions of devices by 2020'.[34] According to the Research Consultancy Mobile World, about 140 million new mobile subscribers were registered in the first quarter of 2009.[35] Trends in this area point to more 'location-aware' mobile devices that may offer personalized shopping suggestions as you walk down the street, or chances for parents to monitor their child's location.[36] As the technology is refined, mobile Internet will look increasingly like fixed line Internet, offering similar but not completely equal services, speed and accessibility. Overall, mobile Internet will mean increased Internet accessibility and faster, more varied communications. At the same time, more location-aware mobile devices have the potential to dramatically reduce privacy and heighten vulnerabilities to cyber attacks.

An extension of the mobile computing phenomenon involves technology known as 'augmented reality', a cousin of the once much-vaunted 'virtual reality'. Augmented reality works by starting with a real environment and then adding to it, overlaying digital information over the real world.[37] For example, a person looking at a mountain landscape with augmented reality might see the names of each mountain imposed over the actual mountain. Augmented reality essentially provides a way to blend data available online with the physical world; it is a bridge between the real and the virtual.[38] Although augmented reality has been around conceptually for some time, the rise of mobile computing and mobile phones equipped with features like satellite-positioning systems, fast Internet connectivity, and a digital compass are making augmented reality much more possible.[39] For now, augmented reality mostly helps a person contextualize their surroundings but, eventually, it may help to make advertisements and other media more interactive.[40] As *The Economist* notes, 'the building blocks of [augmented reality] have arrived and are starting to become more widely available. Now it is up to programmers and users to decide how to use them'.[41]

## Cloud computing

Cloud computing designs the delivery of hosted services over the Internet. The three categories of service included in cloud computing are Infrastructure-as-a-Service, Platform-as-a-Service and Software-as-a-Service. Unlike traditional Internet hosting, a cloud is sold on demand by the minute or the hour, and it is elastic.[42] In other words, a user can have as much or as little of a service as they want at any given time. In slightly more technical terms, cloud computing refers to a style of computing that allows providers to deliver a variety of IT-enabled capabilities to consumers. What makes cloud computing unique and important is the delivery of capabilities as a service, the delivery of services in a highly elastic and scalable fashion, the use of Internet technologies and techniques to develop and deliver services, and designing for delivery to external customers.[43] Because of its elasticity and scalability, the barriers to entering the cloud computing arena are low, and this allows small companies to grow quickly. Overall, cloud computing promises to make accessing the Internet easier and more cost-effective. This is because cloud computing allows businesses a way to increase their IT capacity and capabilities without the expense, time and uncertainty of adding new infrastructure, training personnel or acquiring software licences.[44]

## HashCache

In many parts of the developing world, a major barrier to using ICT is not the lack of computers but the difficulties in obtaining reliable, fast Internet access. Often, in the developing world only low band-width Internet connections are available, and individual users receive a fraction of the speed of the notoriously slow dial-up connections.[45] US computer scientists are working to develop HashCache, a more efficient method of storing frequently accessed web content on a hard drive rather than using bandwidth to repeatedly retrieve the same information. HashCache reduces Random Access Memory (RAM) and electricity requirements by a factor of ten by using a novel hash function that eliminates the need for a RAM-hungry cache index.[46] Although this technology is still in the early stages of development, it is being field tested in Africa and represents a major move forward in the previously stagnant field of caching. As Jim Gettys, the co-author of the Internet's HTTP specification, has noted, HashCache would allow even the poorest schools and most basic computers to cheaply access one terabyte of web content, roughly equivalent to all the coursework that is freely available from colleges such as MIT.[47]

## Computational technology and the rise of supercomputing

Computers are becoming increasingly fast, breaking previous performance records at lightning quick speeds. The world's leading supercomputers can now process information and make calculations at a rate of 1105 quadrillion calculations per second.[48] Since 1961, computers have become faster and the timeline between each increase in speed has become shorter.[49] Two principles in particular highlight this fact: Moore's Law, which states that thanks to ongoing technological advances, computer processing power has been doubling every 18 months for the past three decades (a trend that is likely to continue for at least another decade),[50] and Kryder's Law, which observes that disc memory doubles every 12 months.[51]

Overall, the trend in computational technology is for more functionality to fit into the same amount of space, allowing for more and more powerful computers in smaller and smaller sizes. Smaller dies can also mean less need for power, which in some cases translates to more energy-efficient computers.[52]

In everyday life, faster processors can help process data faster and ensure smooth operations of everything from photo editing to 3D video game playing.[53] However, the real significance of improving computational technology is for science and research. For example, faster computer processors will help astronomers and those searching the universe for extraterrestrial life to measure and process frequencies in space at a much faster rate, allowing more efficient detection of unusual patterns.[54]

Improved computational power will also help in longer range weather forecasting (beyond the current limitations of approximately ten days), better 3D-modelling and, perhaps most importantly, a processor with infinite speed could allow scientists to have a purely numerical and computational model of the universe as opposed to one that relies on rough approximations of continuum mechanics.[55] Some of these latter developments are still hypothetical and may be decades away, but the increasing power of computational technology is already playing an important role in our abilities to process, understand and model information.

### Nano-memory storage

Information storage is a key challenge of the Information Age. New memory chips based on nanotubes and iron particles may be capable of storing electronic data for a billion years. Researchers at the University of California in Berkeley have designed a memory cell by taking a particle of iron and placing it in a carbon nanotube.[56] (For more information on nanotubes, see Chapter 8.) The researchers then placed electrodes at each end of the tube, and when they applied an electrical current, the iron particle shuttled back and forth. In this way, the researchers created a '1' and '0,' the signs required for digital representation.[57] What makes this method of memory storage so durable is the fact that the repeated movement of the iron particle does not damage the walls of the carbon nanotube, meaning the process can be repeated almost infinitely. Researchers still need to design a platform that will exploit millions of these memory storage units instead of just one, but this technology could revolutionize how we store electronic data.[58]

## Regulatory structures

It should come as no surprise that in a multifaceted industry like ICT, there are numerous regulatory structures and competing visions of who should govern what and how. The Internet is a transnational network of

networks, basically accessible to anyone with a computer, and its regulation and governance are particularly thorny subjects, but even more
basic forms of ICT such as telephones are subject to some form of international monitoring and standards setting. An overview of some of the
dominant ICT institutions, issues, and controversies is provided below.

## The Internet

Internet governance is a contested concept,[59] but perhaps the best starting definition is that of the United Nations Working Group on Internet
Governance (WGIG). WGIG has defined Internet governance as 'the
development and application by Governments, the private sector and
civil society, in their respective roles, of shared principles, norms, rules,
decision-making procedures, and programs that shape the evolution
and use of the Internet'.[60] While this is a generally accepted definition
of Internet governance, it is by no means the final word on the debate.
Some have dismissed WGIG's implied calls for a single Internet regulatory structure, arguing that there is more of a need for 'a heterogeneous
and highly distributed array of prescriptions and processes that reflects
the Internet's core features, rather than centralized "one-size-fits-all"
control over a singular system'.[61] According to William J. Drake, a heterogeneous approach will help policymakers to 'evaluate the full diversity of public and private sector practices' relating to the governance of
the Internet and to reflect whether there are cross-cutting issues, gaps
or tensions which have not been properly tackled.[62]

The 'world's most visible Internet governance body' is the US-based
Internet Corporation for Assigned Names and Numbers (ICANN).[63]
Founded in 1998 at the request of the US Government,[64] ICANN is a
non-profit public benefit corporation that coordinates Internet domain
names and address assignments.[65] Due to its history and the fact that it
is based in the United States, ICANN's overall legitimacy has often been
questioned. Yet in 2009, ICANN took a step towards becoming more
internationally accessible and democratic by allowing the creation of
domain names in non-Latin scripts, such as Arabic, Chinese, Russian
or other languages.[66] While ICANN still remains US-centric, this step
opened up Internet access to a range of people who had previously been
restricted in their web usage because of their unfamiliarity with the
Western alphabet.[67]

Overall, the Internet today is managed in a generally ad hoc fashion
with input from both public and private actors. Private management
of the Internet infrastructure is focused on active coordination among
private providers.[68] Meanwhile, multiple Internet industry groups are

currently involved in the development of protocols and technical standards and in improving network interconnection and inter-operability.[69] Many industry organizations participate in the global management of the Internet by establishing sets of rules and regulations on issues such as network security, electronic contracting and digital signatures.[70] Noteworthy structures include non-governmental organizations such as the World Wide Web Consortium (W3C) led by Tim Berners-Lee and the Internet Engineering Task Force (IETF). Some have suggested that this ad hoc regulation and the lack of strong global governance of the Internet have allowed creativity and entrepreneurship to play a central role in the development of the Internet. According to them, it is precisely because the Internet was so open and fluid that new modes of business and operation thrived, thus empowering the Internet to fundamentally change the world and our communication with it.

Yet, this type of regulation has not yet succeeded to the resolve the issue of cyber security. Cyber security refers to 'measures to protect information technology; the information it contains, processes, and transmits, and associated physical and virtual elements (which together comprise *cyberspace*); the degree of protection resulting from application of those measures; and the associated field of professional endeavour'.[71] The main challenge for Internet regulation guaranteeing cyber security stems from the fact that, in order to have the truly transnational cooperation necessary to design and implement effective, global regulations, countries have to be willing to share information about their perceived cyber vulnerabilities and threats. Only by doing this will it be possible to clearly define what the biggest risks are and to establish legal guidelines on how to combat those threats most effectively.[72] Unfortunately, such information is often extremely sensitive and is usually deeply connected with a country's intelligence-gathering capabilities and resources.[73] Thus, when it comes to sharing vulnerabilities and information about a country's cyber infrastructure weaknesses – even if this sharing is purportedly done in order to improve global Internet security – countries rightfully fear that sharing such details could increase their vulnerability. This would be especially true if an enemy country exploited the information that was supposedly shared for the greater good and then used it to attack its rival. Trust between countries is a fundamental requirement if better regulation and security of the Internet is going to be established, but the nature of geopolitics means that such widespread trust is far from realistic.[74]

In an editorial published in 2006, I outlined several other key policy challenges relating to the Internet. These include civil liberties versus surveillance; balancing the potential use of technology by terrorists with the use of technology by governments; and the dissemination of harmful, radicalizing messages versus the spread of positive, peaceful ones.[75] Striking the correct balance between civil liberties and privacy with government monitoring of the Internet for potentially harmful activities is perhaps the most salient Internet regulatory challenge of all, for which the best solution is balanced policies and responsible oversight.[76] Similarly, in trying to ensure freedom of speech, states should look towards the creation of international regulations and organizations that monitor the creation of websites and the types of information they propagate.[77] Even on the Internet, hate-filled rhetoric and messages that incite violence should not be tolerated. Most European countries regulate and aim to eradicate any truly violent or hateful sites.

For its part, the Progressive Policy Institute (PPI) frames the challenge of commercial Internet regulation in terms of jurisdiction and sovereignty, arguing that, at least in commercial transactions, a jurisdiction must be created in which sellers are not subjected to the laws of other countries in their application of new technologies and information systems. The PPI stresses that international rules must protect consumers and create an environment of trust.[78] In terms of sovereignty, PPI maintains that the ideal regulatory framework will 'not let individual nations reach beyond their own borders to control content in cyberspace' but will allow them to 'control both users and Internet hosts within their borders as long as the content controlled is not governed by trade agreements'.[79] Finding a balance between these regulatory recommendations should be a top priority for global policymakers.

### The International Telecommunications Union and the global alliance for ICT Development

The International Telecommunications Union (ITU) is the main UN agency overseeing issues related to ICT. As the second oldest international organization still in existence, the ITU has a history dating back to the advent of the telegraph.[80] Its modern day activities focus on coordinating the shared use of the radio spectrum, assigning satellite orbits, improving the communications infrastructure across the developing world, and establishing worldwide standards to encourage seamless interconnection of a multitude of communications system. Additionally, the ITU aspires to address and ameliorate the effects ICT have on the environment and to contribute to improving global cyber security.[81] Based in Geneva, Switzerland, the ITU's membership includes 191 member states and more than 700 sector members and associates.[82] Since its creation, the ITU has had a foundation in public-private partnerships, and that is one of the key reasons for its longevity and success.

One of the ITU's top priorities is bridging the digital divide, a topic outlined in greater detail below. To this end, the ITU seeks to build information and communications infrastructures and to promote capacity building in the developing world.[83]

The Global Alliance for ICT Development (UN-GAID) is a UN organization that focuses specifically on achieving the ICT-related Millennium Development Goals (MDGs).[84] A platform for policy-dialogue, UN-GAID is committed to improving the accessibility, content, connectivity and educational value of ICT-related technologies.[85]

### National regulation

ICT governance on a national level is a highly complex and challenging task. Although the responsibility for ICT networks lies mostly with governments, these networks are often privately owned.[86] In addition, even if governments adopt national legislation to regulate cyber criminality, prosecution, privacy issues and the like, the transnational nature of the challenge often transcends national borders and therefore eludes the reach of individual governments.[87]

China is a particular case in the sense that it attempts to strictly regulate the Internet in multiple ways. For instance, there are currently 12 government agencies working on censoring the Internet.[88] In 2010, the introduction of new legislation requiring Internet providers and telecommunications companies to inform the Chinese Governments about leaks of state secrets was announced.[89] It is also worth mentioning the ongoing dispute between the Chinese Government and Google. Following Google's decision that it would no longer comply with the censorship rules imposed by the Chinese Government, searches conducted on google.cn were automatically redirected to Google's Hong Kong address, google.com.hk, in order to avoid censorship.[90]

As for the United States, a 2008 report by the Center for Strategic and International Studies acknowledges the importance and the transnational nature of the potential threats posed by the Internet as well as the urgent need to react. More specifically, it finds that: 'cybersecurity is a major national security problem', 'decisions and actions must respect privacy and civil liberties' and 'a comprehensive national security strategy must embrace both the domestic and the international aspects of cybersecurity'.[91] So far, the approach adopted by the United States has been to reinforce collaboration between law enforcement agencies of different countries[92]

In a nutshell, governance requested to face online threats goes beyond national reach. As Buckland et al. argue, private–public collaboration, trans-national collaboration and harmonization of national legislation are key to establishing effective frameworks.[93]

### Other regulatory structures

Outside the realm of the Internet, other ICT-related governance mechanisms have been established in the form of legally-binding agreements within the framework of international organizations as well as informal agreements and consultations.[94] Organizations involved are the OECD, the European Conference of Postal and Telecommunications Administrations, the European Union, Asia-Pacific Economic Cooperation, the Asia-Pacific Telecommunity, the Inter-American Communication Commission, Australian Technology United and the European Telecommunications Network Operators Association.[95]

The United Nations is taking the lead on the creation of an international computer security treaty. In July 2010, a team of cyber security specialist and diplomats addressed a number of recommendations for the creation of such a treaty to the Secretary General. Signed by countries such as the United States, China, Russia, India, Brazil and South Africa, the document produced by this task force can be summarized in five points:

Having more discussions about the ways different nations view and protect their computer networks, including the Internet; discussing the use of computer and communications technologies during warfare; sharing national approaches on legislation about computer security; finding ways to improve the Internet capacity of less-developed countries; and negotiating to establish common terminology to improve the communications about computer networks.[96]

### The role of the private sector

One of the major characteristics of the ICT revolution is that it has pushed the private sector to engage more directly in public policy making. As new technological opportunities in telecommunications merged with computing, multiple facets of the private sector became deeply and irreversibly intertwined, from banking services to commodity markets, and from capital flows to data systems.[97] Because of the increasing levels of global integration, multinational corporations (MNCs) received a new impetus to pressure governments to liberalize domestic and international markets.[98] In fact, the human and financial resources of some MNCs exceed the capacities of many governments and allow them to exercise considerable influence.[99]

Compared to governments, industry has very different priorities in terms of the types of regulation it would like to see over ICT and the reasons for engaging in such regulation. While governments routinely emphasize the risk of cyber warfare and the idea that critical infrastructures could be attacked by a rogue or enemy government, businesses and industry tend to be much more focused on their bottom lines, and they structure their approach to regulation accordingly.[100] In particular, the entertainment industry is concerned with the growing phenomenon of copyright piracy, peer-to-peer file sharing, and illegal downloading of music or films. In general, industry groups will be more willing than government agencies to cooperate with other businesses, the government and in some cases even competitors in order to identify vulnerabilities and to create ways to protect their ICT infrastructures from hackers and other malicious parties.[101]

The question of to whom it is that MNCs ultimately report still arises. It is one thing to say that MNCs and industry in general will look to secure their own Internet and ICT infrastructures as a means of preserving their bottom lines. The questions of who will prevent MNCs with a truly international presence from breaching the security of others and who has the jurisdiction to enforce MNCs adherence to any international guidelines, however, remain sticky ones.[102]

Unlike MNCs, small- and medium-sized firms often lack the means to participate effectively in the public policymaking process. In addition, governments can be selective about which firms they will consult or allow to participate in multilateral forums.[103] Thus, private sector regulatory efforts on ICT are by no means clear-cut.

## Challenges

### Time compression

In the past, the rate of scientific progress has been so predictable that it is treated almost as a law of nature. For example, in scientific research the number of papers published has generally doubled every 15 years, and in astronomy the distance to the furthest galaxy that can be seen from Earth has doubled around every ten years.[104] The ICT revolution has also proceeded at a regular, albeit much faster pace, as evidenced by Moore's and Kryder's Laws, which are mentioned above.

Coupled with the global spread of information through communications networks and the media, this creates what has been called the 'time compression phenomenon'. Essentially, this is a reference to the fact that ICT enables more information to spread more rapidly, which generates pressure for business and policy leaders to react increasingly quickly to new developments. Even if policymakers would prefer to have time for contemplation or analysis of unfolding events, the Internet, blogs, 24-hour news stations and other public forums often create pressure for a quick decision, leaving little time for deep analysis and reflection.[105] Such a situation is made even more dangerous by the potential unreliability of information on the Internet and by the public's tendency to react before all the facts are known.[106]

Time compression is also evident in the economic and financial sectors, where the emergence of a global, active, around-the-clock financial market has resulted in the creation of a trading system that requires computer-dependent decisions and computerized evaluations of market developments for its daily operations.[107] In this environment, businesses and governments alike are forced to react faster to changing developments than they would in the absence of ICT.[108]

People have more access to information and to each other than ever before and, in this context, issues of security and privacy in dynamic wireless networks present themselves. It is important to address the issue of insufficient and sometimes unreliable security infrastructures, as well as the problems related to the security evaluation techniques necessary in the context of emerging wireless networks.[109]

### Death of distance

The Internet allows us to identify, relate to and work with other people, regardless of where in the world they are based. Social networking, business matters or research benefit from these developments since physical proximity is no longer required for interaction. This has important consequences for geopolitics, and not all of them are beneficial. For example, it means that cyber attacks can be launched from any part of the world. Often, the origins of attacks cannot be traced back. For instance, although it is heavily suspected that Russia was responsible for a series of cyber attacks on Estonia in May 2007 (discussed below), there is no conclusive evidence.[110]

### Privacy issues

Privacy issues are a central concern in today's interconnected societies. One possible definition of privacy, as opposed to security, has been proposed by Biggs: privacy refers to 'unwarranted access to private information, but not necessarily breaches in security', whereas security refers to the 'access by non-authorised people to protected sites'.[111] The social networking site Facebook is probably the most prominent case of privacy concerns. In 2009, Facebook changed its terms of service, noting that going forward, Facebook would have ownership of all information and material uploaded to its site forever – even after its users cancelled their accounts. Following uproar from consumer groups, Facebook users and the media, Facebook quickly had to at least temporarily reverse its policy change, but the incident touched the heart of the debate over privacy and who controls personal data in the Information Age. In addition, gaps in privacy resulting in the display of protected information are repeatedly being discovered.[112]

Similarly, Google's recently launched social networking service Google Buzz has been heavily criticized. At the time of launching, Buzz drew on contact information stored in Gmail and used it to automatically create a network of friends. Following public outcry, Google had to change its settings.[113]

As the Internet becomes more and more ubiquitous, the trail of information that individuals leave on the web is growing. Much of this trail comes from search engines, where users' searches are saved, often revealing personal identity clues or information.[114] Other concerns relate to how the government or employers might track or monitor personal emails.[115] For as common as the Internet and other ICT outlets are in everyday life, few consumers have a clear understanding of how websites or companies collect, collate and share data about their personal details or habits. Data collected through different ICTs could be used in a number of ways, from more targeted advertising to potentially tracking individuals and sharing information with third parties (the government, insurance companies or future employers).[116]

With regard to Internet privacy, the distinction between personally identifiable information (PII) and non-PII is important. As is indicated by the name, the tracking, storage and use of the former type of information poses a much greater risk to individual privacy than the latter. Unfortunately, consumers are often naive about the distinction, and the potential for the abuse of PII is strong.

Generally speaking, concerns over Internet privacy are two-fold. On the one hand, there are issues of private companies like Internet search engines or web browsers gathering the information provided by consumers and either using it to better market their products or selling it to third parties to use for unspecified purposes. More menacingly, there are also considerations over governments using ICT as a means to monitor and track their citizens. Information that could once simply be collected can now be digitized and stored on huge interconnected databases, making it easier to build highly detailed profiles of people, their preferences and their behaviour.[117] In order to protect individual privacy in the face of emerging and rapidly evolving ICT, governments must enact strict privacy regulations and rules protecting personal information and regulating how such information can be used.[118] Initially, it seemed as though the dominant paradigm for Internet regulation was going to be industry self-regulation, but such efforts have fallen short and it is becoming increasingly apparent that the government will need to play a bigger role.[119] Priorities for regulation include websites providing notice before collecting information, allowing users' choice over how their information is used, providing consumers with access to collected data, with the ability to contest how such data is used, and the assurance of information security and protection of information from unauthorized use.[120] Companies with privacy policies should not be confused with companies with policies that protect privacy.[121]

Again, such efforts must be broadly multinational, as even if one country protects its citizens' data and Internet privacy, another country may be ready and willing to exploit or sell the same information.[122] Transborder flows of information must be protected, and purely national legislation and guidelines will not suffice on this front.[123] Adding an additional layer of complexity, different nations value privacy and the importance of personal information differently. Indeed, many companies complain about lack of clarity amid competing local, national and international regulatory frameworks.[124] Moreover, many government and private sector leaders worry that overly stringent privacy regulations on the Internet would limit the Internet's commercial potential.[125] Thus, it will be challenging but nonetheless necessary to develop a set of global best practices on this front.[126]

## ICT and the environment

The ICT sector is responsible for about 2 per cent of global greenhouse emissions, meaning there is an increasing impetus to improve the greenness of the industry by introducing power-saving features into the various computing platforms, as well as other, more expansive, initiatives.[127] Green ICT firms are gaining more attention for their potential to both reduce their industry's emissions and help offset overall global carbon emissions.[128] In fact, the Climate Group estimates that reductions in the ICT industry's emissions could reduce total global greenhouse gas emissions by 7.8 billion tons of $CO_2$ by 2020 – more than five times the ICT sectors own carbon footprint.[129] However, the ITU expresses concern about the effects of the financial crisis on eco-ECTs or energy-efficient ICT, warning that there is a risk that the global downturn may affect the investment needed to change to alternative energy sources as well as research and development efforts.[130]

## Geopolitical implications of ICT

### Social implications

ICT is one of the key drivers of social change and plays a key role in empowering global civil society; and new information technologies can affect political systems and stimulate political change. This is especially valid in countries where the print and broadcast media are under state control. New forms of ICT outlets such as blogs or mobile phone text messaging, photography, and video are difficult for governments to control and are therefore accessible platforms for citizens to express dissent. By overcoming the information bottlenecks of state-owned media and bypassing censorship rules, bloggers have the capacity to influence the media and the public at large, and the potential to generate positive and negative change in their home countries.

One prime example of this trend is Iran, where the number of bloggers has increased dramatically in recent years. According to the NITLE Weblog Census, Farsi is the fourth most widely used language in the world of blogging.[131]

In April 2009, the power of the new media and ICT was demonstrated in very high-profile ways in the former Soviet republic of Moldova. In the wake of elections that many people perceived as being rigged in favour of the incumbent Communist party, journalists and young people organized an impromptu protest against the government and the elections. What made this protest unique was the fact that it was organized entirely via social media, in particular through Twitter, a microblogging service that allows users to post updates on their activities and whereabouts. Moldovan journalist Natalia Morar, the mastermind of the so-called Twitter Revolution, explained to the BBC how simply and quickly the protests came together: 'It just happened through Twitter, the blogosphere, the Internet, SMS, websites and all this stuff', she said. 'We brainstormed for 15 minutes and decided to make a flash mob' in order to protest the elections.[132] Within several hours, over 10,000 people had come on to the streets. The events in Moldova show the ways in which technology is increasing the speed and efficiency with which people can respond to events, an extension of the previously described time compression phenomenon. The Moldovan protesters took advantage of new technologies that allowed them to act when their anger and frustration were still fresh, and because of the organic nature of Twitter and other tools used to rally people, the protests were organized and enacted so quickly that the Moldovan authorities did not have time to clamp down on the demonstrations until they were already in full-swing. Although the protests ultimately did not result in a regime change, Moldova did receive a visit from European Union (EU) foreign policy chief Javier Solana who promised to begin EU dialogue and engagement with Moldova's opposition parties.[133]

Impoverished populations may also stand to benefit from ICT developments. For example, a United Nations Economic, Social and Cultural Organization (UNESCO) initiative seeks to provide women in rural parts of the world's least developed countries with access to ICT to improve their information resources and to offset a perceived gender imbalance in the distribution of ICT-related technologies.[134] Through chat rooms and other social networking services like message boards and information sharing websites, isolated or marginalized women now have greater opportunities to share and discuss their experiences in health, agriculture and family-planning and, importantly, they are doing so in their own languages.[135] Initiatives like these are especially important because the gender divide for ICT is quite dramatic. In general, and especially in poor, developing countries, women have much less access to the Internet and communications tools than men. This can reinforce women's marginalization in a culture, which can make it harder for women to get jobs or to access educational resources.

More ominously, new forms of ICT are, in some instances, enabling the news media to perpetuate xenophobia and cultural stereotypes. These discourses, which are often one-sided and separated from fact, can fuel aggression and violence against ethnic and cultural minorities.[136] Governments need to work to establish policies that allow for freedom of speech while still clearly outlining guidelines for responsible journalism.[137] In the short-term, it is important to make sure that journalists can operate independently from government or special interests. In the long run, measures should be taken to promote more inclusive societies via ICT. For example, 'peace radios' should be created and promoted in order to offset the influence of 'hate radios'.[138] Above all, cultural respect should be maximized and xenophobia should be minimized.[139] Governments can help in this process by encouraging media programmes about other cultures, their specific sensitivities, and methods for promoting mutual understanding of different cultures and religions.[140]

Overall, ICT presents great opportunities to generate new scenarios for social relations. New agents from civil society in the field of ICT can have a significant influence on technological, social and political relations on local, national and global scales.[141]

## Military applications of ICT

Information technologies have been widely used in military and national security applications. Not only do militaries use ICT in their day-to-day operations, it is also increasingly likely that, in the future, they will use cyber attacks as a way to weaken their enemies during a more traditional military attack.

According to Michael Vatis, the then Director of the Institute for Security Technology Studies at Dartmouth College in the United States, cyber attacks can be defined as 'computer-to-computer attacks to steal, erase, or alter information, or to destroy or impede the functionality of the victim computer systems'.[142] In terms of 'aggressors' initiating the attack, it is important to distinguish between cyber attacks conducted by private criminals and attacks conducted by governments. Ventre suggests distinguishing between 'cyber criminality type attacks', and what he calls 'information warfare type attacks'. Cyber crimes include tools used by criminals which can be dealt with by national legislation while information warfare attacks, or the use of force and armed attacks, refers to states attacking other states or their own citizens.[143] Yet, not all cyber attacks can be as easily categorized. There are also certain grey zones where an attack may not be conducted by a Government, but where a private agency may have been commissioned to do so, on its behalf.[144]

Cyber attacks may put at risk critical infrastructures, such as telecommunications; government and public health records; food and water supply; agriculture and many areas of production.[145] As the World Economic Forum explains, an attack on or a system failure in one part of a critical information infrastructure 'creates a domino effect, shutting down IT-dependent applications in power, water, transport, banking and finance, and emergency management'.[146]

An enemy government could also use cyber attacks to alter information in order to spread misinformation or propaganda. Such actions could provoke anger or fear, or damage morale among target populations and could contribute to undermining public support for a military campaign. Similarly, such misinformation could cause panic in global financial markets and undermine the economic strength of an adversary.[147] For many smaller countries, cyber attacks are a way of addressing imbalances of power in terms of conventional military strength.[148] Nye, from a geostrategic power point of view, argues that ICT allows smaller actors on the word scene to acquire a disproportionately large share of power. Although resources and geography are still crucial for cyber power, smaller actors do have greater access to ICT power tools than to traditional means. Nye calls this phenomenon a 'diffusion of power', whereby larger powers will dominate the domain less than they dominate areas such as the sea or the air.[149]

Perhaps the most striking case of governments using cyber attacks as part of a more traditional military strategy was in May 2007 when Estonia became the victim of a cyber attack. Although it was difficult to determine exactly who was responsible for this attack, it is widely suspected that the Russian Government played at least an indirect role in the denial of service that wreaked havoc on Estonian businesses and government offices for several days.[150] The attack was initiated in the form of botnets, which are malicious automated computer programs that 'take root undetected in far-flung computers and barrage their targets with useless data'.[151] As a result of the attack, many government and bank websites in Estonia were disabled.[152] The Estonian Government responded calmly and quickly. Although the attacks were disruptive, most sites restored their services in between one and two days. This type of attack – a so-called denial of service – is not the most damaging.[153] Although the Estonian case was very serious, the country's territorial integrity was not compromised, and all damage from the attack was apparently short-term.

The July–August 2008 conflict between Russia and Georgia reflected an escalation of the type of tactics used in Estonia a year earlier. This time, in the conflict over the status of South Ossetia, cyber weapons were openly used as part of Russia's conventional military attack on Georgia.[154] Many government websites, including that of Georgian President Mikheil Saakashvili, were attacked as Russia sought to assert its authority over the former Soviet Republic. Because of this denial of service cyber attack, Georgia's ability to disseminate information to the public during the conflict was severely compromised.[155]

Estonia and Georgia are not the only countries to suffer attacks. In 2007 and 2008 for instance, the United States, Germany, France, the United Kingdom, New Zealand, India, Belgium, China and Russia declared themselves the victim of cyber attacks.[156]

Hopefully, these recent developments will encourage governments to focus more on cyber security laws and policies. So far, most of the victims seem to have been surprised by and unprepared for the attacks. According to Ventre, 'that our security leaders or our governments admit to the world that they were victims [...] is a confession of helplessness, an acknowledgement of vulnerability and a lack of control'.[157]

As is shown by these cases, cyber war can be used as a part of traditional military strategy. There is also an ongoing debate over whether cyber attacks even have the potential to entirely replace the traditional military in the future.[158] Some argue that cyber war will bring a 'second wave of revolution in military affairs'.[159] Cyber warfare presents multiple policy and legal problems that need to be tackled urgently but, even in the light of recent examples, governments seem slow to respond to this emerging challenge. As a case in point, after being a victim of the cyber attack in 2007, Estonia asked the North Atlantic Treaty Organization (NATO) for help with cyber defence. In response, NATO put a renewed focus on improving its common approach to cyber security.[160] The 2008 NATO cyber security policy sets out the general principles underlying the importance of the issue of cyber defence and asking various NATO agencies to create a coordinated and unified approach for resolving the outstanding issues.[161] However, that policy does not tackle important policy issues such as those related to NATO retaliation in case of an attack.[162] A new 'strategic concept' for NATO was to be agreed in 2010.[163]

There are several policy possibilities for tackling weaknesses in cyber security, including improved education and training, improved risk management, adopting the necessary standards and certification, and using benchmarks, checklists and metrics.[164] To successfully address the issue of cyber security, it is important to create global and national cyber security frameworks.

In terms of securing cyberspace, various initiatives have been undertaken at the national and international levels. The European Union has created the European Network and Information Security Agency, a cyber security agency. The G8 has launched a cyber crime response network.[165] In addition, various private organizations deal with cyber security at the national level, including: the Cyber Security Industry Alliance, the Business Software Alliance, the Information Technology Industry Council and the Software & Information Industry Association. Most recently, the United States has established a national cyber command (CYBERCOM) at the US Department of Defense to protect military networks against cyber attacks and provide the Pentagon with offensive cyber weapons.[166]

Connectivity is crucial in this technologically advanced world, which is why the need to protect cyberspace and assets is so important. The rapid advances in ICT require well-defined strategies for tackling potential cyber threats at the national and global levels. Cyber security is a cross-cutting issue across all types of infrastructure, and it is a foundation for many public and private sector operations. Cyber security threats can change very quickly, so they require protective measures to be as rapid. In this context, it is important to establish global and national cyber security response systems as well as awareness-raising and training programmes. It is also important to develop a strong international cyber security cooperation programme.

Interestingly, many states have been reluctant to improve their cyber security. Many potential initiatives are a hard-sell from a political perspective because they can be costly and long term, and many of the end results are invisible to everyday users.[167] Unfortunately, this means that states and international and regional groups often resort to patchwork proposals in response to imminent or just passed threats. A more comprehensive approach is crucial.

### Threats to international cyber security

The US Federal Bureau of Investigation (FBI) has ranked cyber attacks as the third greatest threat facing the United States, second only to nuclear war and weapons of mass destruction.[168] Whether it is the launch of a new virus, an attempt to break into computer systems containing valuable information or something even more sinister, the Internet has many traits that make it both vulnerable and appealing to hackers, terrorists and criminals. The fact that the traits that make the Internet so susceptible to attacks – its interconnectedness and its wealth of information – are the same traits that make it so invaluable to daily life is one of the greatest challenges of Internet security.

Cyber security is extremely difficult to implement, especially in a comprehensive fashion. The Internet has developed at a rapid pace with the constant addition of new systems, and the truth is that the Internet is only as strong as its weakest link. Even a seemingly simple virus can spread rapidly around the world.[169] Adding to the complexity of the threat is the fact that an attack can be launched from anywhere in the world, with surprisingly minimal technical skill or financing required. Moreover, cyber attacks are often difficult to trace to their origin, making it more difficult to prosecute cyber criminals.[170]

Cyber attacks can vary in intention, scope and motive. As is mentioned above, threats may come from individuals, states or even as a result of accidents or infrastructure-damaging natural disasters such as earthquakes.[171] Although the latter factor is a risk, James Lewis stresses that any large-scale cyber catastrophe will have a human element. 'Could a crisis in cyberspace happen without human intervention? This sort of scenario is very doubtful [...] Cyberspace is a human construct and will most likely require human intervention for it to fail.'[172] The motives for cyber attacks can be equally various. From malicious dissidents who unleash a virus to damage the key infrastructures of their enemies to thrill-seeking pranksters simply looking to wreak havoc or even to international criminal groups looking for financial gain, the range of cyber crimes is almost as broad as the Internet itself.[173] Furthermore, new viruses or worms can be released across a wide variety of platforms, facilitating their spread and making it even more difficult to stop them.[174] As we as a society become more dependent on the Internet, the chance that cyber attacks and Internet viruses will move beyond having only a financial impact to having human costs will increase dramatically.[175] In fact, some cyber security experts argue that the main cyber threats come not from attacks that debilitate critical infrastructure such as power grids but from cyber attacks carried out in conjunction with physical terrorist attacks.[176] For example, a cyber attack could debilitate emergency responders' communications tools at the same time as terrorists set off a bomb. Such a coordinated attack would have the dual impact of having a high human cost and a high financial cost. Although the main motive for cyber attacks until now has been financial profit, such technological uses could well be prompted by political, criminal or terrorist motives.

Cyber threats can take different forms, such as hacking, espionage, identity theft or terrorism, spamming, phishing, data leaks, intrusions, site defacements and denial of service attacks.[177] Commonly used strategies include information warfare, stimergy, swarming, open source models as a guerrilla warfare model and psychological manoeuvres such as the propagation of rumours, using blogs, semantic attacks and the use of web applications by insurgents.[178] They might also include attacks on the information infrastructure, conducting hostile information operations or performing reconnaissance for physical attacks. Cyber experts agree that the United States is a prime target.[179] In 2007, the US Department of Homeland Security logged over 80,000 attacks on Pentagon systems and about 37,000 attempted breaches of private sector and US Government computer systems.[180]

The complexity of cyberspace and its related components makes it difficult to predict how different systems would behave in unexpected circumstances, thus making it hard to adequately prepare for a potential cyber attack.[181] The major channels of potential attacks are 'through cyberspace [...] by direct destruction or alteration of physical structure, such as buildings or telecommunications lines, or through intentional or inadvertent actions by a trusted insider'.[182] Various combinations of attacks are possible, and they are not mutually exclusive.[183] Conficker, a malicious software programme or botnet thought to have been designed by criminal gangs in Eastern Europe, is among the more recent, high profile examples of the Internet's enormous vulnerabilities. On its release, Conficker easily infiltrated millions of computers worldwide, often without the computer's owner even being aware of the attack.[184] At the time of its launch, Conficker was quickly identified but its potential ramifications were a mystery. It could send spam from infected computers; capture information typed by users or any other number of possibilities. The uncertainty surrounding Conficker underscores the ambiguity of many cyber threats and, more importantly, the difficulty authorities have in successfully combating them.

Potential improvements of cyber security include: approving the necessary standards and certifications, promoting best practices and guidelines, using risk management, improving training and education, using benchmarks and checklists, building a high degree of security into enterprise architecture and adjusting metrics.[185] More aggressive solutions call for a redesign of the entire Internet, possibly having users give up their anonymity in exchange for more security.[186] In fact, researchers at Stanford University are studying options for ways to introduce a more secure Internet structure without disrupting the Internet we are so accustomed to in our day-to-day lives.[187] Such a structure would have improved security capabilities and the ability to support new, complex, and not-yet-created Internet applications. Security measures would be a much more integral part of this system. While the introduction of a universal redesign to the Internet is far from being fully implemented, the fact remains that most cyber security threats are currently dealt with in a piecemeal fashion that neglects the overall security challenges of the Internet's infrastructure and architecture.[188]

Cyber security, like other forms of security, is a public good, but the Internet is unique because of the high level of private sector involvement. 'In a perfect market, the private sector would purchase adequate security and firms would offer the products needed for it. This has not been the case. While some industry sectors, such as financial services, have moved to increase security, other sectors may not improve without further incentives.'[189] In such a situation, the benefits of government intervention would be great. As both the public and the private sectors become more and more dependent on the Internet and ICT for key day-to-day functions,[190] the need for a comprehensive and coherent policy approach is increasingly apparent. Unfortunately, government regulation of the Internet would be difficult, because the private sector owns many components of the Internet. Additionally, on the Internet, physical jurisdiction is separate from governance.[191] In an ideal world, countries would act to govern the Internet in a more coordinated fashion but, unfortunately, issues of sovereignty are a hindrance in this endeavour.[192]

## The digital divide

The digital divide refers to the global gap between the ICT 'haves and have-nots'.[193] According to the United Nations Conference on Trade and Development (UNCTAD), a person in a high-income country is 22 times more likely to have Internet access than a person in a low-income country.[194] Secure Internet servers are 100 times more likely in rich countries.[195] In 2007, less than 5 per cent of people living in low-income countries had access to broadband networks.[196] Moreover, Internet access in developed, rich countries is much faster and relatively much cheaper than the Internet available to people in poorer countries.[197] As ICT develops and evolves, advances in the technologies tend to be to the benefit of wealthier populations, a trend that has been dubbed the '80/20 factor' (80 per cent of ICT profit is made from serving the richest 20 per cent of the population).[198] Even when the poor do get access to ICT, it is generally a modified version of what was originally designed for the affluent, thus many of the poor's ICT needs end up being ignored or neglected.[199]Although it was once hoped that ICT would help leapfrog developing countries to better technological parity with developed countries, the opposite has been the case. Even as the developing world increases the number of computers per capita and other indicators of ICT progress, the differences in quality and relative price and what is available in the developed world far surpass the developing world's progress. Today, the digital divide acts as 'a potential threat to political stability and economic progress' in countries such as China and India.[200] That is not to say that ICT has not brought any benefits to developing countries; in fact, between 1993 and 2001, spending on IT grew twice as fast in developing nations as in developed ones, and such expenditures have improved the productivity of developing economies – China's in particular.[201] To the extent that digital and ICT technologies have infiltrated the developing world, they have had a positive economic effect. Nonetheless, a gap in the quality and availability of such technologies still exists between the developing and developed worlds.

There are a number of international initiatives under way to shrink the digital divide, including providing Internet kiosks to rural communities, recycling computers, designing affordable computers and providing economic incentives for Internet service providers to operate in poor, rural areas. In spite of these efforts, the digital divide remains a major geostrategic challenge. The United Nations hopes that international efforts to widen access to cell phones, the Internet and other ICT will help to eradicate poverty, and WSIS has increased its calls for donations in order to facilitate reaching this objective.[202]

Advances in mobile technology are also promising on this front. Mobile telecommunications offer the opportunity to avoid the cost and hassle of having to purchase a computer and to pay an Internet service provider for Internet access, a prospect that can be especially costly if Internet services are bundled with telephone landlines and television access.[203] Many people in the developing world already have mobile phones, so now it is just a matter of upgrading that device to a device that is capable of web browsing and adding the Internet plan to that account.[204] In other words, mobile technology is showing promise in areas where other ICT developments have fallen short, specifically in helping developing countries to leapfrog stages of technological advancement in order to be fully connected to the world and the Internet. Admittedly, mobile phones are not a perfect solution to bridging the digital divide, and the challenge of getting Internet-capable mobile phones into the hands of the world's poorer populations should not be understated. Nonetheless, mobile telecommunications do have more potential than earlier forms of ICT to lessen the divide as opposed to expanding it.

## Blogs

For all they offer in terms of freedom of speech and freedom of expression, blogs present a number of geopolitical and geostrategic challenges. They provide easier methods for terrorists to communicate and to develop transnational networks, for example. Furthermore, they can undermine overall security by allowing for the promotion of criminal, violent, racist or dangerous ideas and philosophies.[205] Blogs enable extremist groups to spread their messages more effectively and, importantly, they can make it hard to trace the source of the information or to locate the person spreading such information.[206] Blogs run by military personnel in operations may spread sensitive information.[207] Extensive blogrolls and linking between blogs reinforces this tendency, as it makes it easier to cross-reference and access information related to topics of interest. Case studies on the significance of blogs to geostrategy are touched on in the 'social implications' section of this chapter, but it is nonetheless worthwhile to reiterate their unique importance to global security.

Additionally, there are some blog-specific policy recommendations worth taking into account. For example, it is imperative that anonymous bloggers be discouraged because if bloggers are not willing to take responsibility for their thoughts and postings, the chances of immoral postings or the posting of illegal information will increase.[208] Similarly, it is important to introduce criminal prosecution and liability laws for use against bloggers that incite hatred, violence, criminality, terrorism or other forms of insecurity as well as for bloggers who make unsubstantiated personal allegations or use character assassination.[209] A web of anonymous writers, potentially with connections to terrorists or organized crime, will inevitably undermine global security.

From a geostrategic perspective, the key point is that bloggers need to be held accountable for their writing and actions, and governments need to move to implement the structures that would make such accountability possible. As blogs gain prominence and increasingly play a role in social revolution and political dissidence, it will also be necessary to introduce more quality controls by peers and ethical guidelines for responsible blogging, along with legal remedies in case of unjustified prosecution by authorities or slander and libel.[210] Special priority and preference should be given to blogs that promote peace and transcultural harmony.[211] As blogs become more advanced, technically sophisticated and mainstream, more harmonization and collaboration with the traditional media is also desirable.[212]

## An overview of key trends and developments in ICT

From the Internet to social media to faster processors, ICT has revolutionized our daily lives, the way we access and share information, contemporary business models and even how governments respond to and interact with their populations. Social media and Web 2.0 have facilitated protests against governments from Iran to Moldova, and they have also increased the frequency and quality of dialogues between diverse populations. Faster Internet access through broadband technology and the increasing prevalence of the mobile Internet have also encouraged this trend. Unfortunately, many of the benefits of ICT disproportionately favour the developed world, and poorer countries remain somewhat marginalized from the Internet revolution and other ICT-related developments. Although some initiatives are under way to ameliorate this imbalance, much work still needs to be done. In the end, the hope is that the beneficiaries of the ICT revolution will be developing and developed country populations and industry alike.

Developments in ICT also mean new geopolitical challenges and dangers. Hackers, criminals and even governments are using ICT for malicious reasons, including theft of financial information, denial of service attacks against political enemies and, in some instances, widespread disruptions of Internet servers and Internet access. Even though cyber terrorism is a growing threat to geopolitics, we as a global community are almost universally unprepared to address it. Similar inadequacies exist with regard to ICT regulatory frameworks, and fixing these shortcomings should be a top priority for global policymakers and private sector leaders. Developing appropriate responses to these governance challenges is made even more difficult by the fact that new developments and innovations in ICT happen so quickly that it is difficult, if not impossible, for the slow and often unwieldy policy-making process to keep pace. Multilateral frameworks and an emphasis on global standards and regulation of ICT innovations and applications should be a priority.

# 3
# Energy and Climate Change

## General overview

Climate change and energy security are fundamentally important issues in international politics. The two problems are distinct, but they are still closely intertwined, especially when it comes to their potential technological solutions. This chapter looks at the existing and emerging strategic technologies that are helping policymakers and the global community to better respond to these growing geopolitical challenges.

The mandate for global policymakers is two-fold. On the one hand, they need to act urgently to stabilize and eventually begin to reduce greenhouse gas (GHG) concentrations in the Earth's atmosphere. Working collaboratively, governments must implement policies and incentive-structures that minimize anthropogenic carbon dioxide ($CO_2$) emissions without disrupting economic growth and development. Meanwhile, as global energy demand skyrockets and questions arise about the remaining stocks of fossil fuels, policymakers who are looking to improve their countries' energy security will need to invest in the development of sustainable and renewable energy technologies. In many instances, the strategic technologies that will help the world offset the effects of climate change are the same types of technology that will help individual countries improve their overall energy security and reduce their dependence on fossil fuels. For this reason, it is instructive to look at these two challenges and their potential technological solutions under the same umbrella.

## Climate change: the nature of the threat

Climate change, defined by the Intergovernmental Panel on Climate Change (IPCC) as 'any change in climate over time, whether due to natural variability or as a result of human activity',[1] is a growing threat to global peace and security. Although some climate change can be attributed to natural forces, credible scientific evidence has demonstrated that human activities are accelerating and even altering the Earth's natural cycles. These human-induced changes are already having significant repercussions for humanity.

Tomas Ries of the Swedish Institute of International Affairs has identified three major challenges of climate change. The first is erosion, which alters the foundations of the human habitat and, by extension, the basis of society. Second, Ries maintains that climate change is increasing the rate of 'ecological shocks'. These shocks range from more violent storms to altered weather patterns and ocean currents to higher rates of flooding in coastal areas. Finally, climate change is forcing ecological tipping points, or 'ecological trends that go beyond the point of no return and cause irreversible and critical shifts in the biosphere'.[2] Although Ries' list of the costs of climate change is by no means exhaustive, it is a neat summary of the most salient issues that face global policymakers.

Human activity has always had an impact on the environment, but the problem of anthropogenic climate change really began accelerating in the nineteenth century with the emergence of the Industrial Revolution and the widespread use of coal for fuel. Since the mid-1800s, atmospheric concentrations of $CO_2$, the chief heat-trapping GHG, have risen by more than 35 per cent.[3] In May 2010, $CO_2$ levels in the atmosphere were at 392.94 parts per million (ppm) and continue to rise at a rate of about 2 ppm per year.[4] Many scientists argue that if the dangers associated with climate change are to be successfully mitigated, then GHG emission levels must be stabilized somewhere between 450 and 550 ppm of $CO_2$ equivalent. To achieve this stabilization, the Stern Review argues that current emission levels need to be lowered by at least 25 per cent by 2050.[5]

Unfortunately, current trends point to a steady increase in emissions, not a stabilization or a reduction. The IPCC calculates that if current emission patterns are not altered, global temperatures will rise by 4°C,[6] resulting in harmful levels of precipitation, increased melting of glaciers, dramatically altered global drought patterns and other potentially unforeseen consequences.

Although global climate change is a slow process, it is already influencing geopolitics and geostrategy. The effects of global warming – rising sea levels, extreme drought or coastal flooding – all pose potential security risks. According to the United Nations International Strategy for Disaster Reduction (UNISDR), the past 30 years have seen a threefold increase in the number of devastating storms, floods and droughts worldwide. In 2006 alone, the United Nations estimates that 134 million people suffered from natural hazards that caused over USD 35 billion in damage.[7]

More menacingly, the IPCC predicts that, in the future, the world will face massive food and water shortages as a direct result of climate change. Even a small rise in temperatures above existing norms could significantly disrupt global agricultural production and increase the developing world's vulnerabilities to malnutrition.[8] Changing weather patterns can also have serious impacts on water resources. Already, many countries in the Middle East and North Africa experience water scarcity, and in the future, many more countries are likely to grapple with these problems. Such natural disasters will disproportionately affect already vulnerable populations of women, children, the elderly and the disabled.[9] Speaking in terms of geostrategy, major catastrophic weather events caused by global warming will have an especially strong impact on human and transnational security.[10] For example, lack of access to or disputes over vital resources such as clean water and food will exacerbate or even cause regional conflicts. In the words of one climate group, 'When climates change significantly or environmental conditions deteriorate to the point that necessary resources are not available, societies can become stressed, sometimes to the point of collapse'.[11] As the effects of climate change become more pronounced, many societies will experience a growing degree of tension, with the possibility of escalation to violence.

Even in places where water is abundant, climate change may increase the frequency of heatwaves, potentially compromising the supply of fresh water and increasing the risk of water-borne diseases. Furthermore, the World Health Organization (WHO) cautions that 'changes in climate are likely to lengthen the transmission seasons of important vector-borne diseases, and to alter their geographic range, potentially bringing them to regions which lack either population immunity or a strong public health infrastructure'.[12]

The WHO also raises concerns that rising sea levels will increase the risk of coastal flooding, forcing large-scale population displacement and the accompanying health-related challenges. Rising sea levels could potentially affect the half of the world's population that lives within 60 km of the sea.[13]

It is likely that the developing world will suffer the biggest burden from climate-induced migration, as it is there that 'even a relatively small climatic shift can trigger or exacerbate food shortages, water scarcity, destructive weather events, the spread of disease, human migra-tion, and natural resource competition'.[14] As living conditions change, people are likely to want to move to regions with more favourable envi-ronments and better access to resources. Large-scale migration would present a significant challenge. Moreover, it would reinforce the pros-perity gap within and between many countries.[15] Although some have dismissed the idea of widespread migration caused entirely by climate change (arguing that many other factors contribute to families mak-ing such decisions),[16] the implications of migration or health challenges that are even indirectly caused by climate change are still significant enough to demand attention.

Overall, the security threats posed by climate change have the poten-tial to bring out the best and the worst in humanity. Nobel Prize winner and former US Vice President Al Gore has spoken of the opportunities for cooperation that climate change affords. 'By facing and removing the danger of the climate crisis', he says, 'we have the opportunity to gain the moral authority and vision to vastly increase our own capacity to solve other crises that have been too long ignored'.[17] If global political and business elites recognize climate change as a threat to humankind, they could mobilize to adopt a range of innovative and dynamic envi-ronmental policies at the national and global levels. On the other hand, if they ignore the imminent danger, the German Advisory Council on Global Change warns that:

> Climate change will draw ever-deeper lines of division and conflict in international relations, triggering numerous conflicts between and within countries over the distribution of resources, especially water and land, over the management of migration, or over com-pensation payments between the countries mainly responsible for climate change and those countries most affected by its destructive effects.[18]

Both optimists and alarmists can at least agree that climate change is a multifaceted problem that will require extraordinary and extensive efforts to mitigate and control.

While it is not the only part of the solution, technology will nonethe-less play a fundamental role in helping to reduce global GHG emissions, promoting better fuel efficiency, and possibly eventually contributing to global cooling. Because this chapter is particularly focused on the connection between energy efficiency and climate change, Section C takes a broad look at the challenges facing the global energy industry before examining the strategic technologies that can help tackle both problems.

## The global energy industry: challenges and trends

Global energy demand is growing at a rapid rate. Although the financial and economic crisis has resulted in a drop in demand for the first time since 1981, the International Energy Authority (IEA) predicts a 40 per cent increase in world primary energy demand by 2030.[19]

Of the total global energy mix, fossil fuels will continue to be the world's dominant source of primary energy. Despite a decrease in demand in 2008 and 2009, fossil fuels are expected to cover 77 per cent of the projected overall increase in world demand between 2005 and 2030.[20]

In terms of energy sources, oil will remain the most widely consumed fuel. Coal is expected to see an increase of 53 per cent in global usage between 2007 and 2030,[21] while the respective shares of natural gas and electricity will increase by 42 per cent and 76 per cent, respectively. Moreover, renewable-based electricity generation will increase from 18 to 22 per cent, due to both higher fossil fuel prices and concerns over energy security.[22]

Increasingly, emerging economies are driving the demand for fuels such as gas, oil, coal and uranium.[23] In particular, these economies dominate world demand for coal and natural gas. China and India account for about 53 per cent of annual global coal use.[24] According to the IEA, China is likely to overtake the United States in terms of oil and gas imports by 2025.[25] While these economies are some of the world's fastest growing, they are also the least efficient in terms of their energy use, thus reinforcing their need for extensive new energy resources.[26]

Overall trends in the energy industry indicate that the global centre of the industry's power is undergoing a major shift. From Russia's Gazprom and Rosneft to Brazil's Petrobras to China's National Petroleum Company, the emerging economies' energy companies are equal in scale – and increasingly in influence – to those of the big western oil and gas businesses.[27] Today's world is closely interconnected, and the global energy markets are no exception, a point that was made clear in the aftermath of the 2005 Hurricane Katrina, when energy disruptions in the Gulf of Mexico were felt as far away as the Middle East, Asia and Africa.[28] Taking into account this globalized and interlinked reality, policymakers must seek global and collaborative solutions to combat the shortage of sustainable and secure energy supplies.

The dual challenges of rapid growth in global energy demand and climate change make it highly likely that the next technology boom will be centred on alternative energy.[29] Major corporations are already moving in this direction. For example, Google has initiated a project to make renewable energy cheaper than coal, and General Electric has indicated that it wants to put a new focus on renewable energy solutions such as wind turbines and solar energy.[30]

Today, the European Union (EU) along with South Korea, Japan and China seem to be the frontrunners in the field of renewable energy development. The EU is striving to actively reduce the GHG emissions of its major economies, and the EU leadership is aiming 'to cut greenhouse gas emissions by one-fifth from 1990 levels, use 20 per cent of renewable energy sources in power production and 10 per cent of biofuels made from plants in transport, all by 2020'.[31] In addition to these initiatives, EU member states are already global leaders in carbon reduction practices and policies,[32] and at this moment, there are two EU directives in the field of renewable energy in force – one for biofuels and one for electricity.[33]

## The intersection of climate and energy: relevant strategic technologies

### Renewable energy technologies

There are two main groups of strategic technologies that address the challenges of climate change and growing global energy demand. The first group is technologies that harness and exploit renewable, clean energy resources such as plants, geothermal heat and the Sun. Developing cost-effective, reliable technologies to use these resources is a central element of sustainable global energy security. If adopted on a large scale, these technologies have the potential to greatly reduce and even mitigate global GHG levels.

Renewable energy technology has many benefits. It is cutting edge but also often easily transferred for manufacture within each country. Provided that an adequate market is established, it has the potential to reduce countries' dependence on foreign sources of energy. Additionally, local components and services for renewables create inward investment. Significantly more of a country's investment in renewable energy is spent domestically compared to investments in other energy sources. Renewable energy technology is also often labour-intensive, thus creating several times the amount of employment of equivalent fossil fuel plants. The cost of renewables has already declined and is poised to continue falling. In some markets renewables are competitive with new conventional coal, nuclear and large hydro-generation schemes – even before the external benefits related to health and the environment are taken into consideration.[34] Nonetheless, renewable energy sources face several significant obstacles, including the high cost of storage. Figuring out a way to use renewable resources to produce electricity at a moment's notice, which natural gas and oil already do, will be a key step in making these technologies more mainstream.[35]

Many policymakers see positive reasons to further develop renewable energy sources. For one thing, replacing fossil fuel use with proven renewable energy technology and resources can potentially achieve the deep cuts (60–80 per cent) in GHG emissions that scientists say are required to address climate change. Renewable resources are often abundant within countries and could supply major portions of the energy demand if harnessed. Plus, renewable resources and their harvesting are generally more environmentally and socially benign than fossil fuels. This chapter details some of the most promising renewable energy technologies.

### Biomass, biofuels and biorefineries

Biomass is organic matter that can be used to make biofuels and other forms of renewable energy. It can come from anything from agricultural crops to wood chippings to methane derived from animal excrement.[36] Currently, biomass provides approximately 19 per cent of the world's primary energy supply and generates 1.3 per cent of global electricity.[37]

Certain types of biomass can be converted directly into liquid fuels. These so-called biofuels, such as ethanol and bio-diesel, are renewable fuels that, at least in principle, help reduce global GHG emissions.[38] Alcohol-based biofuels include ethanol, which is produced from corn; P-series fuel, which is made from ethanol, natural gas liquids and ether created from biomass waste; and methanol, which is wood alcohol made from natural gas.[39]

Biomass is processed into biofuels and other viable energy resources at facilities known as bio-refineries. These bio-refineries are essentially to biomass what oil refineries are to petrol. They bring together a variety of technologies in order to integrate biomass conversion processes with the equipment needed to produce fuels, power and chemicals. New bio-refineries are a key part of bringing biofuels into mainstream use and to ensuring that domestic biomass resources can be processed domestically.[40]

The influence of bio-technologies like those available at bio-refineries on alternative energy sources could be significant. It is estimated that over 'one billion tons of biomass could be available in the U.S. to produce biofuels and bioproduct, enough to meet 30 per cent of U.S. demand for transportation fuels and 25 per cent of demand for chemicals'.[41]

Although biofuels seem to have numerous advantages, the sustainability of biofuels has been questioned. Biofuels do burn cleaner than fossil fuels, but questions have been raised about the emissions released during their production and whether the net carbon emissions of the biofuel process are actually more detrimental to the Earth's atmosphere than those of traditional fuels. New developments to address this issue include research into a new generation of biofuel technologies that use algae.[42] According to New Energy Finance, this process is 'a highly efficient converter of sunlight into energy, making it 30 times as productive as current oil seed crops per acre'.[43]

### Solar energy

The Sun is one of the Earth's most abundant renewable resources, and there are several technologies that are being developed in order to harness the Sun's energy. These technologies include photovoltaics (semiconductors that convert sunlight directly into energy), solar heating (solar collectors that absorb the Sun's heat to use for water or space heating) and solar power (the use of reflective materials to concentrate the Sun's heat energy and to use the energy to produce electricity).[44]

Solar thermal technologies are already in use. For example, in 2005, 'world solar hot water/heating capacity totalled 88 GW (63 per cent in China, 13 per cent in Europe, 26 per cent in the rest of the world). Another 13 GW of capacity was added in 2006, 77 per cent of which was in China'.[45] That said, the barriers to solar thermal technologies are numerous, including the need for solar technologies to become cost-competitive with low cost fuels the prices of which do not reflect environmental externality costs.[46]

Another challenge is that solar energy can be difficult to store, although developments in thermal energy storage technologies are slowly being introduced as a way to provide hot water and space heating and cooling. Used in combination with often-intermittent solar energy supplies, energy storage allows the electricity or heat generated to be consumed during peak demand periods.

## Geothermal energy

Geothermal energies take advantage of the hot, hydrothermal fluids below the Earth's surface to produce clean, reliable and home-grown energy.[47] Enhanced geothermal systems extract the natural heat contained in high temperature, impermeable rocks in the Earth's crust.[48] Yet, only a few places in the world are close enough to the hot springs and subterranean magma required to produce geothermal energy. In Iceland, El Salvador, the Philippines, Costa Rica, Kenya and Nicaragua, for instance, geothermal energy accounts for an important share of total energy demand.[49] In these locations, geothermal power can be used to heat buildings, melt snow on roads and pavements and produce electricity.[50] Despite its multiple potential applications, less than 1 per cent of the world's total energy supply currently comes from geothermal sources.[51]

## Wind energy

Wind energy is a type of solar energy that takes advantage of the winds produced from the Sun's uneven heating of the Earth.[52] With the use of wind turbines, the Earth's kinetic energy is converted into mechanical or electrical energy. Wind energy is clean, has no emissions, can be produced domestically and is among the lowest priced of all renewable energy resources and technologies available. On the down side, wind power requires higher initial investment than fossil-fuelled generators, and wind gusts are intermittent – a significant obstacle for a resource whose energy cannot be stored and used later.[53]

In the field of wind technology, US and Chinese companies plan to manufacture wind turbines that use the magnetic levitation (maglev) technology previously envisaged for futuristic monorail systems.[54] Maglev technology would reduce friction, allowing turbines to operate at lower wind speeds and lower cost.[55]

Another emerging technology that could potentially increase the efficacy and adaptation of wind energy is floating wind turbines. In general, wind blows much faster far out at sea than it does near the coast, meaning that turbines in the open ocean will produce more energy than those closer to shore. However, building power plants in such places is a challenge because the water that far out to sea is often too deep for a traditional turbine's tower to be attached to the ground of the sea. Thus, the energy produced cannot be transported back to land.

Some energy companies are seeking to overcome this problem by placing turbines on floating platforms that are tethered to the seabed by cables preventing the turbines from floating away.[56]

The potential of these turbines is enormous. Some even estimate that offshore wind could provide power to all Europe. However, numerous obstacles remain. Chief among them is the fact that connecting offshore wind turbines to the electric grid will be costly and time-consuming. Additionally, maintenance of the turbines would only be possible in good weather. Despite these challenges, offshore wind turbines are one of the most exciting developments in the wind energy field.[57]

## Hydropower

Hydropower has been around in one form or another for thousands of years. It works by taking the natural flow of water and converting its energy into electricity for use in homes and offices. As water falls, it moves the blades of a water turbine, thereby producing energy. The amount of electricity produced depends on the height from which the water falls and the speed at which it is flowing.[58] In recent years, the use of hydropower has significantly increased in absolute terms, especially in China, India and Vietnam. Projections estimate that hydropower generation will increase to 3730 TWh (terawatt-hour) by 2015 and 4810 TWh by 2030.[59] This technology has the added advantage that its energy can be stored[60] and that hydro-resources are widely spread around the world.[61] Moreover, it reduces GHG emissions and its main input, water, is not subject to the price fluctuations seen in fossil fuel markets.[62]

Hydropower's drawbacks include the fact that construction of turbines and dams can damage local ecosystems and cause floods.[63] Certain large-scale projects have even forced the displacement of local populations and raised concerns over increased vulnerabilities to natural disasters such as earthquakes.

## Nuclear energy

Nuclear power is a clean, reliable fuel source that provides nearly one-fifth of the world's electricity.[64] The energy comes from splitting uranium atoms, a process that generates energy, which can be used to make steam.[65] While nuclear power is an emissions-free technology, it does raise a number of serious concerns, including environmental issues over nuclear waste storage and disposal, and questions over nuclear power's risks to health and human security.[66] The fact that the enriched uranium used to generate nuclear power could also be used more maliciously in the production of nuclear weapons is another central concern surrounding this strategic technology.[67]

Nonetheless, nuclear energy offers many benefits, and given that it does not result in any GHG emissions, it must be seriously considered, particularly in the context of the looming energy crisis and climate change. Nuclear fuels can help countries reduce their dependence on foreign fuels, and particularly on regions rich in fossil fuels.[68] In electricity production, nuclear power competes well with more traditional fuel sources such as coal, natural gas, hydropower and solar power, but nuclear's high capital costs and the political and economic hurdles to building new reactors are significant impediments to more widespread adaptation.[69] Moreover, unless caps or taxes are imposed on companies' GHG emissions, coal will continue to be more cost-efficient than nuclear power for the foreseeable future.[70]

One potential method for improving the safety and cost of nuclear power is a travelling-wave reactor, a prototype reactor that requires very little enriched uranium, instead of gradually converting non-fissile material into the fuel it needs to produce energy.[71] In fact, such a reactor could potentially run for decades without refuelling. The travelling-wave reactor envisaged by private sector project leader John Gilleland would run on what is essentially considered waste. Importantly, this would reduce weapons proliferation concerns in addition to being simpler and more cost-effective than traditional nuclear reactors.[72]

*Hydrogen*

Hydrogen is the simplest and most abundant element on Earth, and it has the potential to be an excellent energy carrier. It produces zero emissions when burned in an engine and its only by-product is water.[73] Unlike electricity, the power generated from hydrogen can be easily stored and used later. Although hydrogen is not a renewable energy resource (in fact, hydrogen is often produced using fossil fuels), it is included in this section because once created, it is essentially an emissions-free energy carrier, and, in that sense, it has the potential to help offset climate change.

For all its advantages, hydrogen has one major drawback: its cost. Because of its simple atomic structure, hydrogen does not exist independently in nature. Instead, it binds with other elements, and in order to produce hydrogen atoms for energy, the atoms must first be separated from natural gas,[74] water and biomass, among other things.[75] Hydrogen can be obtained in a number of ways including from fossil fuels, electricity, and renewable and nuclear energy. Yet, these processes for hydrogen isolation can be extremely costly, which offsets some of hydrogen's perceived benefits.

Hydrogen's growth has been rather slow. Its use in generation grew by only by 2 per cent from 1970 to 2000.[76] Major obstacles to future

development of hydrogen-related power include the extensive financial requirements, numerous social and environmental challenges, and
water- and food-related conflicts.[77]

If these challenges could be overcome, then hydrogen would be one
of the world's best solutions to the growing demand for energy. Clean
energy from renewable and nuclear energy sources could be used in
direct combustion to power motor vehicles.[78] In addition, hydrogen
could provide electricity for several applications in the industrial, transport and residential sectors.[79]

According to the IEA, the level of investment required to supply
hydrogen to the world's transport sector alone would be 'in the range
of several hundred billion dollars over several decades' or 'USD 0.1–1.0
trillion for pipelines and USD 0.2–0.7 trillion for refuelling stations'. It
regards this level of investment as achievable in the long term, but suggests that 'building infrastructure now would be premature because key
$H_2$ technologies are still under development'.[80]

### Tide and wave power

The transformation of ocean energy currently provides a minor share
of global energy production. In 2008, it accounted for less than 1 TWh
worldwide. Yet, while technology is still at quite an early stage, various
governments, including France, Canada, China, Russia, Norway and
the UK, have expressed strong interest in developing it further. The
IEA's Reference Scenario estimates that global tidal and wave power will
provide 14 TWh in 2030.[81]

## Technologies to improve the efficiency of non-renewable resources

The second type of strategic technologies important to the energy
industry and to the global climate change issue is the technologies that
make traditional forms of energy, such as fossil fuels, less polluting and
more efficient. As scientists and researchers work to make renewable
energy technology mainstream, this second type of technology acts as
an important interim step for helping to reduce GHG emissions without
disrupting economic productivity.

Many of these second types of technology focus on reducing emissions from coal because, 'If there was a most-wanted list for climate
change culprits, coal-fired power stations would be number one'.[82]

### Wet scrubbers and coal

Wet flue gas desulphurization (FGD) technology is one of the world's
most widely used methods for reducing the sulphur dioxide emissions

of coal.[83] The technology works by passing coal's flue gas emissions through a spray tower or absorber where the gas is exposed to water slurry made up of about 10 per cent limestone or lime sorbent.[84] During this process, the sulphur in flue gas reacts with the lime, creating calcium sulphate and calcium sulphite deposits.[85] Not only does this process have the potential to reduce the sulphur dioxide emissions of coal by 80–95 per cent,[86] but it can also produce gypsum, a material that can be used commercially to make products such as dry wall and cement.[87]

Wet scrubber technology, as FGD is commonly called, is commercially established in developed countries, and it has great potential to be widely used in the developing world, especially since it is easier to construct wet scrubbers from scratch than to retrofit the technology to existing power plants in the developed world. However, many power plant owners in poorer countries are reluctant to adopt this technology because although wet scrubbers are one of the best options for improving the emissions quality of an existing power plant, the installation process is expensive and would have a notably negative effect on the company's bottom line or force the firm to raise prices, thereby negatively affecting consumers. An existing power plant would have to close from three to six weeks to construct and install the scrubber.[88] Once installed, the scrubbers would demand 1–2 per cent of the factory's energy output to run them. Since wet scrubbers do not produce any revenue, firms and power plants have little incentive to install and use them. Air quality is a public good and most firms prefer to be free riders rather than proactive environmental warriors.

### Coal gasification

The conversion of coal to gas is another potential technology that may reduce the negative environmental implications of coal while also transforming coal into a more versatile and efficient fuel. During the gasification process, a gasifier applies heat and pressure to coal under the presence of steam. The coal is broken apart by the gasifier's heat and pressure, ultimately producing a hydrogen-rich gas known as syngas.[89] Minerals from the coal that are not transformed into gas settle at the bottom of the gasifier, meaning that only minimal waste matter is emitted into the atmosphere.[90] Chief among its many advantages is the fact that syngas burns as cleanly as natural gas.[91] Furthermore, under the right circumstances, syngas may also provide the hydrocarbon ingredients necessary for gas and diesel fuel. As a second step to gasification, some firms are currently experimenting with the liquefaction of syngas, a process that would make it easier to transport from the fuel factories and mines to cities.[92]

Although gasification has a number of positive environmental benefits, its major drawback is that it only allows factories to preserve about 55 per cent of coal's original energy potential.[93] This may make it a feasible technology for large firms or government-sponsored corporations, but for private power plants and smaller businesses current gasification technology is an unmanageable financial burden.

### Integrated Gasification Combined Cycle (IGCC) technology

IGCC technology takes coal gasification to the next level by combining it with a carbon capture and storage (CCS) plan. It is potentially a major technological breakthrough because it burns the more efficient gasified coal and then takes those already reduced emissions, sifts out $CO_2$ components and deposits these carbon wastes in underground reservoirs and aquifers. The potential of IGCC is promising, and the benefits are notable. Emissions from IGCC technology are lower than those of pulverized coal, and the process consumes less water than traditional burning methods and generates less solid waste.[94]

To describe the process in more technical terms, when syngas burns, it is not emissions free, but its gas streams are much more concentrated than those of traditional coal. Thus, as gas is emitted from a syngas burning factory or plant, it can be passed through turbines, which are able to separate soot, $CO_2$ and other pollutants from the emissions.

Unfortunately, IGCC plants cost 10–20 per cent more than pulverized coal-fired plants,[95] and power companies and heavy industry around the world still lack the financial or political motivation to implement this technology. Furthermore, highly integrated IGCC plants have long start-up times and require significant capital investment.[96] That said, 'the emissions of particulates, $NO_x$ and $SO_2$ from IGCC units is expected to meet, and possibly to better, all current standards',[97] and therefore IGCC technology is presently one of the hottest areas of clean coal technology research and investment.

### Carbon capture and storage

Toxins from processes like IGCC can be transported and sequestered in underground saline aquifers and aging oil reservoirs, a process known as carbon capture and storage. Should this technology become mainstream, there would be enough capacity in old oil fields around the world to handle carbon waste storage needs for hundreds of years. Obviously, this technology offers enormous potential to reduce emissions from carbon consumption (the IEA estimates that it could reduce emissions from coal-fired power plants by up to 85 per cent),[98] but the technology is still extremely new and its benefits do not come without cost. So far, carbon capture and storage is not being employed on a large enough scale to make it cost-efficient or practical.[99]

*Hybrid vehicles*

Hybrid vehicles, powered by a combination of traditional internal combustion and electricity, are an excellent starting point for reducing vehicle emissions while technologies like hydrogen fuel cells are in development.[100] Additionally, they can be considered a stepping-stone to completely electric cars. With both an internal combustion engine and an electric one, hybrid cars automatically select the most efficient fuel source, depending on driving conditions.[101] Although they can be 10–20 per cent more expensive than traditional vehicles,[102] they have increased fuel efficiency and they release GHG emissions at a reduced rate.[103] Today, the majority of car producers are working on hybrid or fully electric cars, although their market share is still under 3 per cent and the evolution of consumer demand is hard to predict.[104]

*Catalytic converters*

Although hardly new technology, catalytic converters play an important part in reducing toxic vehicle emissions. By taking harmful toxins from vehicle exhaust and transforming them into less hazardous gases, catalytic converters are capable of improving air quality and minimizing the harm incurred by petroleum-fuelled vehicles.[105] While catalytic converters are standard on cars used in the United States, developing – but highly polluting – countries like China and India have yet to adopt them on a widespread level. If this existing, affordable technology could be employed on a wider scale, it could have dramatic effects on air quality.[106]

*Sensor tyres*

A new type of tyre that is equipped with built-in sensors can help improve fuel efficiency. The so-called cyber tyre measures acceleration and deceleration and transmits this information to the car's braking and control systems.[107] Such monitoring helps to reduce fuel consumption by optimizing suspension and braking. The sensor tyres could be combined with low-rolling resistance tyres, which reduce fuel consumption by reducing the resistance between the tyre and the road. A tyre that helps improve fuel efficiency is an important emerging strategic technology that will help reduce GHG emissions.

## Governance and regulation

As the effects of climate change become more pronounced and the world experiences more frequent global energy price fluctuations, international governance and regulation of emissions and energy are becoming increasingly important. According to Ries, 'the key issue is to reduce mankind's ecological footprint'.[108] There are two ways to do this. 'The first is inward looking: to modify our behaviour, transcending current patterns of consumption and pollution. This is ultimately a spiritual approach requiring a change in attitude [...] the second is outward-looking: to try and modify our environment [...] They are not mutually exclusive'.[109]

Speaking of climate change, the UN Secretary-General Ban Ki-moon recently noted that 'business as usual is no longer an option'.[110] Policymakers must accurately analyse and carefully consider the potential geopolitical and societal implications of climate change. This analysis should lead to effective and innovative policymaking based on increased cooperation in the fields of energy- and environment-related technology research. Even if there is disagreement regarding the extent of future changes, it is important to assess how climate change might eventually affect global political and security structures.

### The United Nations and climate change initiatives

Concern over various climate-related issues has been on the global policymaking agenda for many decades. Among the earliest large-scale initiatives was the 1972 UN Conference on Human Environment (the First Earth Summit).[111] The Governing Council of the United Nations Environment Programme (UNEP) was established there, designed to be the 'voice for the environment within the United Nations'. UNEP presents studies and reports, coordinates international environmental regulatory efforts, educates, and encourages cross-sector partnerships on behalf of the environment.[112]

In 1989, UNEP and the World Meteorological Organization established the International Panel on Climate Change to examine the greenhouse effect and the pace and scope of global climate change.[113] In 1992, the Second Earth Summit was held in Rio de Janeiro, Brazil. Its Agenda 21 and the Rio Declaration set a new framework for the development of international agreements to protect the integrity of the global environment. At the same time, the United Nations Framework Convention on Climate Change (UNFCCC) was opened for signature.[114] The UNFCCC aims to stabilize atmospheric concentrations of GHGs at a level that does not damage the climate system.

One of the most successful international environmental regulatory efforts was the 1987 Montreal Protocol, which was created to phase out ozone layer depleting gases. Over the past two decades, the Montreal Protocol has led to significant reductions in chlorofluorocarbons (CFCs), chemicals that were once common in products such as hairsprays and that are linked to climate change.[115] Without the successful implementation of the Montreal Protocol, it is estimated that CFC levels would be five times higher than they are today.[116]

The Montreal Protocol was followed by the Kyoto Protocol, which formally entered into force on 16 February 2005. Another major step in the fight against global climate change, the Kyoto Protocol aims to cut overall emissions of the major GHGs (including $CO_2$ and methane) to at least 5 per cent below 1990 levels in the commitment period between 2008 and 2012.[117] Although the Kyoto Protocol was a major achievement, its success was undermined by the failure of the United States – the world's largest emitter of GHGs – to ratify it.

The first commitment period of the Kyoto Protocol will end in 2012 and a new legally binding treaty is urgently needed. The 2009 UN Climate Change Conference in Copenhagen in 2009 did not succeed in negotiating such an agreement. Attended by the largest number of world leaders ever present at a summit, including US President Barack Obama representing a shift in US environmental policy, as well as numerous civil society participants, and with the world looking on, the conference did not result in a 'new deal to save the climate'. Despite desperate efforts, no committing accord was signed at the conference.[118] The resulting 'Copenhagen Accord' was neither legally binding nor supported by all the participating countries. Instead, it acknowledged the fact that global temperature rises should be kept under 2°C, without fixing commitments for $CO_2$ reduction. Furthermore, an annual amount of USD 100 billion annually by 2020 was pledged by developed countries to be used for adaptation and mitigation.[119]

The reasons for failure are multiple and complex. Dvorsky of the Institute for Ethics and Emerging Technologies suggests a summary in five points: nation states are too self-serving and reluctant to take a lead, democracies are ill-equipped to deal with crisis, China adopted an unilateral approach, the powerful corporations do not necessarily think in the long term and there was a weak consensus on the reasons for global warming.[120]

Since the Climate Change Summit, the need to take action has become even more urgent. The next UN Climate Change Conferences attempting to produce a single global framework was planned for Bonn, Germany, and Cancun, Mexico, in 2010, and were thus even more loaded with expectation.

The problem of climate change is global and can only be solved by a global and coordinated response. This is true in terms of both emissions regulation and fostering technological innovation. This response requires clear vision and a long-term approach. It is important to focus on stabilizing GHG concentrations in the atmosphere because, unfortunately, 'a technological breakthrough that would lead to a decisive, near-term reduction in the concentration of carbon dioxide in the atmosphere remains far away'.[121]

## The Stern Review: recommendations

For many, setting a global price for carbon and creating an international carbon market are crucial steps in controlling GHG emissions and encouraging the development of clean, renewable energy resources. *The Stern Review: The Economics of Climate Change* is a study commissioned by the British Government. It proposes four strategies for dealing with global climate change: emissions trading, technological cooperation, action to reduce deforestation, and adaptation.[122] The review proposes that the global emission trading schemes should be linked and expanded as 'a powerful way to promote cost-effective reductions in emissions and to bring forward action in developing countries: strong targets in rich countries could drive flows amounting to tens of billions of dollars each year to support the transition to low-carbon development paths'.[123]

So far, the EU has overseen the world's most established carbon markets: the EU Emissions Trading System (ETS), but many obstacles remain. For example, the EU issued too many permits in its initial allocation, and thus the price of permits has fallen, making them seem like a bad investment to the companies that made the original investments.[124] Additionally, many perceive the ETS as a regulatory burden rather than a profit-making opportunity. Because of this, many smaller firms have indicated that they would prefer to pay fees rather than deal with the administrative burden of actively trading emissions permits.[125]

Another recommendation brought forward by the Stern Review calls for an increase in the degree and effectiveness of global investment in innovation through active support for energy research and development.[126] It also emphasizes the important role of forests in climate change, noting that the problem of deforestation adds more to worldwide emissions each year than the transport industry.[127]

Because developing countries are the most vulnerable to climate change, policymakers must focus on the issue of adaptation. In this context, it is important to gain a better understanding of the impacts of climate change.

### The private sector: the business of climate change

In the future, new climate change-related regulations on businesses will increase the cost of emissions and, consequently, the cost of doing business.[128] Accordingly, if companies 'persist in treating climate change solely as a corporate social responsibility issue, rather than a business problem, [they] will risk the greatest consequences'.[129] Firms are increasingly being asked to evaluate emission costs as well as their 'vulnerability to climate-related effects such as regional shifts in the availability of energy and water' and the reliability of their infrastructures and supply chains.[130] In order for a company to establish its position on the issue of climate change, its leaders must understand 'how the impact of the firm's activities on the climate (in both its physical and its regulatory manifestations) may affect the business environment in which the firm competes'.[131]

The Global Reporting Initiative has developed the most respected framework of reporting principles for business, setting out guidance and standards for disclosures on environmental, social and economic performance.[132] Such disclosures are increasingly important to businesses, and failure to disclose can result in strategic disadvantage. Environmental performance reports serve as a key accountability mechanism.[133]

Some of the leading corporate environmental initiatives include the Chicago Climate Exchange, the Pew Centre's Business Environmental Leadership Council, the Global Roundtable on Climate Change and the World Business Council for Sustainable Development.

To cite a specific industry example of the intersection of business, climate change and regulation, in July 2008 the European Parliament approved a measure to include air traffic in the EU's ETS from 2012. Airlines will be issued with most of their carbon permit allocation free of charge, but will have to buy 15 per cent of their quota at auction: a proportion that may rise.[134] As this development indicates, climate change will become a central part of a business's bottom line decision-making processes.

Indeed, the airline industry is one of the areas most responsible for and affected by global climate change. Some in the industry are responding proactively in the face of potential additional regulations. According to the *Financial Times*, 'airlines are being forced to accelerate efforts to improve the efficiency of their fleets in response to [...] increasing environmental pressures to reduce emissions and noise'. They are also under pressure to demonstrate that 'improvements in technology can at least mitigate some of the sector's future growth'.[135] The director general of the International Air Transport Association (IATA), Giovanni Bisignani, has called for a 'zero emissions' aircraft within 50 years', in response to the 'reputation crisis' faced by the industry.[136] According to IATA, 'the airline industry contributes about 2 per cent of global carbon emissions at present, and [IATA] forecasts that traffic growth will lift this to 3 per cent by 2050'.[137] This is disputed, however, as environment specialists argue that 'greenhouse gases from aeroplane engines have a more damaging effect on the climate when emitted at a high altitude'.[138]

As is noted above, the ICT industry is another highly polluting industry that is responsible for approximately 2 per cent of global carbon emissions, a similar share to that of the aviation sector.[139] Among this industry's biggest regulatory challenges is how to limit the amount of electricity used by computers and data centres, both of which are heavy power consumers. It is estimated that there are about 28 million servers in the world, and this figure could rise to 43 million by 2010 as our demands for information and data continue to grow.[140] In order for these data centres and servers to function, they must be kept cool, a requirement that can demand tremendous amounts of power and electricity for air conditioning. A number of companies have already begun self-regulating on this front, implementing things like server virtualization, which uses software to increase the efficiency of a computer by 'allowing several applications to be run from a single machine'.[141] Power management uses software to switch machines off or put them on standby when they are not in use.[142] Such solutions on their own, however, are not sufficient without a broader focus on renewable energy solutions.[143] One leading example of a company at the cutting edge of this trend is Google, which has voluntarily installed solar panels at its headquarters in Mountain View, California. These panels generate some 30 per cent of the company's energy needs.[144] As ICT continues to be a significant part of daily life in the twenty-first century, such private sector self-regulation will need to be complemented by more comprehensive international rules and frameworks for reducing industry emissions.

**Other regulatory issues and suggestions**

In the long run, unmitigated climate change could have a direct impact on the ability of global policymakers to adapt. It is essential to develop:

> A portfolio or mix of strategies that includes mitigation, adaptation, technological development (to enhance both adaptation and mitigation) and research (on climate science, impacts, adaptation and mitigation). Such portfolios could combine policies with incentive-based approaches, and actions at all levels from the individual citizen through to national governments and international organizations.[145]

Citizen initiatives will be important because they play a key role in raising public awareness of climate change and technological development.

The International Task Force on Global Public Goods has suggested that an International Consultative Group on Clean Energy Research be established, including both developed and developing countries, to 'collaborate and exchange information on research and development of more efficient and cleaner energy technologies'.[146] For his part, Michael A. Levi of the Council on Foreign Relations has downplayed the potential role of the United Nations in the climate change process. Instead, he has suggested that the G20 would be a more effective forum, as its members consist of many of the world's most significant polluters and the G20 works at the head of state level.[147] Interestingly, in the Leaders' Statement released after the April 2009 summit of G20 countries, the G20 members committed to build an 'inclusive, green and sustainable recovery' in response to the 2008 global financial crisis.[148] How this commitment will play out on a more practical level remains to be seen.

## Geopolitical implications of technology for energy and the environment

### The food for fuel debate

Summer 2007 witnessed what has been dubbed the perfect storm for global food production.[149] Mediocre harvests in the United States and Europe coincided with drought in Australia, and the price of grains began to spike. At the same time, places as far apart as the Ukraine and Argentina became concerned over rising prices and began limiting their own food exports in order to protect the domestic food supply.[150] From early 2005 to early 2008, international food prices rose by nearly 80 per cent.[151] Not surprisingly, the world's poor were disproportionately affected by these price rises.[152] For many observers, the increase in global food prices was dramatically worsened by the international biofuels industry, which uses food crops such as corn to produce fuel. According to a study by the World Bank, biofuels pushed global food prices up by as much as 75 per cent.[153]

The problem lies in the fact that the production of biofuels diverts crops and farmland that would otherwise be used for food production to the production of the often more profitable biofuel crops. Although numerous factors drive global food prices, biofuels are a central part of the recent increases and there is an ongoing debate over their relative value. The food for fuel debate positions two of the world's greatest challenges, climate change and global hunger, against each other. Proponents of biofuels argue that most biofuels are currently made from corn, an agricultural product that is more commonly used to feed animals than humans.[154] As such, they argue, the impact of biofuels on global food prices has been dramatically overstated. Critics of biofuels argue that 860 million people are hungry or malnourished worldwide, and that the use of biofuels is essentially a victory of the world's car-driving population over the world's poor and hungry.[155]

Yet, overall, the increasing use of biofuels is just one of several forces driving the increases in global food prices. Policymakers need to approach this debate in a broad manner. Investment in the research and development of non-food source biofuels is an important first step. Governments will also need to evaluate the proportion of land that can be used for harvesting biomass. In this assessment, the focus should be on long-term sustainability and on the preservation of natural landscapes.

### The Kyoto Protocol and developing countries

A second geostrategic challenge involving energy, the environment and strategic technology deals with the Kyoto Protocol and its emissions targets (or lack thereof) for developing countries. Currently, non-industrialized countries have no binding emissions reduction targets under the Kyoto Protocol, the logic being that they have lower per capita emissions than their industrialized counterparts and that imposing emissions caps on countries in the process of developing would be an unfair disadvantage – especially compared to the industrialized nations who have been free to emit unlimited amounts on their path to development.

Thus, one of the key geostrategic challenges is that developing countries, despite having low per capita emissions, are still responsible for 47 per cent of global $CO_2$ emissions.[156] If the global community is going to effectively deal with climate change, these emissions, which are likely to grow dramatically in the future, cannot simply be ignored. The Kyoto Protocol has a few mechanisms in place for offsetting this challenge. 'Clean Development Mechanisms' (CDM) allow developed countries to pay for emissions reductions in developing countries. In return, the developed countries can receive credits against their own emissions.[157] The first commitment period of the Kyoto protocol will end in 2012. Since the 2009 UN Climate Change Conference failed to produce a legally binding agreement, expectations for the 2010 UN Climate Change Conferences in Bonn, Germany and Cancun, Mexico, were high.

### The looming global energy crisis

According to the US Energy Information Agency, the world's total energy consumption is expected to grow by 50 per cent between 2005 and 2030,[158] with countries such as China and India responsible for the most significant portion of this increase. As the world's economies and populations grow over the next half-century, energy scarcity will become an increasingly widespread and persistent problem.[159] Already, the world's total energy demand is far outpacing discovery of new energy resources. To put the problem into perspective, the IEA found that in 1964, the amount of new oil resources discovered worldwide was 48 billion barrels while total global consumption for the year was 12 billion barrels. By 1988, global oil consumption and discovery were roughly equal at 23 million barrels. However, in 2005, the world consumed a total of 30 billion barrels of oil but it discovered only five to six billion barrels of new reserves.[160] From these statistics, it is clear that the question of whether the world will face a global energy shortage is a matter not of 'if' but of 'when'. Although the problem of a global energy shortage is seemingly inevitable, the solution is far from obvious. The energy company Royal Dutch Shell aptly summed up this challenge in the title of its 2008 world energy scenarios report: TANIA, or 'There Are No Ideal Answers'.[161]

To respond to this looming energy crisis, policymakers must strongly encourage the development of clean, renewable energy resources. Tax credits for companies willing to invest in research and development in these technologies is a natural first step and one advocated by Steven Chu, Nobel Laureate and Energy Secretary for President Barack

Obama.[162] Some of Chu's more lofty propositions reflect the urgency around the world's finite energy resources, and he strongly advocates the development of an Apollo-like research programme in which revolutionary energy technology is promoted and pursued without regard to cost.[163]

Although Chu's recommendations are intended for the United States, they could easily be applied to all countries around the world, since, in some way or another, all countries will be affected by energy supply shortages. Speaking from a global perspective, Mohamed Al-Baradei, the head of the International Atomic Energy Agency (IAEA), has strongly pushed for the creation of a comprehensive global energy organization. Indeed, given the fundamental importance of energy to virtually every facet of modern life, it is somewhat remarkable that there is not yet a globally inclusive, authoritative body to coordinate and direct efforts on this front. Al-Baradei argues that a world energy organization would be able to exert government influence in areas where market forces fail, for example in the long-term development of new energy technologies. Additionally, such an organization could coordinate and compile global energy data, provide comparative risk and environmental assessments, accelerate the transfer of energy technologies to poor and developing countries, and help countries oversee energy supplies in the face of crisis or natural disasters.[164]

Governments and the private sector should also have contingency plans prepared for unexpected energy supply disruptions. Having diverse forms of energy coming from multiple sources is thus a central component of a strong national energy security policy.

### Nuclear energy: a debate over values

Nuclear energy is gaining increasing popularity as an efficient energy form that has few waste products, abundant inputs (specifically uranium) and the ability for most countries to produce such energy domestically. It also decreases dependence on traditional fossil fuels, and can be cost-competitive with things like natural gas, coal and solar-powered energy.[165] Unfortunately, nuclear energy generation associated with isotope separation technologies increases the risk of nuclear proliferation.[166] In some ways, the issue of pursuing nuclear energy presents a challenge to global values: how best to balance concerns over increasing energy shortages and global warming with fears over nuclear proliferation and an unthinkable attack by terrorists armed with a nuclear bomb.[167]

Nuclear energy has numerous benefits, but the technology is not without costs. In particular, the security and safeguards that need to be put in place around nuclear facilities can be expensive to build and maintain, and they require well-trained manpower to operate and oversee them.[168] The storage of highly radioactive nuclear waste and the risks of nuclear meltdown, demonstrated in the 1986 Chernobyl disaster, are also daunting challenges.

Currently, nuclear energy is responsible for producing 14 per cent of the world's electricity.[169] Interestingly, even if the world's nuclear power capacity increased threefold between now and 2050, nuclear power would still only account for 20 per cent of the world's electricity needs, due mostly to the rapidly growing demand for electricity. Although nuclear power is often touted for its environmental friendliness, the extent of the impact it can have on global climate change is notably limited.

While the full scope of nuclear energy's impacts on the environment might be overstated, the risks it may pose to nuclear proliferation are not. With the end of the Cold War and the dramatic spread of manufacturing and technological capabilities, the former mechanisms used to prevent nuclear proliferation (most specifically the Nuclear Non-Proliferation Treaty (NPT)) are becoming increasingly obsolete.[170] A new framework for the NPT must be developed that better reflects the changing status of our world, specifically acknowledges the growing role of transnational actors and addresses the increasingly dual-use nature of certain nuclear technologies (for example the use of isotope separation for medical or industrial purposes).[171]

Overall, from a geopolitical perspective, it is important to consider that any country capable of generating nuclear power using enriched uranium is capable of generating nuclear weapons, or at least the fissile materials required for a nuclear bomb. Thus, in order to ensure that terrorist organizations do not gain access to nuclear technologies, it is imperative that any country wishing to pursue nuclear power agrees to submit itself to strict international inspections that will ensure that nuclear facilities are being used for exclusively peaceful purposes. Similarly, countries could restrict their use of uranium in nuclear power generation to natural, unenriched uranium that is unable to produce weapons-grade material. Canada has already taken this step, and other countries should be encouraged to follow.

## An overview of key trends and developments

As policymakers look to reduce and mitigate the anthropogenic forces driving climate change and to improve their energy security, they are focusing on the development of sustainable, renewable energy resources that can reduce and eventually eliminate dependence on highly polluting fossil fuels like coal and petroleum.

Clean, renewable and sustainable are the main buzzwords for energy and climate change-related strategic technologies in the twenty-first century, and there are two ways for policymakers to approach these objectives. The primary way should be by investing in the development of sustainable and renewable energy resources that are viable and widely accessible alternatives to traditional fuels. Many such technologies already exist, including solar power, wind power and hydropower, but, unfortunately, these technologies are not yet reliable enough or affordable enough for mass application. While scientists, businesses, and researchers work to develop new, clean energy resources, policymakers should look to more immediate steps to reduce or offset the emissions of existing, widely used technologies. Whether it is catalytic converters in cars or the adaptation of clean coal technologies, the world is faced with the challenge of taking immediate steps to minimize human effects on climate change.

Scientific research related to existing practices, innovative processes and new materials is also important for developing new energy sources and improving long-term energy sustainability.

From an energy perspective, some key geopolitical challenges and trends characterizing the global energy market include the growing demand for energy, linked to the rapidly growing economies of China, India, the Middle East and Eastern Europe; emerging regulation resulting from the need to address climate change; and geopolitical tensions related to supplies of natural gas and oil, and threats to the oil supply from potential terrorist attacks. Energy has a major impact on our environment and is therefore closely associated with a number of environmental and social issues.[172]

Climate change and energy are complex global problems that require coordinated, global responses. It is crucially important to integrate technology development with environmental energy security. Translating scientific knowledge into policy recommendations is one of the biggest challenges in these areas of science. But I believe that the degree of global awareness of the current environmental challenge means that special attention needs to be paid to the development of new environmentally sustainable technologies as well as to finding further scientific solutions.

# 4
# Health Care

## General overview

When it comes to the geopolitics of health, today's developments in strategic technologies present both challenges and opportunities. Improvements and breakthroughs in pharmaceuticals, medical devices, vaccines and patient care have meant improved standards of living and longer life expectancy for much of the world. In certain parts of the world, many infectious diseases have been contained or eliminated, and improvements in sanitation and food safety have reduced the spread of common illnesses, especially in poorer, developing countries.

Nonetheless, the advancing field of health science also leaves much to be desired. Advancements in health-related strategic technologies are improving life in parts of the world, but elsewhere these advances are merely serving to emphasize the ever-growing gap between the world's haves and have-nots. Medical care in the United States and Europe is a marvel of modern technology and innovation, yet much of the world's population still suffers from basic and often treatable diseases. Globally, around one billion people still live in abject poverty with no access to proper health care. What is most frustrating about this situation is how frequently policymakers regard health and security challenges as divergent institutional spheres.[1] Traditionally, health 'is not immediately perceived as a security issue' but 'recent potential health threats, particularly within the context of food production, travel, and migration, force us to think of health as a paramount security concern'.[2]

Health-related technology depends on the expertise, knowledge and innovation generated by multiple strategic technologies, including biotech, nanotech, and genomics. The types of relevant health technologies available today are seemingly as diverse as the list of existing ailments and illnesses. Major health technologies fall into the categories: diagnostic, preventive, therapeutic, auxiliary and rehabilitative.[3] Some of the most essential health technologies include: diagnostic imaging; medical devices and equipment; blood transfusion; transplantation services; hospital and surgery-related technologies; and laboratory services.[4] Remarkable achievements in terms of medical devices in the diagnostic domain include imaging technologies (such as Magnetic resonance imagining (MRI) or echography) and electrophysiological methods (such as electrocardiography (ECG) and electromyography (EMG)).[5] Therapeutic technology examples include lasers, defibrillators, cardiac pacemakers, auditory and visual prostheses and ionizing radiation devices.[6] Assistive and rehabilitative technological innovations include respirators, braces, hearing aids and eyeglasses.[7]

From a policy perspective, the health care focus needs to be on raising global awareness over transnational health threats, taking preventative measures, and sensitizing people to the increasingly global dimensions of health risks.[8] A multilateral treaty on disease control requirements and protocols would be a good first step in this direction.[9]

## The global health care industry

In developed countries, costs of health care are rising at a rapid rate. In 2010, total health care expenditure outpaced GDP growth in all OECD countries, reaching an average ratio of 9 per cent of GDP, as opposed to 7.8 per cent in 2000.[10] Health care spending is expected to increase even more in the foreseeable future.

One of the largest sectors of the industry is the global prescription drug market, which was valued at USD 550 billion in 2006.[11] Other major sectors include medical equipment and supplies, biotechnology, alternative medicine and health care services. Some of the defining forces of the modern health care industry include increasing the productivity of drugs, hospital equipment and other supplies; the rising costs of health care and the growing number of patients from industrialized countries who are seeking treatment in cheaper, developing countries; a rise in ailments among the elderly, resulting in demand for new types of drugs; and new drug discovery and development especially with regard to genetic sciences.[12]

With global health care costs on a steady upward trend, the overall inclination is for the industry to try to offset its costs, particularly through privatization, higher returns on investment and better management. Governments will maintain their regulatory function in the health care industry but, increasingly, they will look to offload the practical provision of health care to more professional organizations. Another industry trend especially important to governments and public health officials is the globalization of disease and addressing the fact that health developments in one corner of the globe can have dramatic impacts on far away populations.[13] This is true from the perspective of spreading infectious diseases but also in terms of more mundane issues such as the global price of pharmaceutical drugs.[14] A final industry challenge is the brain drain of medical professionals from developing countries. These talented professionals often leave their home countries in search of better education and opportunities in the industrialized world, thus undermining the quality of medical care at home.

## Key technological drivers in health care and medicine

Genomics, information technology, antimicrobial drugs and nanotechnology are likely to be the key technological drivers in the fields of health care and medicine over the next ten years. Overall trends for the twenty-first century point to a gradual shift to early detection and prevention and to more targeted interventions.[15] Health science and research will be predetermined by the technical understanding of processes and pathways and the molecular vision of disease.[16] It will also be more evidence based.[17] Yet, despite major progress in health science and geopolitics, much work remains to be done. According to the World Health Organization (WHO), in 2030 the key causes of death will be cerebrovascular disease (stroke), ischaemic heart disease, chronic obstructive pulmonary disease and HIV/AIDS.[18] Major improvements in health technologies will have to be made to face these challenges.

### Genomics

In 2003, scientists finished identifying and sequencing the nearly 25,000 genes in the human genome.[19] Since then, genomics has rapidly gained momentum and influence in the field of global public health. In many ways, genomics and genetic-related research play a pivotal role in the geopolitics of health care.

For example, in the field of pharmaco-genomics (the process of adjusting therapeutics to the patient's genetic make-up)[20] and in the related field of pharmaco-proteomics (the use of proteomic technologies in the development of drugs), recent research has focused on tailor-making pharmaceuticals to suit specific genetic profiles. Using knowledge gleaned from genomics, researchers have been able to shape drug design around specific disease-causing proteins expressed by genes.[21] These types of advancement could have positive results for cancer patients, among others. By introducing genes that can stimulate the patient's immune system or make toxins to eliminate tumour cells, genomics will make it possible to fight cancer in a more effective way.[22]

When speaking of genomics, it is important to note that genomicists have sequenced more than just the human genome. In the past decade, researchers have also sequenced the genomes of at least 50 microbial pathogens, and the results of this research may assist in the control of communicable diseases,[23] helping in the treatment and diagnosis of diseases and potentially limiting the outbreak of new diseases.[24]

Thanks to genomics, many old, expensive health care procedures can now be replaced by cheaper, more accessible ones. Additionally, recombinant vaccines, which bring together biotechnology and genomics, have the potential to be safer and cheaper to produce and store than their traditional counterparts.[25]Finally, genomics, in combination with information technologies, can make a substantial contribution to health care and medicine. One recent example is the Bio-Linux Computing Platform, a computer program specifically designed to facilitate environmental genomics research.[26]

**Information technologies**

Advances in digital and information technologies are playing a key role in spreading scientific research and medical information. Better access to information and improved sharing of medical records and research are some of the most powerful tools for improving global health care. For example, electronic health records make it easier for hospitals to track patients and for non-primary care physicians to access full patient histories.[27] Static reference information systems like these can be transformed into dynamic decision support systems in hospitals, in private and public sector clinics and at home.[28] Similarly, the Internet can serve as both an informational resource and a tool to allow doctors to consult with each other in real time, regardless of geographical distances.[29]

Data sharing is an increasingly important component of health care and evidence-based medical research. Current health databases are often small and inadequate for large-scale research, and data is acquired on a project-by project basis. This results in databases that are small, unique, specialized and difficult to navigate and use on a large scale. By combining all this data into streamlined national databases, the health care industry could benefit from rapid learning, a health care technology rooted in IT that would help increase the value of health care, monitor health care costs, and track the success of treatment based on the experience of tens of millions of patients being updated in real time.[30] One of the most successful instances of this is the US National Cancer Institute's cancer Biomedical Informatics Grid (caBIG), an information network that enables members of the cancer community, including researchers, physicians and patients, to share data and knowledge.[31] The hope is that caBIG will help to accelerate the discovery of new approaches to diagnosing, treating and preventing cancer.[32] Such technology could easily be applied beyond the realm of cancer therapy, and it has promising potential for improving health care in the twenty-first century.

Information technology is essential to advancing health care, reducing and containing diseases, and streamlining patient care. There are several existing global communications initiatives designed to manage global health crises such as the outbreak of a new disease. For example, the WHO Global Outbreak Alert and Response Network (GOARN) partnership, created in 2000, uses advanced electronic communication networks and computer applications to 'enhance the network's power in global surveillance and response'.[33] GOARN, which is maintained by Health Canada, regularly and systematically searches websites, news wires, local online newspapers, public health email services and electronic discussion groups for key words that could signify disease outbreaks.[34] When a new disease outbreak is identified, GOARN has the resources and contact lists to globally disseminate information to doctors and public health officials.

Overall, the global market for health care information technology is expanding rapidly. It is expected to grow at an annual rate of 13 per cent between 2009 and 2014 and at an annual rate of 16.1 per cent by 2014.[35] ICT and what has been called 'e-health' have large potential to confront the financial and sustainability challenges the health sector is currently facing. However, many OECD countries still lag behind and government intervention is essential for this potential to be realized.[36]

## Antimicrobial drugs

Antimicrobial drugs are a major technological challenge for global health care. The complexity of pathogens such as HIV and malaria, the slow pace of the development and approval of new antimicrobials and the lack of commercial incentives for new drug development mean that new vaccines and drugs are not keeping pace with fast-evolving, drug-resistant pathogens.[37]

The central challenge with antimicrobial drugs is their over-prescription. Indiscriminate use of antibiotic drugs facilitates microbial resistance to these drugs. As pathogens mutate in response to antibiotics, drugs that were once at the cutting edge of technology quickly become obsolete. For this reason, it can be said that the antibiotic revolution is facing a new range of challenges.[38] As Professor George Fidas explains:

> The first line drug treatment for malaria is no longer effective in over 80 of the 92 countries where the disease is a major health problem. Penicillin has substantially lost its effectiveness against several diseases, such as pneumonia, meningitis and gonorrhea, in many countries. Even vancomycin, the last defense against hospital-acquired infections, is losing its effectiveness. Influenza viruses, in particular, are essentially efficient in their ability to survive and sometimes change into deadly strains.[39]

Ensuring that research and development of new antimicrobial drugs keep up with the rapid evolution of diseases and pathogens will be one of the biggest challenges of global public health in the twenty-first century.

## Nanotechnology

Nanotechnology has the potential to facilitate health care in a number of ways. For example, surfaces enhanced by nanotechnology may eventually be completely bacteria-retardant, thereby facilitating the sterilization of hospitals and other medical facilities. Similarly, nanotech filtration systems could help hospitals improve air quality at a lower cost.[40] Also, certain medicines could potentially be released into the bloodstream on the nanoscale, thus improving their speed and efficiency.[41]

Public acceptance of the use of nanotechnology, as will be discussed in more details in Chapter 7, is likely to depend on the public's perceptions of its risks and benefits. For this reason, it is important to create a strong legal and ethical framework for the development and application of such innovations.[42]

## International regulation

The WHO is the world's pre-eminent health organization. It is responsible for providing leadership on global health issues, shaping research agendas, setting standards and making policy recommendations, and it sets the tone for global health initiatives.[43] Although most regulatory issues like drug approval process or compliance issues are dealt with at the national level, the WHO is the one organization poised to respond to the outbreak of a global epidemic and to comprehensively address transnational health issues such as strengthening the world's health infrastructure and promoting development and health security, especially in the developing world.[44]

To this end, the WHO has established a set of International Health Regulations (IHR), a set of legal principles that are binding on the 194 participating countries around the world. The IHR aims 'to help the international community prevent and respond to acute public health risks that have the potential to cross borders and threaten people worldwide'.[45] Specifically, the IHR requires countries to inform the WHO about specific disease outbreaks and public health events. This, in turn, sets off a series of preventative measures, such as alert and response. Because of its vast infrastructure, its experience and the diversity of its member countries, the WHO is especially well poised to act in this capacity.[46]

Although the WHO oversees and coordinates global health issues, it has no particular mandate or philosophy that relates specifically to health care technology.[47] Again, the trend is for these issues to be dealt with at the national or, in some cases, the regional level.

## Geopolitical implications of health care technologies

### Infectious diseases

Naturally occurring diseases have always been a huge threat to humanity, both from a health perspective and an existential one.[48] In recent years, naturally occurring diseases have been responsible for 25 per cent of deaths worldwide and for 45 per cent in developing countries.[49] Since the 1970s, 20 maladies have spread or re-emerged in more aggressive forms on the world scene, including highly contagious and deadly diseases such as tuberculosis and malaria.[50] In addition, at least 30 previously unknown diseases have been discovered (for example Ebola and SARS). The reasons behind the emergence of new diseases and the spread of existing ones are many. For example, the overuse of antibiotics to enhance the growth of chicken and cattle has led to the emergence of more drug resistant microbes.[51] Additionally, there is increasing human settlement of formerly remote areas such as jungles, and humans have no established immunities to the viruses and bacteria in these areas. Other reasons for the increasing number and growing rates of infectious diseases include the development of megacities in countries with poor health infrastructures, an increasingly globalized world with greater opportunities for travel and, consequently, greater numbers of microbial 'hitchhikers' that transfer disease from one region to another, and climate change which is leading to a new distribution and evolution of pathogens.[52]

Obviously, infectious diseases and their spread are a major threat to global public health, but they also influence many other elements of security and geostrategy, including economic strength, national and regional stability and social fragmentation. Sub-Saharan Africa, a region that is unfortunately often cited in this context, has an HIV/AIDS infection rate of nearly 5 per cent of its population between the ages of 15 and 49,[53] and the area is reflective of the profound influence that infectious diseases can have on the security and stability of a region. Emerging technologies have the potential to offset some of the gravest challenges posed by these diseases and help to offset their spread. For this reason, emerging strategic health technologies will contribute greatly to global peace and security.

Infectious diseases can take many forms, and whether it is haemorrhagic fevers, AIDS or drug-resistant iatrogenic infections, these are all major health challenges. It is also important to stress that the fight against these infectious diseases must go hand-in-hand with tackling other deadly health problems such as cardiovascular disease, cancers and major chronic, hereditary and non-communicable diseases.[54]

To put infectious diseases into their broader context, it is worth looking at some statistics. Every year, almost two-thirds of child deaths and between a quarter and one-third of all adult deaths worldwide are linked to infectious diseases.[55] The global incidence of AIDS, tuberculosis and malaria continues to rise, and AIDS and tuberculosis are 'likely to account for the overwhelming majority of deaths from diseases among adults in developing countries by 2020'.[56]

Strategic technologies are crucial for the purposes of prevention of major infectious diseases. They can help, for example, by preventing HIV transmission due to unsafe blood transfusions.[57] If the prevention of health care-associated HIV transmission is not improved, unsafe medical injections and blood transfusions could cause up to four million new infections.[58] Antiretroviral drug technology can change an incurable, fatal disease such as AIDS into a condition that can be managed and dealt with almost like a chronic illness.[59] This helps infected people to live longer and stronger, meaning less disruption to society. New advances in medicine and medical technology are minimizing the risk of parents spreading HIV to their children, and making these technologies cheaper and more widely available should be a top priority for global policymakers.

In the case of an epidemic or pandemic, strategic technologies can play a fundamental role in diagnosing the illness, developing new vaccines, and slowing or stopping the spread of new or evolving viruses. Some of these technologies are medicinal – for example, new vaccines or antiviral medicines – but others merge health technologies with other forms of strategic technology. New technologies played a key role in the detection and fight against the global outbreak of the H1N1 virus in 2009. Airports for instance used infrared heat sensors to monitor the body temperature of all travellers arriving from abroad. Anyone with an abnormal reading had to submit to a medical evaluation to ensure that they were not infected with the potentially deadly virus.[60]

Another example is Veratect, a Seattle-based bio surveillance firm, was able to send early warnings of a potential outbreak to the US authorities.[61] It had identified the potential disease thanks to its information gathering technology tools, which combine artificial intelligence, bio surveillance and the Internet among other things.[62] Early warning and rapid response systems in general are fundamental elements of preventing global pandemics and ones that are dramatically facilitated by new technologies. Historically, disease surveillance systems have been passive and disease-specific. These numerous individual systems often lack information sharing capabilities and unknown diseases are only detected haphazardly.[63] Information and communication technologies can help in the surveillance and notification of the spread of infectious diseases.

Genomic technologies can also provide viable future solutions to increasing resistance to viruses. For example, researchers from Geneva and Lausanne, Switzerland, have united their efforts to pioneer 'a technique identifying the gene variants that offer resistance against the [HIV] virus and are passed from generation to generation'.[64] They managed to locate a place where the virus interacted with the immune system of an individual and, as a result, 'to map the susceptibility of one part of the human genome and uncover a family of proteins that had not been previously associated with HIV'.[65] Technologies like antimicrobial drugs that fight infectious diseases will be invaluable in the fight to improve global public health security.

## Global initiatives dealing with public health issues

Although emerging strategic technologies have great potential to limit the spread and outbreak of disease and illness, getting these technologies to the people who need them most is an ongoing global challenge. A number of initiatives exist specifically to match technologies with the neediest populations, and this section describes a few of these in greater depth.

Health care is a strategic challenge that is uniquely suited to global collaboration and public-private cooperation. This fact is reflected in the number of global public health initiatives that encompass representatives from the public and private sectors as well as governmental and non-governmental groups. Some of the most prominent strategic partnerships in the field of global health are the Global Fund to fight AIDS, Tuberculosis and Malaria; the Global Alliance for Vaccines and Immunisation (GAVI) Alliance; the WHO; the Bill & Melinda Gates Foundation and the World Economic Forum Global Health Initiative.

The Global Fund came into being in January 2002. Its principle mission is to finance national programmes that fight AIDS, tuberculosis and malaria. Funded by over 50 countries from around the world (the largest proportion of the funding comes from the United States),[66] it has so far committed over USD 18 billion to 136 countries.[67] The Global Fund is a multi-stakeholder partnership, and although the largest contributors are states, organizations such as the International Olympic Committee and individuals such as the former United Nations Secretary-General, Kofi Annan, also contribute money to the cause.[68] The Fund's governing bodies include donor and recipient governments, intergovernmental organizations, representatives from non-governmental organizations (NGOs), and the private sector. There are four non-voting members: the government of Switzerland, the Joint United Nations Programme on HIV/AIDS (UNAIDS), the WHO and the World Bank.[69] The private sector is represented by the Bill & Melinda Gates Foundation and the McKinsey consultancy.[70]

The diverse nature of the Global Fund allows for unique initiatives that would be unimaginable for a purely governmental group. For example, Bono, the lead singer of the rock band U2, proposed a series of (RED)™ products designed to engage commercial partners in the fight against disease by getting them to design some of their products with special Global Fund branding.[71] A portion of the proceeds from these products goes to the fund, thus engaging the public and the private sector in the fight against AIDS, tuberculosis and malaria in a unique (and fashionable) way.[72]

The GAVI Alliance is a public-private partnership, the main objective of which is to improve access to immunization for children in poor countries, thus improving the overall health and social and economic capital of some of the world's most impoverished countries. The main members of the alliance are the United Nations Children's Fund (UNICEF), the WHO, the World Bank, the Bill & Melinda Gates Foundation, developing and donor country governments, research and technical health centres, pharmaceutical and vaccine industry representatives and NGOs.[73] Interestingly, in contrast to the Global Fund, civil society is significantly outnumbered in the GAVI Alliance by the private and commercial sector.[74] The GAVI Alliance's 'resources are designed to accelerate the development and introduction of new and underutilized vaccines, enhance injection safety and strengthen routine immunization services as well as the health systems that support immunization and child health care services'.[75] GAVI works closely with the WHO to assure as much continuity in these vaccines as possible.

In addition to supporting projects like the Global Fund and the GAVI Alliance, the WHO is extremely active in monitoring and developing new health technologies as part of its own organizational mandate. Its Essential Health Technologies division works in the areas of AIDS, malaria and tuberculosis to support the expansion of access to necessary technologies; to provide diagnostic and laboratory support, including prevention and best practices; to promote the strengthening of capacity building to produce positive results at country level, through a dedicated country focus strategy; and to support 'the Millennium Development Goals in reducing child mortality, improving maternal health and combating HIV/AIDS, Malaria and other diseases'.[76] As the group's website notes, 'access, including in primary health care, to safe and effective health technologies relies on policies for selection and management based on scientific evidence and best practice for organization of their management and use'.[77]

Another major player in the global health scene that is committed to further developing and increasing global access to health technologies is the Bill & Melinda Gates Foundation. Founded by Microsoft chairman Bill Gates and his wife Melinda, this organization has a wide mandate that covers everything from increasing access to education to finding a cure for AIDS. Overall, the foundation has invested some USD 20.5 million in the study of the genomes and immune systems of the so-called elite controllers, people able to control HIV naturally without drugs.[78]

The World Economic Forum's Global Health Initiative (GHI) was launched in 2001 in Davos, Switzerland: explicitly designed to promote and strengthen private-public partnership in global public health, the initiative supports CEOs from various sectors in their efforts to fight

HIV/Aids, tuberculosis and Malaria.[79] According to the World Economic Forum's own figures, GHI and partner initiatives provide access to information, prevention, testing and treatment services for over nine million people.[80]

**Health, development and poverty**

Human health is a top security priority and a crucial part of global development and stability. This reality is demonstrated by the fact that the nations that are most vulnerable to illness are often the ones that also suffer from the damaging interaction of poverty and conflict.[81] Disease reinforces underlying social problems and can sometimes escalate tendencies towards violence. Health problems fuel or foment all sorts of other grave issues, and what is particularly frustrating is that, in most cases, failures in global public health come more from policy shortcomings rather than from a lack of scientific knowledge and resources. The twentieth century was marked by great progress in medical science and human health, but regional inequalities of access to preventive health care and treatments for disease remain and continue to widen. This sad truth is reflected statistically: in developed countries, life expectancy is approximately 80 years or over, whereas in certain sub-Saharan nations it is 40 years or below; and a woman living in a least developed country is 300 times more likely to die in pregnancy or childbirth than a woman from a developed country.[82]

Several initiatives are under way to address this dramatic discrepancy in global health care. In 2000, the then WHO Director-General, Gro Harlem Brundtland, launched the Commission on Macroeconomics and Health.[83] It was chaired by Jeffrey Sachs and had a mandate to study the links between macroeconomic problems and health issues.[84] The Commission's aim was to examine the connections between health and poverty and to show that health investment can prompt and accelerate economic growth.[85] The Commission found that the poor experience greater levels of preventable suffering and death. The lack of clean water, malnutrition and poor sanitation make them more susceptible to disease, and they often lack adequate access to health care.[86] The epidemics of HIV/AIDS, tuberculosis and malaria greatly affect the socioeconomic structures of most developing countries.

Medical science and knowledge are most often targeted towards the needs of rich nations.[87] The Global Forum for Health Research estimates that just 10 per cent of global health research resources is spent on the diseases that affect 90 per cent of the world's population: the so-called

'10/90 gap'.[88] As the International Task Force on Global Public Goods notes:

> Only a small fraction of the worldwide expenditure on health research and development is devoted to the major health problems of the majority of the world's population. Of the 1,233 drugs that reached the global market between 1975 and 1997, only 13 were for tropical infectious diseases that primarily affect the poor.[89]

The private sector of the health care industry is primarily focused on the potential for profit in wealthy countries.[90] More money is spent per year on finding a cure for baldness, a problem of particular concern to rich, Western men, than on developing a vaccine for malaria.[91]

Health disparities within countries are of just as much concern as inequalities between countries. In the United States, for instance, until recently, the private-public health care system resulted in insufficient access to health care and insurance for large parts of the population.[92]

The health reform legislation adopted by the Obama Administration is expected to lead to significant improvements. On the one hand, the health reform is likely to trigger more regulation and more investment in health technology. On the other hand, it promises to 'make insurance more affordable [...] bring greater accountability to health [...] and end discrimination',[93] thereby covering the basic needs for dignity of many more US citizens. Additionally, it will reduce the cost of health care because more preventive medicine will help impeding diseases.

Over the past decade, the trend in international technology transfers was for 'an increasing role for the private sector and a more static role for the public sector'.[94] Of the USD 73.5 billion spent on health research and development funding in low- and middle-income countries in 1998, only 3 per cent was public sector funding, and a further 8 per cent was private non-profit funding.[95] By contrast, in the developed world research is mostly financed by the private sector. In 2000, about 68 per cent of all US research was carried out by the commercial sector.[96]

One challenge in increasing the developing world's access to health care and drugs is concerns over intellectual property rights. Researchers, scientists and private sector firms would be reluctant to invest heavily in new product development if they could not cover the initial investment in their work. As a solution to this problem, the International Task Force on Global Public Goods recommends balancing the developed countries' concerns over intellectual property rights with a multilateral agreement on open access to basic science and technology (ABST).[97] According to the Task Force, the ABST would:

Help resolve the free-rider problems that reduce investments in science and technology relative to a global optimum. Second, it could restrain the tendency of governments to restrict access and to encourage privatization of basic knowledge. This rebalancing of technology development norms in favour of expanding the public domain could help vitalize scientific research in many countries while promoting applied innovation. Third, the agreement could provide an important plank for the construction of modern technological capabilities in poor countries while sustaining access to information for education purposes.[98]

Through an agreement of this type, rich countries would agree to support developing countries in their efforts to generate, assimilate and diffuse knowledge while still profiting from their investments and intellectual capital.[99]

It is important to focus on the health problems affecting the most disadvantaged, poorest people. The list of key challenges includes improving vaccines, creating new vaccines, controlling insect vectors, improving nutrition, limiting drug resistance, curing infection and measuring health status.[100] Resolving these challenges would require substantial work aimed at understanding which immunological responses ensure protective immunity, creating single-dose vaccines for newborns, creating vaccines that do not need to be refrigerated and can be delivered without needles, controlling insects that are transmitting agents of disease, creating optimal bioavailable nutrients in a single staple plant species and conceiving immunological methods to cure chronic infections.[101] Visionary scientific research and technological innovation will have to be combined to meet these global challenges.

### An overview of key trends and developments in health care technologies

Across the globe, the challenge of providing quality and affordable health care is a daunting one. Numerous emerging strategic technologies have the potential to help world leaders and communities meet this challenge. Advances in fields such as genomics, ICT, nanotechnology and drug delivery coupled with better medical tools and equipment are gradually making it easier for doctors to detect, treat and prevent many illnesses. In many cases, this is true even for diseases that were previously thought to be incurable. Technologies are making it easier to control and stop the spread of infectious diseases, and improved health is closely correlated with stronger development for poorer countries. Nonetheless, while technological progress is accelerating across many facets of the health care industry, problems of inequality are rampant. Developing countries have little or no access to many health care technologies, and often cost is one of the key prohibitive factors. Genetic testing, prescription drugs and antibiotics can be an economic strain even for wealthy countries, and this reality is only magnified for poor countries. Many international and public-private initiatives have been launched to improve the quality of global health care and to correct imbalances in access and research to health care and related technologies. Their investments will hopefully result in further discoveries and developments that will improve global health, an issue that affects literally every person on this planet.

# 5
# Biotechnology

## General overview

From stem cell research and genetically modified crops to anthrax and the manipulation of disease-causing agents, the science of biotechnology is truly at the centre of politics, religion, ethics and innovation. Biotechnology has enormous potential to improve life expectancy, food production and the treatment of illness and disease. It also has possible applications for the environment, manufacturing industries and in the creation of alternative energies, making it of huge interest to both the public and private sectors.[1]

Against this backdrop of promise and potential, biotechnology poses very real dangers, ranging from bioterrorism and the creation of new, potentially lethal, viruses to the unanticipated health consequences of genetically altered foods and viruses.[2] Overall, the field of biotechnology presents a unique set of ethical and cultural questions. How these questions are answered, both by individual societies and by the international community at large, will have enormous geostrategic consequences.

In the broadest sense, biotechnology is 'the application of biological techniques and engineered organisms to make products or modify plants and animals to carry desired traits'.[3] It is a combination of various technologies that focus on the manufacturing capabilities of cells and on using the advantages of biological molecules such as proteins or deoxyribonucleic acid (DNA).[4]

Currently, biotechnology is most prominent in the fields of medicine and health care. Biotech drugs treat everything from psoriasis to non-Hodgkins lymphoma, and genetic testing and other diagnostic tools made possible by developments in the field of biotechnology can facilitate early detection of cancer and disease in high-risk patients. Additionally, biotechnology has allowed for more efficient production of new drugs such as human insulin and vaccines against hepatitis B and rabies.[5]

Outside health care, biotechnology has important applications in agriculture, and energy. One of the most high-profile (and controversial) examples of this is genetically modified crops. Thanks to developments in the field of biotechnology, farmers can now raise crops that are more resistant to pesticides and diseases, that ripen more slowly, or that have added nutritional value, among other traits.[6] Although scientists have raised valid concerns over the unintended consequences of genetically modified agriculture,[7] the potential for biotechnological advances to lead to innovations that reduce world hunger, increase crop productivity in the developing world and reduce post-harvest losses are enormous.[8]

Biotech advancements in agriculture also contribute to the energy industry, specifically in the context of biofuels. As discussed in Chapter 3, although biofuels burn more cleanly than conventional fuels, their widespread use remains surrounded by controversy. The diversion of potential food resources for fuel production has contributed to global food shortages and sparked an international food-for-fuel debate.[9]

Many other industries also benefit from biotechnology applications, albeit on a smaller and more specialized scale. Often, these biotech innovations also help to offset the emissions of some of the world's dirtiest industries. For example, the chemical industry uses biocatalysts to produce novel compounds, reduce waste by-products and improve chemical purity. The plastics industry has used biotechnology to reduce the use of petroleum in its production processes by making 'green plastics' from renewable crops such as corn and soy. Meanwhile, the heavily polluting paper industry has improved its manufacturing by reducing the toxic by-products of pulp processes. Similar advances have been made in the textiles industry, especially since the discovery of new enzymes that make fabric detergents more effective and less polluting.[10] Even the food industry has improved its baking process thanks to biotechnology, applying fermentation-derived preservatives and analysis techniques to food safety. Finally, the livestock industry has added enzymes to increase nutrient uptake by animals and to decrease phosphate by-products.[11]

Like other emerging strategic technology fields, the pace of change and development in biotechnology is becoming increasingly quick. Similar to Moore's Law that was discussed in Chapter 3 on ICT biotechnology

– and specifically the capacity of sequencing technology – has been said to be governed by Carlson's Law which states that the number of sequenced and/or synthesized base pares per person per day doubles every 12 months.[12] Whereas starting in 2002 it took a team two years to synthesize the polio virus, a similar process, when started in 2005, only took two weeks.[13]

## The biotechnology industry

Biotechnology, in its simplest sense, is an age old science. Thousands of years before our era, Sumerians were brewing beer, and ancient Egyptians were using fermentation techniques to make wine and to leaven bread.[14] Although we now take these processes for granted, they are actually everyday examples of biotechnology in action. The modern biotech movement dates back to the 1970s,[15] when the pioneering scientists Herbert Boyer and Stanley Cohen discovered how to 'clone genetically engineered molecules in foreign cells'.[16] Since then, the biotechnology industry has developed over 200 new vaccines and therapies aimed at treating diabetes, cancer and HIV/AIDS, among others, and over 400 products targeting hundreds of diseases are currently undergoing clinical trials.[17] By 2008, the total global value of publicly traded biotechnology companies was in excess of USD 360 billion.[18]

The United States is the current world leader in the global biotech industry, followed closely by the countries of the European Union, and Canada.[19] The industry is dominated by small to medium-sized companies, but the revenues from biotechnology are enormous. The revenues of the world's publicly-traded biotech companies reached USD 73.4 billion in 2006,[20] and, until the onset of the 2008 global financial crisis, growth in the biotech industry had been strong and consistent.[21] Between 2005 and 2006, for example, the total amount of capital raised by the global biotech industry rose from USD 19.7 billion to USD 27.9 billion.[22] In the same time frame, research and development expenditure increased by 31 per cent to USD 5.2 billion.[23]

A particularly notable trend is the shift in the pharmaceutical industry towards biotechnology. Attracted by the profitable market and threatened by competition from cheap generics, traditional drugs are increasingly complemented or substituted by biotechnical drugs.[24]

In 2006 the biotechnology industry achieved more than USD 50 billion in biomedical product sales, grew more than 250 million acres of crops, raised more than USD 20 billion in new investment, made 341 new deals between biotechnology companies and 490 new deals between biotech and pharmaceutical companies, and negotiated 157 mergers and acquisitions.[25] Even the financial crisis could not seriously affect growth trends in the biotechnology industry. In the United States, for example, employment grew 1.4 per cent in 2008, despite it being the first year of the recession.[26]

## The different areas of biotech research

Biotechnology is a vast, multifaceted science with many areas of specialized research that enable biotech to influence a spectrum of different industries and political issues. Major biotechnology-related specialities and technologies include bioprocessing, recombinant DNA, monoclonal antibody technology, cloning, protein engineering, biosensors, biomarkers, nanobiotechnology, DNA finger printing, cantilevers, stem cell research, biomimetics, biomedicine and synthetic biology. A brief overview of each of these specialities is provided below.

### Bioprocessing

Bioprocessing, the oldest form of biotechnology, is the science of converting biological materials into other practical forms.[27] Using biomaterials such as DNA, single-celled organisms (bacteria and yeast) and enzymes,[28] bioprocessing today is an integral part of the manufacture of vaccines, antibiotics, vitamins, amino acids, pesticides, food-processing aids and biodegradable plastics.[29]

### Recombinant DNA

Much of modern biotechnology is centred on recombinant DNA (rDNA), produced by combining different pieces of DNA from different sources in order to create new genetic variations that can be used in everything from drug research to industry. Because of advances in this field, it is now dramatically easier for scientists to isolate a specific gene or segment of DNA, allowing researchers to better study and manipulate DNA and ultimately to reinsert the modified sequences into a living organism.[30] Innovations in rDNA assist in processes such as developing new drugs and vaccines, increasing agricultural yields, improving the nutritional value of food and reducing air and water pollution.[31]

### Monoclonal antibody technology

Monoclonal antibody technology is a primarily medical speciality that uses immune system cells to make antibodies. These antibodies can in turn be used to help the human body destroy foreign invaders such as viruses or bacteria.[32] Monoclonal antibodies are also important for locating environmental pollutants, detecting harmful micro-organisms in food, distinguishing cancer cells from normal cells and diagnosing infectious diseases in humans, animals and plants more quickly and more accurately.[33] These antibodies also offer concrete therapeutic compounds for treating cancer and preventing the rejection of transplanted organs.

## Cloning technology

Cloning technology focuses on the development of genetically identical cells, molecules, plants and animals. Gene cloning creates 'genetically identical DNA molecules' and is an important tool of biotechnology.[34] Key cloning techniques include embryo splitting and somatic cell nuclear transfer.[35]

## Protein engineering

Protein engineering is essential to developing proteins that do not exist in nature and to improving existing proteins. This form of biotechnology is essential to designing ecologically sustainable industrial processes. Moreover, protein engineering has been used successfully in medical research to develop new proteins that can deactivate tumour-causing genes or viruses. It is also important in food production.[36]

## Biosensor technology

Biosensor technology combines advances in microelectronics and biology. Biosensors are made of a tiny transducer linked to a cell, antibody, enzyme or DNA strand. They are used to measure the nutritional value, freshness and safety of food; measure vital blood components; locate and measure environmental pollutants; and detect and quantify explosives, toxins and biowarfare agents.[37]

## Biomarkers

Biomarkers are biochemical agents that can be measured or tracked as an indicator of biological processes.[38] They can be used to track the progress of a disease or monitor the evolution of a treatment.[39] They have significant potential for identifying the presence of cancer before it has advanced too far, thus making it easier to treat the disease in its early stages when treatment is more effective.[40] Biomarkers have similar environmental applications, as they can monitor environmental contamination or help track how a toxin affects a biological organism.[41]

## Nanobiotechnology

Nanobiotechnology combines molecular biology and nanotechnological breakthroughs. Some of the most relevant nanobiotechnology applications are 'increasing the speed and power of disease diagnostics; creating bio-nanostructures for getting functional molecules into cells; improving the specificity and timing of drug delivery; miniaturizing biosensors by integrating the biological and electronic components into a single, minute component; and encouraging the development of green manufacturing practices'.[42]

## DNA fingerprinting

DNA fingerprinting is a particularly interesting biotech development, especially in the context of international security. Unique to every person, DNA fingerprints use people's genetic differences as a means of identification and distinction.[43] Currently, there are no known ways to alter DNA fingerprints, making this technology an important part of forensics in criminal investigations, diagnostic medicine and personal identification.[44]

## Cantilevers

Cantilevers are a new type of assay that can test for disease markers more efficiently and cheaply than the enzyme-linked immunosorbent assay, which is currently being used in the diagnostics field. The assay is a cantilever that 'bends in response to protein markers at levels 20 times lower than what is needed for diagnostic purposes'.[45] This may help doctors to detect cancer and many other diseases earlier.

The width of cantilevers is about 50 microns, the length 200 microns and they are about half a micron thick. For genetic tests, doctors and researchers can coat a cantilever with antibodies or a single strand of DNA. Subsequently, when matching proteins or DNA bind to the coatings, the lever bends downwards, indicating a positive match. In addition to detecting the amounts of target markers in a piece of DNA, cantilevers also have the advantage that they can search for multiple markers in a single reaction. As researcher Thomas Thundat notes, 'the primary advantage of the microcantilever method originates from its sensitivity, based on the ability to detect cantilever motion with subnanometer precision, as well as the ease with which it may be fabricated in to a multi-element sensor array. No other sensor technology offers such versatility'.[46]

## Stem cell research

Stem cells are the building blocks of all organs and tissues in the human body:[47] 'they are like a blank microchip that can ultimately be programmed to perform particular tasks'.[48] Research in the field of stem cells provides insights into how living organisms grow and develop, as well as how cancers form and spread.[49] From a medical perspective, embryonic stem cells have numerous potential applications, as they are capable of generating all types of cells in the body. Embryonic stem cell research is therefore an important area of research for genetics and biotechnology but, because it often depends on harvesting stem cells from human embryos in the early stages of development, it is also one of the most controversial in ethical terms.[50]

## Biomimetics

Biomimetics is a field of research aimed at developing robots based on biological systems. These robots may demonstrate greater robustness in performance in unstructured environments than non-biologically inspired robots.[51] Combining existing knowledge of biology with developments in materials science, fabrication technologies, sensors and actuators, biomimetics can be applied to everything from reconnaissance to demining.[52]

One practical example of biomimetics occurred in Japan in 2007 when scientists used a fish tank that housed an electric eel to power the lights of a Christmas tree. By placing metal plates at the ends of the tank, the scientists were able to channel the eel's electricity and power the Christmas bulbs.[53]

The eel is not the only animal to use electrically charged atoms to do work. Even humans rely on electrical impulses to release calcium ions. The key emerging technological developments in this field involve studying the operation of these electricity-producing cells and membranes in order to mimic that power in batteries. Such a development could potentially create an efficient, cheap source of electricity.[54]

## Biomedicine: bone-repairing glue

Biomedicine is an admittedly broad term that covers the processes and systems that create, sustain and threaten life.[55] Advances in this field influence our understanding of health and disease and create better opportunities for the research and development of more innovative therapeutic strategies. In the long term, advances in this field may help to offset the negative effects of aging or to manage chronic disease.[56]

Recently, researchers in biomedicine have used an adhesive produced by a maritime worm to begin treatment for fractures in human bones. Through research and analysis of the secretions of the sandcastle worm, a sea creature that lives in a mineral shell built from sand and its sticky secretions, scientists have developed a polymer that solidifies in response to changes in acidity and temperature. This glue sticks bits of bone together in watery environments, and it is exceptionally strong. Eventually, the applications of this glue could mean a much easier healing process for compound fractures.[57]

### Synthetic biology

Synthetic biology is one of the most recent fields of biotechnology. An expert group of the European Commission defines synthetic biology as follows:

> The engineering of biology: the synthesis of complex, biologically based (or inspired) systems, which display functions that do not exist in nature. This engineering perspective may be applied at all levels of the hierarchy of biological structures – from individual molecules to whole cells, tissues and organisms. In essence, synthetic biology will enable the design of 'biological systems' in a rational and systematic way.[58]

Synthetic biology has huge potential applications in many fields ranging from energy generation to medical uses. A major breakthrough was achieved in May 2010 by Craig Venter, the same researcher who led the completion of the Human Genome Project (HGP) (further discussed in Chapter 6). Venter and his team claim to have created an entirely synthetic cell, which according to Venter is 'the first self-replicating species we've had on the planet whose parent is a computer'.[59]

However, as promising as synthetic biology may be, its potential downsides and the ethical implications of the capacity to 'create synthetic life' have started a vivid debate. According to Steven Benner, there are three main potential dangers from synthetic life forms: (i) they have the capacity to evolve; (ii) they are self-sustaining; and (iii) they are made from 'standard terran biochemistry', that is, they are 'made out of the same stuff as we are'.[60] The cell created by Venter fulfils all these criteria. Mirroring the importance of these recent developments, President Obama has asked his Commission for the Study of Bioethical Issues to undertake a study of the implications of the advances of synthetic biology.[61]

## Biotechnology, the dual-use dilemma and research governance

In virtually every field of biotechnology, the equipment and knowledge used to produce beneficial and cutting edge biotech products and research can also be used for more sinister purposes, including the production of biological weapons, weaponized diseases, or new, potent toxins.[62] So far, the most high-profile biological weapon attacks have been relatively small in scale. In 2001, two people died in the United States after anthrax spores were distributed in letters sent through the US Postal Service.[63] Numerous other lower-level or failed attacks have occurred around the world in the past few decades.[64] Although these biological attacks have been small or amateur in nature, the potential damage from a large-scale, more coordinated biochemical attack is enormous. While the chances of a successful, large-scale attack are relatively small, governments still need to strive to make such attacks even less likely. Strong international regulation across all facets of the biotech industry is a key component of achieving this goal. Since the recent growth in the global biotechnology industry has resulted in a rapid expansion of dual-use technologies and research, the impetus for global regulation of and standards in this field of science is high. Currently, there are two general approaches for dealing with and regulating biotechnology: international frameworks and agreements, and information and research controls. Unfortunately, both approaches have major shortcomings, and neither is nearly as developed as it should be.

The most prominent international framework governing dual-use biotechnology is the Biological and Toxin Weapons Convention (BTWC). Signed by 162 states and ratified by 159, the BTWC, which entered into force in 1975, prohibits the development, production, acquisition, transfer, retention, stockpiling and use of biological and toxin weapons.[65] The BTWC states parties also convene annual meetings, where states and experts can discuss how to promote common understanding and effective action on improving biosafety and biosecurity.

Although seemingly impressive in its scope and mandate, the BTWC does not prohibit defensive research programmes. Nor does it have provisions for verification. Furthermore, there are 23 states that are not yet signatories to the convention and 16 states that have not yet ratified it, including 'competing' states such as Syria and Israel, both of which are believed to be developing their own biological weapons programmes.[66] Equally problematic is the fact that the BTWC was negotiated and agreed in the early 1970s, at the height of the Cold War. The field of biotechnology and biological warfare has changed dramatically over the past three decades, but the BTWC has not been updated accordingly, thus making the need for improved regulation in this area increasingly urgent.[67] The United States, despite the fact that it introduced a new Biological Weapons Security System in late

2009, emphasizes that it will not take the lead on the development of a new legally binding verification protocol. As Ellen Tauscher, then-Undersecretary of State, put it, such a protocol would 'not be able to keep pace with the rapidly changing nature of the biological weapons threat'.[68]

A final shortcoming of the BTWC is that it has no provisions on and no authority over private sector organizations and individuals, despite the fact that these groups have increasingly easy access to biotechnological research data and equipment. For instance, many biotech machines are available on popular websites such as eBay for a mere fraction of their original cost.[69] This flaw has reinforced concerns about terrorist groups gaining access to dual-use biotechnology. Aware of this vulnerability, more and more states have begun investing more heavily in biodefence measures such as better training for public health officials, increased protection of food and water supplies, and preparing the military to respond to biological threats and attacks.[70] However, the BTWC does prohibit the transfer, by states, 'to any recipient whatsoever, directly or indirectly [...] [of] any of the agents, toxins, weapons, equipment or means of delivery prohibited by the convention'. Moreover, United Nations Security Council resolution 1540 of 28 April 2004 obliges all states to 'adopt and enforce appropriate effective laws which prohibit any non-state actor to manufacture, acquire, posses, develop, transport, transfer or use [...] biological weapons and their means of delivery, in particular for terrorist purposes, as well as attempts to engage in any of the foregoing activities, participate in them, as an accomplice, assist or finance them'.[71]

At the national level, many states are taking their own precautions to limit the dual-uses of biotechnology. The US Patriot Act and Bioterrorism Response Act are two of the more contemporary examples of national legislation, with the Patriot Act prohibiting any individual from possessing a biological agent for any inappropriate reason and restricting the transfer of certain biological agents to specific persons.[72] The Bioterrorism Response Act establishes requirements for the registration of possession of specified biological agents or toxins and also requires the Secretary of Health and Human Services to establish rules regarding the physical security of listed biological agents.[73] Other countries, such as India and Australia, have similar national guidelines and regulations for the biotechnology industry and for the handling and possession of biological agents and toxins. The European Union has also developed a whole set of common regulations on biosafety and biosecurity in addition to member states' regulations.[74]

The governance of research represents a second method for dealing with the dual-use dilemma of biotechnology. Scientific knowledge, especially in biotechnology, is advancing at a fast pace, which leads policymakers and members of the scientific community to raise legitimate concerns about whether certain types of information should be made publicly available. In biotechnology, as is the case with many other scientific fields, the central question is who controls what information and how important is freedom of information. There have been a number of instances where scientists have published information on their biotech innovations and discoveries only to be met with accusations that publishing such material could inadvertently aid terrorist groups.[75]

The scientific publishing industry is a vast one, and largely unregulated. In 2002, Morgan Stanley calculated that publishing of science, technology and medical journals was worth USD 7 billion globally,[76] and it is estimated that commercial publishers and academic societies publish on average 1.2 million articles each year in some 16,000 journals.[77] Therefore, it is increasingly necessary for all major actors – governments, industry, international organizations and academia – to be actively involved in developing appropriate research governance measures for these potentially dangerous research publications. Recommendations on this front begin with self-censorship by scientists but also include suggestions for more formal action.[78] A related issue concerns potentially dangerous accidents in laboratories. Employees are exposed to substantial risks while experimenting with deadly pathogens, infectious agents or genetically modified cells. Current safety rules are outdated and focus mostly on chemical rather than biological risks.[79] Similarly, the unintended outcomes of research constitute an unregulated risk of biological research.[80]

The US National Research Council of the National Academies has published two reports that tackle issues central to biotechnology and security.[81] These reports recommend establishing a number of guidelines and monitoring mechanisms for high-risk research, which should be the subject of careful scrutiny at the funding, research and publication stages. These 'experiments of concern' include those that could make pathogens resistant to antibiotics and vaccines and those that aim to turn biological agents and toxins into weapons.[82]

The University of Maryland's Centre for International and Security Studies has suggested a global system of mutually agreed rules for the oversight of potentially dangerous research into pathogens.[83] Such a system would aim to oversee certain high-risk research and its publication. In addition, there have been numerous suggestions and attempts to create codes of conduct to safeguard high-consequence scientific research.[84] For example, my book *Global Biosecurity: Towards a New Governance Paradigm* proposes a new biosecurity governance model that is network-based, integrative, flexible and open, and that actively engages all the actors' concerned (private industry, governments, international organizations, science and academia).[85] For its part, the United Nations University has launched its *Biodiplomacy* initiative for a global negotiation process to construct new policy and legal frameworks to tackle the emerging challenges.[86]

## Geopolitical implications of biotechnology

Biotechnology is a broad term that encompasses multiple scientific disciplines and industries. When it comes to the geostrategic implications of biotechnology, the list is just as diverse and wide-ranging as the science itself. Whether it is health care, agriculture, industry or manufacturing, the effects of biotechnology are wide-ranging. This section focuses on biotechnology and the geostrategic questions surrounding stem cell research, drug discovery and bioterrorism. Other issues related to biotechnology, such as genetically modified organisms and their role in the global food crisis, are explored in greater depth in Chapter 6 on genomics. Similarly, biofuels and the food-for-fuel debate are addressed in Chapter 3 on energy and the environment.

### Stem cell research

As is noted above, stem cell research is an area of biotechnology that has huge potential to change the way illnesses from cancer to Parkinson's disease are treated. One of the most promising forms of this research, embryonic stem cell research, depends on using special cells from human embryos that are three to five days old.[87] Because this research currently requires the disaggregation of the embryo in order to acquire the desired stem cells, many people have raised concerns that stem cell research devalues human life and that such research may lead down a slippery slope to even more controversial practices such as human cloning.[88] The intensity of this debate is reinforced by strong religious overtones.

In 2001, US President George W. Bush signed a law that dramatically restricted the available stem cells that federally funded researchers working in the United States could use.[89] Fuelled by concerns from the religious right in the United States over the sanctity of human life even in the embryonic stage, Bush allowed religion to take precedence over science. He even banned the use of embryonic stem cells left over from in vitro fertilization treatments that were already tagged for destruction.[90] Many religious leaders in the United States praised Bush's decisions, drawing parallels between embryonic stem cell research and abortion. US President Barack Obama reversed this directive in early 2009, allowing federal funds to be used for research on new stem cell lines.[91] Yet the legality of this executive order was challenged, and in July 2010 a federal district judge declared it a violation of federal legislation, thereby blocking all scientific projects working with federal money.[92] At the time of writing, the Obama Administration had announced that it would appeal the court ruling.[93]

While the United States has spent much of the past decade resisting new inquiries into stem cell research, other countries have taken more proactive stances. South Africa and India, for example, allow the therapeutic cloning of embryos for research purposes.[94] Countries such as Singapore and Belgium are known for their support of stem cell research, investing considerable sums of money in attracting top research talent to their facilities.[95]

Other ethical issues pervade the stem cell research and biotechnology debate. For example, somatic cell nuclear transfer (SCNT) is a procedure that produces stem cells that are genetically matched to the donor organism. SCNT can result in the creation of a human embryo and, by extension, the creation of a human life. For this reason, bioethicists around the world have raised concerns over how such technology could affect children, families and society as a whole,[96] and some have speculated on the dangers of 'using the seeds of the next generation as mere raw material for satisfying the needs of our own'.[97] Worries over SCNT range from the possible reduction of diversity in the gene pool to the physical and psychological risks that an SCNT-derived clone would have on the cloned child.[98] Reproductive SCNT provokes an especially strong moral reaction for many people, and many countries are moving to ban human cloning. Whether the international community can agree universal standards for stem cell research and cloning will have important effects on geopolitics. Standardized procedures and norms will dramatically reduce the potential for abuse of this emerging technology. However, if one country or region has less stringent standards than the rest of the world community, there is likely to be a shift of human and financial capital to that area. This country or region may gain a competitive advantage over the rest of the international community, but it may also single-handedly undermine international norms and values on the sanctity of human life.

## Health care

Biotechnology is fundamental to improving global access to and the quality of health care. Drug discovery, for example, is another major component of biotechnology that potentially has global implications. Like biotechnology and related fields, companies are able to research and develop drugs in a more systematic fashion. The increases in the efficiency of the drug discovery process are having far-reaching effects. For example, certain drug discovery advances are helping to revolutionize the way people in poor countries receive vaccines. Currently, vaccines are mostly delivered by injection, requiring cold storage and special handling before they reach their destination. Certain advances in biotechnology could lead to vaccines delivered through food, making it cheaper and easier for poor countries to gain broader protection against diseases.[99]

Biotechnology applications also enable more accurate diagnostic tests, as well as drugs and vaccines with reduced side effects. Biotechnology-based tests allow the diagnosis of a number of cancers, including ovarian and prostate.[100] Related technologies also help provide quicker, less expensive and more accurate diagnoses of Parkinson's disease, Alzheimer's disease, type I and II diabetes, osteoporosis and emphysema.[101] They also play an important role in improving therapeutic practices and treatments for rheumatoid arthritis, leukaemia, haemophilia, transplant rejection and anaemia, among others.

Some biotechnology firms focus on developing small-molecule therapies, while others focus on living systems therapies using cells, proteins and genes.[102] Pharmacogenomics, 'the use of information about the genome to develop drugs' or 'the study of the ways genomic variations affect drug responses',[103] are biotechnology applications that are important in the field of personalized medicine.

Biotechnology-related applications also play an increasingly important role in the field of regenerative medicine. For example, tissue engineering, by focusing on the progress made in materials science and cell biology, offers an opportunity to develop semi-synthetic organs and tissues. Collagen and synthetic polymers are used for tissue engineering. In recent years, good results have been obtained with a renal-assist device – a biohybrid kidney that helps patients who suffer from acute renal failure for the period needed for the injured kidney to restore itself.[104]

As alluded to earlier, biotechnology also plays an innovative role in vaccine production. Some of the key discoveries in this area include antigen-only, as opposed to actual microbe-only, vaccines able to tackle such diseases as meningitis or hepatitis B. Biotechnology has also focused on designing DNA vaccines that are able to offer immunization against those microbes for which vaccines have not yet been developed.[105]

Overall, the potential for these technologies to reduce the costs of health care, improve access to health care for poor and rural populations, and minimize the occurrence of global disease is very promising.

### Bioweapons and bioterrorism

As is touched on in the regulatory section of this chapter, bioweapons and bioterrorism are two forms of the misuse of biotechnology that pose grave risks to geopolitical security. This is especially true because even when biotechnology research is conducted with only the purest, most benevolent intentions, it is highly likely that these technologies will also have military or weapons applications – even if such uses are completely unplanned or unintended.[106] In other words, all biotechnology research could potentially be applied in a dual-use fashion. As the Institute of Medicine and National Research Council notes, 'human history seems to suggest that as technology advances, malevolent use is the rule, not the exception'.[107]

On the one hand, bioterrorism as a risk to global security should not be overstated as there are certainly more wild card factors for a group looking to execute a biological weapons attack compared to conventional, chemical, or radiological attacks. Bioterrorism relies on the cultivation and dispersal of living organisms that are, by definition, difficult to produce and disseminate. On the other hand, if a group was successful at developing a new pathogen, it could be deadly and spread rapidly simply because it has no natural foes.[108] In this context, novel, bioengineered pathogens could lead to global pandemic along with growing fear and massive financial consequences.

Biological weapons, it is worth stressing, differ in terms of proliferation potential and effectiveness from other weapons such as nuclear weapons. Although both biological weapons and nuclear weapons are often categorized as 'weapons of mass destruction', such a label 'fails to capture the disparate future trajectories of the technologies underlying biological, chemical, and nuclear weapons'.[109] Biotechnology and bioweapons are inherently harder to regulate and oversee than nuclear weapons because so many of the key technologies, materials and knowledge are much more widely dispersed. From hospitals and pharmaceutical companies to large research universities, much of the research and technology of biotechnology is part of the day-to-day work of researchers and scientists around the world. These issues make the prevention of the proliferation of bioweapons difficult for even the strictest of states or regulatory bodies to enforce. Moreover, while it is relatively easy to monitor a country's nuclear weapons progression through satellite technology and other techniques, biological weapons development can be done out of public view. There are few concrete 'observables' when it comes to biotechnology.[110]

Overall, the biggest risk posed by biotechnology in the context of biological weapons is the creation of a class of new, highly virulent biological agents capable of attacking distinct biochemical pathways and eliciting specific effects.[111] These dangers come from those who intentionally manipulate biological agents in order to create deadly pathogens, and from those who accidentally create a new deadly agent in the course of scientific research. Another potential threat comes from the accidental release of an old illness that has been mostly eradicated, such as polio or measles.

To cite just two examples of existing biological weapons and how they came into existence, one could look at the Australian researchers who inadvertently discovered that the integration of a standard immunoregulator gene could greatly increase the lethality of smallpox or anthrax, or the team of biology researchers who were recently able to recreate the polio virus from scratch.[112]

Should terrorist organizations with the proper capabilities decide to launch a biological attack, there are many potential pathogens they could use. According to the North Atlantic Treaty Organization, there are as many as 39 potential bacteria, viruses, rickettsiae and toxins that could be used as biological weapons.[113] However, acknowledging the abundance of potential agents should not result in neglecting the difficulties of distributing biological agents.

Containment of a biological warfare agent is even more difficult, and it presents what is perhaps the single biggest obstacle to smaller players being able to effectively use biotechnology to launch biological warfare.[114] Because bacteria and viruses do not discriminate between 'friend and foe', they often end up contaminating the person or agent who seeks to release them. This so-called boomerang effect is a built-in method of deterrence that helps prevent biological attacks.[115] Moreover, releasing a biological agent into the air for distribution would be affected by changes in wind and weather patterns, making dispersion impossible to control and leaving open the possibility of many unintended consequences. All in all, to launch a high-profile, 'credible' event, a biological warfare agent must have many or all of the following properties: 'the agent should be highly lethal and easily produced in large quantities. Given that the aerosol route is most likely for a large-scale attack, stability in aerosol and capability to be dispersed [...] are necessary. Additional attributes that make an agent even more dangerous include being communicable from person to person and having no treatment or vaccine'.[116] In the context of existing technologies, smallpox and anthrax best fit this profile. Bioterrorism is a major geopolitical challenge, but the immediate risks of a widespread attack, particularly one with dramatic, rapid and wide-ranging effects, is minimal. For the time being, the US Centers for Disease Control (CDC) predict that existing public health systems should be able to respond effectively to any possible attempts to release biological pathogens.[117]

## An overview of the key trends and developments in biotechnology

Biotechnology offers unique opportunities for addressing global challenges related to health, aging, the environment, sustainable development, the growing world population and global food supply. Not surprisingly, the biotechnology industry is experiencing a period of exceptional growth. Health care is one of the most prominent areas for biotechnology applications, and over 200 biotechnology drug products currently exist to target diseases, cancers, diabetes and HIV/AIDS. Advances in major biotechnology-related tools and technologies such as bioprocessing technology, recombinant DNA technology, monoclonal antibody technology, cloning, protein engineering and biosensors will only facilitate progress in these and other areas. Stem cell research is another promising area of biotech research and may revolutionize the way common diseases are treated. However, religious and ethical issues are limiting progress in this area, and it is necessary for the global community to establish standards and guidelines for this promising but controversial technology.

Beyond health care, biotechnology has several other key geopolitical implications. Chief among them are the dangers of the hostile use of biology. Globally, the rise of the biotech industry has also meant a growing likelihood that innovations in biotechnology will be dual-use. Such technologies, especially if applied to bioweapons or bioterrorism, would constitute a serious global security threat. Small-scale bioweapon attacks have already been launched, with mixed success, but rapid progress in biotechnology could make a large-scale attack more likely to occur. Striking a balance between scientific innovation and strong, prudent regulation is perhaps the most significant biotechnology-related challenge for global policymakers.

# 6
# Genomics

## Genomics: a general overview

Genomics is the study of genes and the specific characteristics that make people, plants, bacteria and animals who or what they are. In some senses, genomics is the science most focused on the unique traits of the individual. Although genetics can be an intensely personal field of science, its applications and the applications of related sciences such as proteomics affect human health and agriculture on a global scale.

Genomics has the potential to revolutionize the way we understand the causes of diseases and how we treat them.[1] The list of current and prospective applications of genome research is impressive. For example, in the field of molecular medicine, potential applications include improving the diagnosis of disease, using gene therapy and control systems as drugs, detecting genetic predispositions to disease and designing customized medical treatments based on individual genetic profiles. With respect to microbial genomics, genome research can help with the timely detection and treatment of pathogens, the development of new energy sources, such as biofuels, and environmental monitoring. Genome research can also be used for risk assessments, DNA identification and bioprocessing.[2]

Additionally, advances in genomic research could also establish cures or therapies for numerous diseases that have thus far been considered incurable, such as HIV, malaria and tuberculosis. Genomic technologies may help to speed up drug development, making it more effective and less costly. Moreover, technological progress has accelerated the speed at which genes are identified, characterized and manipulated. This new wealth of information may lead to further developments in the above-mentioned areas as well as in areas that are beyond genomics' current applications.

As is the case with all the major strategic technologies, genomics presents numerous challenges, ethical questions and potential for abuse. Some key geostrategic concerns include the dangers of genetic selection and a growing 'genomics divide' between rich and poor countries.[3]

## The genomics industry: past, present and future

In 1953, biochemists James Watson and Francis Crick discovered the double helix structure of DNA. This discovery opened the door to understanding DNA replication, protein synthesis, gene expression, and the exchange and recombination of genetic material, all building blocks of modern genomics.[4] Over the past 50 years, genomics has advanced rapidly. By the 1970s, scientists were able to manipulate genes directly, and they were then able to precisely modify the 'nature' of an organism within a single generation.[5] In 1998 RNA interference was discovered,[6] making it possible to manipulate gene expression and to explore gene function on a whole-genome scale.[7] As early as the mid-1990s, scientists were sequencing entire genomes of eukaryotic cells. A year later, they sequenced the first animal genome; and by 2001, they had the first draft of the human genome.[8]

The latter discovery was part of what is probably the most significant undertaking in the history of genomics: the Human Genome Project (HGP). Started in 1990, the HGP was a 13-year endeavour to discover all the genes in human DNA and, more broadly, to understand how genes work, how genes code for cell-building proteins, and how genomic knowledge can be used to prevent, diagnose and treat human disease. Scientists from over 18 countries worked together to identify the 20–25,000 genes of human DNA as well as DNA's three billion chemical base pairs. On completion of the sequencing, the results were made public so that researchers from the public and private sectors around the world could use the information for further studies.[9]

The HGP has laid the foundation for numerous genomics-related research projects, including projects that seek to describe the common patterns of human DNA sequence variation and ones that look at how genetics influence humans' response to environmental factors, among other areas.[10]

The HGP has also inspired a handful of individuals to pay fees of hundreds of thousands of dollars to have their personal genomes mapped. Despite the preliminary nature of most genomics research, these individuals hope that having their personal genetic blueprint will ultimately enable them to identify their predispositions to genetic diseases.[11] Although very few individual genomes have been sequenced, each completely sequenced genome adds an important new dimension to genomics data and research.[12]

The overall market for genetic data and technology is estimated to be worth tens of billions of dollars,[13] and it will continue to grow as research and innovation advance.

Yet almost ten years after the successful completion of the HPG, critical voices argue that its benefits have been fewer and slower to materialize than initially expected. The translation of research insights and knowledge into effective therapy remains a major challenge. The return on the major investment in genomics research is yet to appear, which has prompted some researchers to dub the HGP a 'social bubble' or a 'Genomics bubble'.[14]

Naturally, a number of controversies surround genome research, including the ethics of embryo research, particularly as it relates to the use of embryonic stem cells; cloning issues; DNA databases and issues of 'open' versus 'specific' consent; the privacy of personal genetic data; sharing of tissue samples and genetic data; and issues relating to genetic testing and insurance.[15] These issues are explored in greater depth below.

## Different areas of genetic research

The sequencing of the human genome has resulted in the development of new technologies and bioinformatics tools that are helping researchers 'to study expression of genes, the function of proteins, metabolism, and genetic differences within populations and between individuals'.[16] Overall, genomics-related scientific research will have a great impact on the prevention, diagnosis and treatment of various diseases.

There are two types of genomics: structural and functional. Structural genomics focuses on 'the construction and comparison of various types of genome maps and large-scale DNA sequencing' and on discovering new genes.[17] Some examples of structural genomics include the HGP and the Plant Genome Research Programme. Functional genomics focuses on the structural, chemical and functional differences that underline the composition of a gene. Using existing data, researchers analyse the functionally essential parts within genes, learn to recognize gene expression patterns and use this knowledge for medical and other purposes.

The results of structural and functional genomics analysis can be applied to the diagnosis of diseases; to identify genetically modified food products; for antisense molecules to block gene expression; to test for microbial contaminants in food products or donated blood; to test for drug-resistant strains of HIV and other pathogens; and to develop gene-based therapeutics, such as DNA vaccines and gene therapies.[18]

More specific areas of genomics-related research and related technology developments include:

### RNA interference

RNA interference uses long, double-stranded RNAs (dsRNA) to silence the expression of certain genes in cells and organisms.[19] By turning off selected genes, scientists may be able to inhibit the function of disease-causing genes, thus creating new treatments for degenerative diseases or leading to the development of new products such as tear-free onions.[20]

### DNA shuffling

DNA shuffling technology takes related versions of the same genetic sequence (for example genes from the same or similar species), breaks them apart and then recombines or shuffles these genes to produce a new, possibly enhanced version of the original sequence.[21] DNA shuffling is a very powerful tool for breeding and hybridization, as it allows for the simultaneous mating of numerous species. Early experiments in this field indicate that DNA shuffling may produce DNA sequences that have as much as 540 times the activity of the original genetic sequence.[22]

### Genome Wide Association Studies (GWAS)

GWAS aims to relate an individual genetic variation to disease susceptibility.[23] The huge reduction in the cost of discovering genetic variation has allowed scientists to scan the DNA of people who suffer from certain diseases, analyse genetic differences and undertake a comparative analysis.[24] As scientists gain a greater understanding of the whole human genome, this information can be combined with clinical and other data to increase understanding of the biological processes influencing human health. GWAS will also help in the prediction and understanding of the occurrence of disease as well as in the personalization of medicine.[25]

### Genetic mapping

Genetic mapping involves identifying where a gene lies on a specific chromosome, either through linkage analysis (a determination of the relative location of a gene compared to other genes on the chromo-some) or through physical mapping (a process for locating a gene's exact location).[26] Overall, genetic mapping can help in the identification of genes that cause rare disorders. Progress is also being made towards mapping the location of genes that cause more common ailments, such as asthma and diabetes.[27]

## Genetic profiling

Genetic profiling, or identifying a person based on their unique DNA traits, is an important part of criminal investigations, paternity tests and fraud cases. Such technology can also be used to give individuals access to secure systems such as weapons or computers, making this technol-ogy important from a security point of view. Additionally, genetic pro-filing is offering insights into the nature and behaviour of stem cells, making it likely that scientists will someday be able to induce stem cells to reproduce human tissue on a large scale.[28]

## Genetic engineering

Genetic engineering allows gene modification by transferring the genetic material of one organism or species to another. The technol-ogy makes it possible to design organisms with traits that can make those organisms more valuable to researchers or more marketable to consumers.[29]

Using the recombinant DNA technique, genetic engineers manipulate DNA and transfer it from one organism to another, making it possible to introduce traits of almost any organism to a plant, bacteria, virus or ani-mal.[30] The creation of transgenic plants, for instance, may result in food with higher nutritional quality.[31] Although the potential downsides are widely discussed, genetic engineering may offer plausible solutions to problems in world agriculture, the global environment and health.[32]

## Genetic testing

Genetic testing tests blood or tissue to find genetic disorders.[33] This technology can be used to identify genetic diseases in people, unborn babies and embryos. It can also confirm the diagnosis of disease or iden-tify the presence of a genetic disease in an adult even before that person begins to show symptoms of the disease.[34]

## Gene therapy

Gene therapy is an experimental method of addressing a genetic problem or defect at its source.[35] More specifically, it attempts to fix flawed genes by replacing them with healthy ones, thereby enabling the cells, tissues or organs affected by the genetic mutation to work properly.[36] Although much research still needs to be done in this field, gene therapy may, in the future, allow doctors to treat a disorder by gene replacement instead of drugs or surgery.[37] Other possible approaches to gene therapy include deactivating a malfunctioning gene or introducing a new gene into the body in order to help fight disease.[38]

## Cloning

Once primarily the purview of science fiction writers, cloning has become a mainstream area of genetic research. Through a variety of different processes, cloning allows scientists to create genetically identical copies of a biological entity.[39] These clones can be used to test new drugs and treatment strategies, as well as to make copies of animals with desirable genetic traits such as high milk production or lean meat.[40] Therapeutic cloning is valuable in the production of stem cells that are genetically identical to the donor cell. These stem cells may be used to help grow healthy tissue that can be used to replace diseased tissues.[41]

## Pharmacogenomics

Pharmacogenomics looks at how an individual will respond to certain drugs based on their genetic make-up.[42] Eventually, this field may allow for drugs that are tailor-made to a person's genetic disease profile, thereby improving the quality and efficiency of drugs while also reducing their side effects and speeding up patient recovery time.[43]

## Bioinformatics

Bioinformatics is a cross disciplinary field merging biology, computer science and information technology into one discipline. In its earlier days, bioinformatics involved the creation and maintenance of a database storing biological information.[44] Now, the field is evolving to focus more on the analysis and interpretation of the collected data. Bioinformatics offers the possibility of 'a more global perspective in experimental design' and 'the ability to capitalize on the emerging technology of database mining'.[45] As scientists continue to gather more and more data on the human genome and other sequence databases, there is a need for methods to store, organize and index this information. Bioinformatics is the solution to this challenge.[46]

## The 100-dollar genome

In 2009 the Massachusetts Institute of Technology's *Technology Review* identified a nanofluidic chip designed by the firm BioNanomatrix as one of the top ten emerging technologies of the year.[47] The firm is in the process of developing genome sequencing technology, fast and cheap enough for the entire human genome to be read in eight hours for roughly USD 100, technology that could be ready within five years. Company founder Han Cao bases this projection on the nanofluidic chip that allows researchers to image very long strands of individual DNA molecules.[48] By neatly aligning DNA, Cao's chip allows scientists to sequence DNA in single strands, reducing many of the complexities associated with sequencing double-stranded DNA. If this technology can be successfully refined, a physician 'could biopsy a cancer patient's tumour, sequence all its DNA, and use that information to determine a prognosis and prescribe treatment – all for less than the cost of a chest X-ray'.[49] In short, this emerging strategic technology offers the potential for quicker genome sequencing and, as a result, improved medical care across the board.

## Proteomics: a complementary science

When the HGP released its results in 2003, many scientists were stunned to learn that the human genome, initially believed to have over 100,000 genes, actually had only about 20–25,000.[50] Immediately, these scientists began speculating on other sources of human diversity, theorizing that proteins – the output of genes – are the real building blocks of human life and the main drivers of both individuality and disease.[51]

The study of how these proteins produced by genes fit together is called proteomics, and many scientists and researchers now believe that proteomics may be more promising than genomics when it comes to the diagnosis and treatment of disease. Proteomics is an important complement to genomics because genes express their qualities and effects through proteins. The proteome, the protein counterpart of the genome, is a combination of proteins in a cell or in an entire organism.[52] A core characteristic of the proteome is that 'the cellular complement of proteins changes throughout the cell cycle in every cell, is different in different tissues, and can alter in response to environmental changes'.[53]

Because we have a full map of the human genome, mapping of proteins can be done in a very methodical and comprehensive fashion,[54] and this research may have applications in the pharmaceutical, bio-technology and diagnostic industries.[55] Current molecular research and technologies focus on targeted interventions for prevention and treatment. Most existing drugs were based on and developed from 500 known molecular targets, but as proteomics provides a deeper understanding of proteins, diseases, disease pathways and drug-response, thousands of additional targets may be discovered and exploited.[56] This would expand the potential for revolutionary drug developments.

Proteomic techniques are already being used: (i) to identify disease markers related to Alzheimer's, rheumatoid arthritis and heart disease; (ii) to analyse and discover new drugs; (iii) to develop new diagnostic kits; (iv) in plant biotechnology; (v) in cellular interaction studies; and (vi) to detect pathogens.[57] Despite the potential of proteomics, some scientists and policymakers have raised concerns that the data related to proteomics are being collected faster than scientists' capacity to understand and integrate them with existing data.[58]

Future research objectives in the field of proteomics are likely to include: '(a) cataloguing all of the proteins produced by different cell types; (b) determining how age, environmental conditions and disease affect the proteins a cell produces and discovering the functions of these proteins; (c) charting the progression of a process, such as disease development, the steps in the infection process or the biochemical response of a crop plant to insect feeding, by measuring changes in protein production; and (d) discovering how a protein interacts with other proteins within the cell and from outside the cell'.[59]

## Metabolomics

Related to genomics and proteomics, the science of metabolomics deals with a new field of research that allows the biochemical consequences of changing a proteome to be observed and analysed more directly than with proteomic analysis alone.[60] Metabolomes are low molecular weight molecules that are produced at different times in the lifecycle of a living thing. Metabolomics attempts to categorize and quantify these metabolomes and the conditions under which they occur.[61] By doing this, scientists can more precisely follow the effects of mutations, changes in the environment and treatment with drugs.[62] This may contribute to the development of new drugs and expand our understanding of how drugs work, interact with the body and how side effects are caused.[63]

## Epigenetics

The most recent breakthrough in genetics is the field of epigenetics. This science maintains that the behaviour of genes can be modified through environmental factors and, most surprisingly, that these changes can be passed down through generations.[64] As John Cloud explains, the epigenome 'sits on top of the genome, just outside of it [...] It is these epigenetic "marks" that tell your genes to switch on or off, to speak loudly or whisper. It is through these epigenetic marks that environmental factors such as diet, stress and prenatal nutrition can make an imprint on genes that is passed from one generation to the next'.[65] The implications of this science are potentially huge, as it implies, for example, that people who smoke cigarettes in their youth may alter their epigenetics. As a result, their children and grandchildren may go through puberty earlier than average.[66] This is just one example of the many potential implications of epigenetics.

Epigenetics is not evolution. It does not change DNA, but instead represents a biological response to an environmental stressor. If the environmental stressor is removed, then, over time, epigenetic marks will fade and DNA will revert to its natural state.[67] While it is important to note that epigenetic changes do not last forever, it is also important to stress that its transience does not mean it is not hugely powerful.

Scientists are working to determine how they can manipulate their growing knowledge of epigenetics. For example, they are working to compare the epigenomes of diseased cells with healthy ones in order to determine exactly how epigenomes may lead to cancer and other diseases.[68] In 2004, the US Food and Drug Administration approved the first epigenetic drug. Known as Azacitidine, the drug treats patients with myelodysplastic syndromes, a group of deadly blood malignancies. Azacitidine targets epigenetic markers to tone down genes in blood precursor cells that are essentially overexpressing themselves.[69] Since 2004, the FDA has approved three more epigenetic drugs that work, at least in part, by stimulating tumour-suppressor genes that have been silenced by disease.[70]

The thought of mapping the entire human epigenome is an enormous and potentially daunting task. It is likely to contain millions and millions of markers, and a full map of it would require major advances in computing power. That said, the potential of a mapped epigenome is overwhelming as it implies the ability to modify genetic predispositions at will.[71]

## International governance and regulation

In contrast to other emerging technologies, genomics governance has been widely discussed and regulated at the national and international level. The main trend within the international community is to adopt a 'human rights-based approach' to governance of the human genome.[72] Relatively early as compared to other emerging strategic technologies, The United Nations perceived genomics as an issue requiring regulation and it has adopted a consistent stance on the question. The 2005 UN Declaration on Human Cloning[73] and the 1998 Universal Declaration on the Human Genome and Human Rights[74] both emphasize the principles of human dignity and human rights.

At the national level, many countries have adopted legislation to regulate the use of genomics. For instance, the United States has introduced a number of bills on topics as diverse as research issues, privacy issues, employment non-discrimination and health insurance non discrimination.[75]

## Geopolitical implications of genomics

### Genetic engineering, food security and food safety

Global population growth, the financial crisis and the damaging effects that climate change has had on crop production and agricultural yields make global food supply and security major geostrategic challenges. The United Nations Food and Agriculture Organization (FAO) estimates that the number of undernourished people in 2009 has increased to 1.02 billion, which represents an aggravation of previous trends.[76] Moreover, the world's population is likely to double in the next 20 years. For the future food supply to match the same relative levels as today, it too would need to double. In order for the food supply to truly eliminate hunger, the current global food supply would need to triple in response to a doubling population.[77] Genomics and genetically modified crops (GMCs) could substantially improve food supply, food security and food safety, and, for these reasons, genomics and related technologies will have a dramatic influence on geopolitics and international security.

In recent years, scientists have used transformational technologies to construct seeds and plants that add various yield-enhancing traits to existing crops. Such traits include shorter growing periods, greater pest resistance or better overall quality. Researchers have also studied how to improve qualities such as herbicide and stress resistance through various genetic modifications. Importantly, GMCs may also be produced at lower cost than their non-modified counterparts – an important element of making food, and especially healthy foods such as fruits and vegetables, more available to impoverished and hungry populations.[78]

Other genetic modifications will mean that crops and livestock can be modified to be immune to certain regional diseases or viruses that would otherwise ruin entire seasons of production.[79] Similarly, plants can be made to better weather cold snaps and other meteorological events.[80] Genetic modifications can improve the taste of fruits and vegetables and increase the amount of nutrients in a product. Also important, genetically modified crops offer the chance for these better, tastier, more nutritious plants to be grown on less land. Further into the future, it is likely that we will see plants with modifications that enable them to act as human vaccines against infectious diseases such as hepatitis B, animals that are resistant to deadly diseases, and fruit and nut trees that yield years earlier than their non-modified counterparts.[81]

Corn has one of the most complicated plant genomes to have been deciphered. It took over 150 researchers and several years to understand corn's genetic make-up well enough to manipulate it. In the future, researchers hope to develop corn and other crops that require less water to grow and that meet specific demands of the biofuel industry, among other developments.[82]

The potential benefits of genetically modified crops are real, but they are not without controversy or difficulty. For example, the risks of GMCs include the possibility of toxicity and the creation of new allergens, alteration of the nutritional quality of foods, antibiotic resistance, the probable creation of new viruses and toxins, and threats to the genetic diversity of crops.[83] Nor are GMCs a blanket solution to global food problems and shortages. Issues such as infrastructure, governments' detrimental trade policies, and lack of more basic yield-improving technologies such as fertilizer also contribute to the problem.[84] Moreover, global hunger issues are not just a question of food supply, but also of food distribution and access to food for the poor. On a similar note, it is worth pointing out that many of the advantages gleaned from genetically modified organisms could also come from other techniques, such as hybridization or tissue culture methods that propagate virus-free root stocks.[85]

In the global debate, the European Union has been especially reluctant to embrace GMCs and, to a large extent, it has been a trendsetter for attitudes in the developing world.[86] The European distrust of GMCs can be partially attributed to the fact that there is a disconnect for European consumers between the possible dangers stemming from the genetic alterations to food and the benefit they could gain from such alterations. Food scares in the European Union and questions about food safety regulations during the 1990s also increased European qualms over GMCs, even though these earlier crises were unrelated to biotechnology.[87]

While GMCs certainly pose some risks and the EU is right to have a certain level of concern, many of the dangers associated with these foods are overstated or at least not yet fully understood or confirmed. Unfortunately, this fear is paralyzing potentially major agricultural players from getting more involved in cultivating GMCs.[88] Safeguards are in place to offset some of the problems, for example, modifying plants with genes from commonly allergenic foods such as peanuts is discouraged, unless it can be demonstrated that the genes being transferred are not responsible for the allergy.[89] Similarly, genes transferred into GMCs may potentially be adverse to human health if they transfer to the cells of the body. To minimize this risk, the WHO encourages GMC manufacturers to use genes that are not resistant to antibiotics so that any subsequent infection can be quickly fought.[90]

Another risk from cloning and the genetic modification of plants and animals is the potential for eliminating biodiversity. The implications of such a development are hard to anticipate but could potentially be quite serious.[91] Finally, although some concerns over genetically modified foods are valid and more research on health and safety implications is needed, plants bred through traditional techniques may also raise concerns about food safety. In fact, these plants are screened for safety and health issues in a much less careful way than their genetically modified counterparts.[92]

Overall, the safety of genetically modified foods should be assessed on a case-by-case basis, not as a judgement on the genre of foods as a whole. Certainly, some genetic modifications will result in unintended consequences or challenges, but others will be hugely beneficial to the causes of world hunger and the environment. A multifaceted approach is key. Appropriate regulation, industry transparency, monitoring and research will be necessary in order to avoid the potentially negative effects of genetically modified food technology.

### The global genomics divide

As is the case with other strategic technologies, one of the major concerns over genomics is the growing divide between rich countries, which can afford to invest large amounts of money in research and development of new technologies, and those poor countries that cannot. This issue is especially evident in the field of genomics, where developed countries, and particularly the United States, account for the vast majority of global investment in genomics research. The United States spent an average of USD 2.9 billion between 2003 and 2006.[93] Many public health officials are concerned that genomics may focus on the health needs of people living in industrialized countries at the cost of the widespread health problems that are prominent in developing countries. Such an imbalance will potentially cause even more disparities between the quality of health care in developed and developing countries.[94]

Another challenge for genomics and creating an environment of equal access is that of intellectual property and research incentives. Because of factors such as intellectual property rights and the desire of private drug companies to make big profits, people in poorer nations often cannot afford drugs that would treat widespread health problems. While this is not exclusively the result of for-profit pharmaceutical companies, these business models do present certain impediments to providing greater equality of access to the world's poor. There are few, if any, incentives for the companies with the technological and intellectual capabilities to develop the kinds of genetic treatments that would specifically benefit poorer populations. Subsidies from the government and public-private partnerships would be positive steps towards ameliorating this discrepancy.

On a similar note, a central challenge for genomics and other related areas of research is that discoveries in this field have characteristics of both public and private goods. Genomics research and knowledge of the human genome can benefit all of humanity, making it, on the one hand, a classic public good. UNESCO, for example, has declared the human genome part of the 'heritage of humanity'.[95] On the other hand, new developments in this field can lead to marketable patents and medicines with characteristics of private goods, which lead to the intellectual property problems outlined above.[96] This dual nature of genomics makes it particularly hard to regulate it efficiently and distribute its benefits equally. Without patentability, companies would be discouraged from investing in research. By the same token, such patentability reduces the poor's ability to access basic life-saving drugs and technologies. Ultimately, this growing genomics divide could augment social and political unrest in at-risk areas.[97]

Equitable economic investment in genomics research, more even distribution of and access to clinical research, services and technologies globally are key steps towards bridging the global genomics divide.[98]

Some have also suggested more inclusive and in-depth global dialogue over the genomics divide. Such discussions could set an agenda and lay the groundwork for a global genomics governance framework.[99] Singer, Daar and Dowdeswell also argue for 'financial investment models that direct resources to undercapitalized genome-related biotechnologies [that provide] both a social and an economic return on investment'.[100]

## Privacy issues

Mapping the human genome was a tremendously important scientific breakthrough that has already led to new understandings of disease and human individuality. Nonetheless, such data, particularly when culled on an individual level, raises questions of privacy and of who should have access to what information. The advent of the Internet, e-commerce and social networking mean that our traditional notions of privacy are coming under attack. To date, most international privacy initiatives have focused on information privacy and audio-visual surveillance but, slowly, concern is also growing over issues such as the collection, storage and use of personal data in public and private databases, both within individual countries and across international borders.[101]

This latter concern is particularly relevant to the field of genetics where there are concerns that insurance companies may begin to use genetic mapping as a basis for, among other things, denying coverage to at-risk individuals, thus limiting individuals' access to health care.[102] For some people, just the idea of genetic information being collected and used for scientific research is a cause of discomfort, and some even view such information collection as a violation of their human rights. 'Thus, the mere *existence* of information about genetic make-up is sufficiently bothersome to make some people anxious. While some people see the so-called genetics revolution as the holy grail of modern medicine, others approach the entire area with trepidation'.[103] Such trepidation comes from the fact that information could also be used as a basis for discriminating against certain ethnic groups, exacerbating social and international conflicts.[104] For example, if a certain ethnic group is predisposed to certain diseases that can be triggered by environmental factors, a group or an individual looking to undermine that ethnic group could work to alter the environment to the minority's detriment, potentially provoking health and reproductive problems for an entire group of people.

The WHO is working with the Geneva International Academic Network and the Bioethics Unit of the Medical Faculty of Geneva University to study how genetic databases can be established and maintained in an ethically responsible way.[105] As population genomics becomes an increasingly popular and promising area of study, standards and norms for the protection of privacy will become paramount.[106] It should be noted that many issues surrounding genetic privacy are similar to those surrounding the privacy of medical records and other sensitive information. These areas could provide a good starting point for discussions on how doctors, hospitals, patients, insurance companies and governments should handle individual genetic information.[107]

## Evolutionary metabolism and health security

Looking at the course of human history over the very long run (millions of years), it is clear that for the vast majority of prehistoric and historic times, the human diet consisted primarily of animal meat, offal and select fruits and vegetables. Indeed, 'the general primate dietary pattern is ancient',[108] and over the course of about 99 per cent of our human evolution, the human diet has been firmly rooted in the hunter-gatherer model.[109] Based on the different foods available at different points in history, the human body has been able to adapt to changes in diet, but it is important to stress that the base of our genetic metabolism is the foraging and scavenging diet that our prehistoric ancestors survived off for generations. This genetic legacy affects both how we process our food and our overall health today.

The traditional human diet continued through the Palaeolithic period nearly 10,000 years ago when humans were still hunter-gatherers and were forced to put substantial effort into securing food supplies. The scarce food supply consisted primarily of animal meat, leaves and root vegetables. At the time, these food items were the most accessible and, importantly, they were safest to eat (grains and beans are unsafe to eat unless they have been cooked, a process which was not available to these hunting nomads).[110] Because of this diet, conditions such as cardiovascular disease, cancer, stroke, hypertension, depression and diabetes were extremely rare.[111] Such a diet would also indicate a high muscle density and the prime period for the growth of human cranial capacities during this period of human development.[112]

With the advent of agriculture in approximately 7000 BC, our ancestors saw a dramatic change in their diets and, even thousands of years after the agricultural revolution, the human metabolism is still adapting to these changes.[113] For example, the agricultural revolution led to a much greater integration of grains and cereals into the human diet.[114] The introduction of these foods had an enormous impact on human health, in many cases making it possible to more than double daily calorie consumption as well as making it easier to store and transport food.[115] Consumption of dairy products also increased, something that, from an evolutionary perspective, human bodies were not really conditioned to digest – especially after early childhood.[116] In many instances, diet shifts were regional. Aleutian migrants coming to Alaska thousands of years ago had less access to agriculture, and their diets continued to be centred around protein and fat (from seals, birds, whales, caribou and fish).[117] However in the area of ancient Rome, there was a greater integration of fruits, vegetables, rice and wheat into the daily diet.

Over time, these variations in diet led the health of different ethnic groups and of people from particular regions to change in particular ways. For example, today, Eskimos and Native American populations are dealing with the rather abrupt and dramatic introduction of carbohydrates and processed foods into their traditionally fat- and protein-rich diet. One of the results of this is that they are seeing higher rates of diabetes and cancer compared to their ancestors.[118] Such community-wide proclivities to a disease such as diabetes are a huge strain on local and regional health resources and the quality of life of these populations.

The human genome evolves very slowly in response to changes in diet, and the evolutionary features of our metabolisms that were established thousands or even millions of years ago have a lasting impact on human health security today. This can be seen in a variety of ways around the world, but Mexican populations are a prime example. Prior to the agricultural revolution, diseases such as diabetes and obesity were very rare in Mayan and indigenous Latin American societies. This was largely because food supplies varied across seasons, and therefore humans would have faced long periods of scarcity. Such situations would have favoured 'thrifty' genotypes, or ones that increase the storage of fat and give their phenotypes a competitive edge.[119] With the emergence of the agricultural revolution, the thrifty gene was no longer as valuable or essential to survival as it had been. The slow evolution of the genome in relation to the change in diet helps to explain the increased incidence of obesity, even among malnourished individuals, and this trend is particularly notable in Mexicans and Mexican Americans.[120] Today, evolutionary metabolism is defining the health problems of entire regions of the world, from increased rates of obesity and diabetes to higher incidences of and predispositions to cancer. In response to this problem, states need to promote healthier lifestyles and to make health screening more widely available. Fighting obesity and improving physical fitness are also important priorities.

## An overview of key trends and developments in genomics

Progress in genomics technology has accelerated the speed at which genes are identified, characterized and manipulated. The sequencing of the human genome was perhaps the most notable achievement in this emerging technology's history, and it has resulted in the development of new technologies that allow the study of genes, proteins and the metabolism. Although scientifically promising, the HGP has raised a number of ethical and social issues related to embryo research, cloning, genetic testing, gene therapies, the privacy of personal data, the existence of the DNA databases and so on.

Outside HGP, other branches of genomics research focus on improving the diagnosis of disease, detecting genetic predisposition to disease, designing customized treatment based on individual genetic profiles, timely detection of pathogens, monitoring the environment, developing new sources of energy, DNA identification, risk assessment and bioprocessing.

From a health perspective, genomics will be key in addressing multiple global health issues. Genome sequencing and blood tests will be able to assess the probability of developing certain diseases. Such information will be valuable in preventing diseases, but will also raise privacy issues and questions over equality of access to health care. Other issues that genomics will influence include the global food supply, particularly in the context of genetically modified foods. Although we have the technology available to manipulate and modify foods to improve nutrition and crop yields, we must be wary of possible dangers in these foods and the unintended consequences of manipulating plants on a large scale. Finally, the evolution of our metabolism and our prehistoric genetic disposition to certain diets also affect regional and ethnic health issues. While genetics may seem like an intensely personal science, discoveries and advances in this field are influencing global health care, food supplies and the overall quality of life for many populations.

# 7
# Nanotechnology

## General overview

Nanotechnology is a 'fundamental enabling technology': the science of engineering objects that are less than one micrometer in size.[1] Even though, or precisely because, nanotech products are infinitesimally small, the potential for revolutionary discoveries and innovations in this field is enormous. As William Sims Bainbridge has put it, nanotechnology is a 'region where many technologies meet, combine, and creatively generate a world of possibilities'.[2] An interdisciplinary science involving disciplines such as physics, robotics, chemistry, biology, and electrical and mechanical engineering,[3] nanotechnology already strongly influences fields as diverse as medicine, energy and the environment. It is expected to have an even deeper impact across additional realms of science, technology and society in the future.

In technical terms, nanotechnology refers to the science and engineering involved in manipulating and/or manufacturing objects with features in the 1–100 nanometer range, that is, smaller than one micrometer. As the US National Nanotechnology Initiative explains, nanotechnology is:

> The understanding and control of matter at dimensions of roughly 1 to 100 nanometers, where unique phenomena enable novel applications. Encompassing nanoscale science, engineering, and technology, nanotechnology involves imaging, measuring, modeling, and manipulating matter at this length scale. At the nanoscale, the physical, chemical, and biological properties of materials differ in fundamental and valuable ways from the properties of individual atoms and molecules or bulk matter. Nanotechnology research

and development is directed toward understanding and creating improved materials, devices, and systems that exploit these new properties.[4]

In short, nanoscience and nanotechnology are based on the understanding that all materials posses certain properties at the nanoscale level and that these properties define those materials' actions. Working at the nanoscale, scientists can begin to exploit, control and predict those properties, thereby gaining the ability to manipulate the materials and their functions.[5] Nanoscale materials such as carbon nanotubes and silicon nanowires are finding interesting applications in electronics, giving rise to the emerging field of nanoelectronics. Moreover, since biological materials such as DNA, amino acids and proteins are in the nanometer range, the convergence of nanotechnology with biotechnology opens extremely interesting horizons of research and development.[6]

Nanotechnology is a particularly important field of science because it brings together a number of revolutionary technologies in order to create and assess new models for everything from converting solar energy into electricity to producing a new generation of materials that could be up to 20 times stronger than steel.[7]

Although nanotechnology is still at an early stage of development, significant advances in this field are already being made.[8] For example, nanowires and carbon nanotubes (molecular-scale tubes of graphitic carbon with remarkable properties, not to mention the strongest fibres known to man)[9] are already showing tremendous potential to generate and store energy, and researchers are excited about nanotechnology's potential to render objects invisible and to improve the lifespan of batteries, among other developments.[10] The Project on Emerging Nanotechnologies estimated that, by April 2008, there were at least 600 nanotechnology-related projects going on around world.[11]

Nanotechnology has the potential to fill basic human needs and improve human quality of life across a variety of sectors and industries. As the field evolves, major breakthroughs are expected in such areas as energy and the environment, global and national security, medicine and health care, manufacturing and materials, biotechnology and agriculture, and information technology and electronics.[12] Although some have warned of a 'romantic mythology' surrounding nanotechnology,[13] it is still safe to say that as nano and related technologies become increasingly mainstream, the whole way in which materials and systems are created and employed will undergo significant transformation.

## The nanotechnology industry and key areas of research

In 1959 the Nobel Laureate in Physics Professor Richard P. Feynman gave a lecture entitled 'There's Plenty of Room at the Bottom'.[14] This speech, widely regarded as the conceptual birth of nanotechnology, raised the idea of the direct manipulation of atoms and considered the scientific potential of altering matter on an atomic scale.[15] In the decades since Professor Feynman's speech, and particularly in the past 10–15 years, research and development in nanotechnology have accelerated rapidly across numerous sectors. Between 2003 and 2008, global public and private investment have increased by 27 per cent annually.[16]

Although still a relatively new industry and still in the early phases of discovery and innovation, nanotechnology has tremendous promise, and expectations for the industry's near and distant futures are huge. Four billion USD was invested in nanotechnology research in 2005; by 2015 the expected aggregate market revenue for nanotechnology is expected to reach one trillion USD.[17] The global market for nano-products is expected to grow to USD 2.5 trillion over the next four years.[18] Over the next two decades, nanotechnology's multi-industry potential growth and increases in profits are staggering.

The United States is the clear global industry leader, but China, South Korea and the European Union are striving to catch up by investing large sums in research and development. In the United States, the importance of nanotechnology to the economy is widely acknowledged. President George W. Bush gave an additional boost to the nanotech industry when he signed the 21st Century Nanotechnology Research and Development Act in 2003,[19] which emphasized nanotechnology's ability to create revolutionary new processes and products as well as to improve existing ones. At the initiative of John Kerry, the National Nanotechnology Initiative Amendment Act 2009 was introduced and is being discussed in the US Congress. It aims to further enhance research and development and to improve the translation of research findings into practical applications. Furthermore, it proposes to put more emphasis on nanoscience at both the college and the university level.[20]

Giovanni De Micheli of EPF Lausanne, Switzerland, has given an admittedly optimistic prediction about the future of nanotechnology in terms of the following trajectory. According to him, in 2000, science was focused on the development of passive nanostructures, reinforcing fibres in new composites, and carbon nanotubes in electronics. Today, we are witnessing the rise of 'active nanostructures' capable of changing their own properties – materials that change shape, for example.

By 2015, we will likely have self-assembling nanosystems and materials that can essentially construct themselves. By 2020, De Micheli predicts that we will have heterogeneous nano networks that will come together and construct themselves.[21] At this point, we would conceivably be able to build any kind of material from scratch. Herein lies some of nanotechnology's greatest potential. 'Fully realized, nanotechnology promises dramatic changes to all aspects of facilitating structure'.[22] That said, most of nanotechnology's potential is still quite a distant prospect, and caution should be exercised, as it can be 'easy to exaggerate the opportunities for technological dynamism offered by this radical new technology'.[23]

## Regulating nanotechnology: risks and opportunities

For all its potential and early successes, nanotechnology is not without its dangers. By definition, nanotechnology deals with particles at the atomic level, and its products can be as small as one red blood cell or one molecule of DNA.[24] Thus, the same characteristics that make nanotechnology products valuable and useful to humans are the same traits that may also render them extremely dangerous, both from a human health perspective and an environmental one.[25] Especially worrying is the fact that many of the dangers of nanotechnology are unknown, not well researched or merely speculative.[26] As Vicki Colvin, a professor of chemistry at Rice University, has said, 'If you fund five teams to understand nanotube toxicity, and they get five different answers, your research investment hurts you, because it creates uncertainty. The bad news is that we have way over five different opinions about carbon-nanotube toxicity right now'.[27] In sum, not only does this mean that nanotechnology potentially has many unidentified risks, but it also makes the industry extremely difficult to regulate.

Nanoscale technologies and products could be the source of unintended or unforeseen risks,[28] such as the development of weaponized pathogens. While nanotechnology could provide an opportunity to develop tests that would reduce the threat from virtually every known infectious disease, the same technology could also increase the risk of catastrophic global pandemics.[29] Similarly, nanotechnologies have the potential to reduce greenhouse gas (GHG) emissions and improve the efficiencies of all industrial and manufacturing processes but, at the same time, we also have to take into consideration the fact that certain nanotechnology-related products and processes may create their own versions of emissions – the presence of man-made nano-particles in the atmosphere may have unintended consequences for the environment.[30]

Many of nanotechnology's potential risks stem from the fact that nano scale particles may behave differently than larger particles, even if the particles have the same chemical structure. Therefore, nano-particles and nanotechnology products may cause unpredictable and unintended problems.[31] Nano-particles are already being used in products such as sunscreen and cosmetics, and scientists and consumer groups have raised concerns about the possible dangers and negative chemical reactions caused by absorption of nano-particles by the skin.[32] Similarly, nano-particles produced or used in industry and manufacturing may, if inhaled, penetrate more deeply into lungs than larger particles of the same material. Once in the body, these particles would be small enough to penetrate the blood-brain barrier, potentially damaging the central nervous system.[33]

From an environmental perspective, nanotechnology products and by-products could contaminate soil and groundwater[34] or lead to inadvertent depletion of the ozone layer.[35] Some of the nanotechnology industry's risk assessments read like science fiction, with concerned observers warning of a '"Grey Goo" catastrophe in which self-replicating microscopic robots the size of bacteria fill the world and wipe out humanity'.[36]

While the grey goo scenario can be written off as an almost ridiculously alarmist assessment of the risks of nanotechnology, the key problem remains that many of nanotechnology's dangers are unknown or poorly understood. This is reinforced by the fact that nano-particles are not a homogenous product with uniform behaviours. Thus, each nano-product has to be individually assessed for its own potential dangers.

The uncertainty surrounding the risks of nanotechnology makes the industry extremely difficult to regulate. Moreover, the industry is evolving so fast that any existing rules or guidelines run the risk of becoming obsolete before they even take effect.[37] Defining nanotechnology adds another difficult dimension to the debate. Even people within the nanotechnology industry have sparred over the creation of a universal, legal definition of their industry's scope. Defining nanotechnology is not mere semantics. Any acknowledged definition of nanotechnology will have serious implications for both the public and private sectors and for how the industry is regulated.[38] The impact of this proposed definition on regulation remains to be seen. As it currently stands, nanotechnology does not exist as a category in industry- and patent-related classifications, and it is not yet possible to speak of only one identifiable nanotechnology sector or industry but only of a number of various applications related to a number of different sectors and industries.[39]

International regulation of the nanotech industry should begin with standardization of nanotechnology testing and research,[40] a process already begun by the International Standards Organization.[41] Additional regulatory frameworks are being debated and evaluated by the OECD, the members of which are some of the world's largest and most significant nanotech players.[42] In March 2007, the OECD established the Working Party on Nanotechnology, with the goal of fostering international cooperation in the research, development and commercialization of nanotech products.[43]

De Micheli recommends that, given the ongoing growth of nanotechnology sensor systems, governments put in place good practices for nanotechnology. Such practices would include verification that ensures correct operation under all possible scenarios.[44] According to De Micheli, nanotech systems must also 'be secured against possible malicious attacks that may force them into undesirable modes of operation'.[45] Data encryption and hierarchical access to system control are two steps towards achieving this goal.[46]

More specifically, De Micheli advances the following policy recommendations:

(a) use common sense and see if regulation would hold in the *meter* range; (b) use toxicity standards that are well understood in other branches of science; (c) help stakeholders identify the highest quality nanotoxicology studies and (d) section nanotechnology into a number of narrowly defined subfields for regulation purposes.[47]

Because nanotechnology has a wide variety of possible military applications (discussed in greater depth below in this chapter), some have called for international laws to prevent a nanotechnology arms race. Within existing international frameworks, there are already guidelines that regulate areas relevant to nanotechnology developments. As a starting point, these agreements could be explicitly expanded to include potential nanotechnology advances. For example, Jürgen Altmann and Mark A. Gubrud suggest 'limits on military autonomous vehicles and robots, in particular with a combat function'.[48] Additionally, 'small mobile systems should be mostly prohibited in the military as well as in civilian society'.[49] These broad-brush recommendations fall generally within the parameters or mandates of existing international regulations, and such rules would cover certain elements of nanotechnology, if not specifically then at least by extension. Yet, Altmann also questions the very capacity of the international system to deal with nanotech weaponry. Given the fact that these tools will be used by not only states but also non-state actors, and that the complexity of the military uses of nanotechnology is increasing at a high speed, he poses the question whether the international system needs a fundamental, democratic modification in order to cope with this challenge.[50]

Current international efforts to design regulations or best practices for nanotechnology include the International Dialogue on Responsible Research and Development of Nanotechnology and the International Council of Nanotechnology.[51] However, these forums remain mostly a place for dialogue among various nanotechnology professionals. The groups have no legislative power or authority. In addition, there is some ad hoc regulation of the nanotech industry at the national and regional levels. For example, in the United States the Food and Drug Administration must approve the use of all drugs made, which naturally includes those drugs designed with the help of nanotechnology or that involve the use of nano-products. At the regional level, China, Japan, South Korea and Taiwan have begun centralizing their nanotech research and development.[52] Meanwhile, NGOs such as the Erosion, Technology and Concentration (ETC) Group have tried to politicize the nanotechnology issue by calling for a moratorium on the use of nanoparticles until a set of 'best practice' guidelines has been established.[53]

With so little public sector regulation of the nanotech industry, there is increasing pressure for the private sector to engage in self-regulation.[54] In this regard, comparisons are often drawn between the nanotech industry and the more established biotech industry. As the international community looks for ways to regulate the nanotechnology industry and calls on the private sector to oversee its own actions, it would be well advised to begin by looking at the lessons learned from biotechnology, information technology and other related emerging technologies.

## Nanoconvergence: combining nanotechnology with other technologies

As is the case with other technologies discussed in this book, nanotechnology has even wider geopolitical implications when used in conjunction with other strategic technologies than it does on its own. In fact, some argue that the real technological revolutions of the next century will not occur in one specific field, as has been the case in the past (for example the ICT revolution) but across disciplines. Looking exclusively at the pace of development in one individual area of scientific research may give the impression that innovation is slowing after a period of high growth and, thus, the new way forward in technological and societal development is through synergy, collaboration and convergence. Such convergence is, at its core, based on 'material unity at the nanoscale and on technology integration from that scale'.[55] In short, nanotechnology is the field that is capable of bringing all other fields together in one way or another.

A report by the US National Science Foundation put it best when it said, 'The sciences have reached a watershed at which they must combine in order to advance most rapidly'.[56] Indeed, technological innovation and society's progress will be dependent on convergence and collaboration between fields previously viewed as separate from each other. As ICT, biotechnology, cognitive science and artificial intelligence technology converge, nanotechnology will be the common thread pulling these seemingly distinct sciences together. Convergence promises to dramatically increase the rate of scientific progress; this will be easier if we begin thinking about science from a holistic perspective rather than a strictly specialized one.[57]

Already, there are numerous cases that show exactly how convergence is revolutionizing everything from medicine to space exploration. For example, nanotechnology combined with biotechnology is allowing for the development of pre-engineering nanostructures. Certain genes in aquatic micro-organisms can now be re-engineered with the coding for various microstructures, such as microtransistors or silicon chips. This will have important commercial value.[58] At the University of California, scientists have combined bio- and nanotechnologies to attach rat heart cells to silicon-based or plastic artificial muscles to create mini-robots that are 'alive' and able to grow and assemble.[59]

Other examples abound. A system that brings together nanotechnologies, biotechnologies, cognitive science and ICT 'would remove barriers to communication caused by physical disabilities, language differences, geographic distance and variations in knowledge'.[60] This 'Communicator' system would improve the effectiveness of cooperation in schools, government agencies and businesses around the world. Moreover, it could reduce language barriers making communication easier across cultures and borders.

The US National Aeronautics and Space Administration (NASA) has been a leader in converging technologies, using the production and scale-up of carbon nanotube-reinforced polymer matrix composites that can be applied in a variety of structural applications as a starting point.[61] Such nanotubes might also have high-temperature applications and may be able to reduce the weight of launch vehicles by up to 80 per cent.[62]

Along with space technology, nanotechnology could help to create new robotic systems able to successfully operate in complex environments such as outer space while also allowing the elimination of expensive human inputs.[63] Nanotechnology could also lead to greater functionality and smaller size of space-related products such as launchers and payloads. Eventually, combinations of nano and space technologies could lead to deeper and more profound exploration of the solar system.[64] On a related note, the application of advanced medical nanosystems' could eventually make long-term survival in space possible.[65]

The fast pace of manufacturing in high-performance nanoscale products can inspire innovation, research and development in other technologies. For example, 'the ability to rapid-prototype finished test hardware at relatively low cost may allow significantly more aggressive experimentation'.[66] The aerospace, health care and ICT sectors may all benefit from such experimentation.

With regard to combinations of energy and nano technologies, carbon nanotube technologies possess a high degree of tensile strength, and they offer promising opportunities for the storage of hydrogen for transportation or power generation applications, as well as computer-related switching and innovative high-tech composites.[67] Given all this potential, governments should increase their funding of convergence-focused research and development. As the National Science Foundation puts it:

> A vast opportunity is created by the convergence of sciences and technologies starting with integration from the nanoscale and having immense individual, societal and historical implications for human development [...] Science and technology will increasingly dominate the world, as population, resource exploitation, and potential conflict grow. Therefore, the success of this convergent technologies priority area is essential to the future of humanity.[68]

We must prepare ourselves for these potentially rapid changes in our social and technological existences. One key way to do this is to improve education and training in issues specifically related to convergence.[69] Our education system as it stands is still too rigid, and someone who studies biotechnology or computer science has few tools for applying his expertise to another industry or field. Other recommendations for the advancement of nanoconvergence include the reorganization of universities and curricula to reflect the changing nature of technical study and to better train the labour force for the future; fostering partnerships between manufacturing, biotechnology, information and medical service companies in order to increase investment in production facilities based on the new principles of nanoconvergence; and government funding for more national and international research and development on converging technologies.[70] The development of an ethical and legal framework for converging technologies is another priority, and one that should be addressed in an international forum such as the United Nations. These governmental initiatives should be reinforced by private sector initiatives to create professional societies focused specifically on interdisciplinary pursuits and convergence.[71] NGOs should also follow this lead in order to maximize the benefits of convergence to their diverse constituencies.[72]

## Geopolitical implications of nanotechnology

### Social

According to Bainbridge, who has extensively studied the interaction between nanotechnology and society, 'the social and economic consequences of nanoscale science and technology promise to be diverse, difficult to anticipate, and sometimes disruptive', but the science nonetheless 'provides a unique opportunity for developing a fuller understanding of how technical and social systems affect each other'.[73]

One key question surrounding the future of nanotechnology is whether advances made in this industry will narrow or expand the gap between developing and developed countries.[74] 'On the one hand, countries with the personnel and resources to discover, patent, and use nanotechnologies would have advantages that could widen the gap between nations. On the other hand, the potentially inexpensive nature of nanotechnology could lead to a narrowing of the gap, as is the case in communication and information technologies'.[75] If nanoscience and nanotechnology-related knowledge from developed countries could be applied at a lower cost in developing countries, it could contribute to improvements in agriculture and food production.[76]

Nanotechnology also poses similar societal challenges with regard to intellectual property rights and concerns over the protection of personal privacy.[77] Indeed, as is alluded to above, questions have already arisen over how to define nanotechnology and related concepts in a legal sense. The complexity and interdisciplinary nature of nanotechnology means that the patent issues surrounding innovations in this field are equally complex.[78] As was demonstrated by one of the earliest nanotechnology innovations, the scanning tunnelling microscope, nanotechnology often does not conform to existing classifications of intellectual property.[79] Because of its multidisciplinary nature, the microscope did not classify under normal headings such as diagnostic tool or pharmaceutical. Ultimately, each element of a nanotechnology product or tool may require multiple patents from all sorts of different disciplines.[80] For example, in the United States, the patent and trademark authority has technology centres for biotechnology, organic chemistry, and chemical and materials engineering (each of which has various subsections), but there is still no specific nanotechnology section.[81] Because of this, nanotechnology innovations run the risk of either having their patents rejected because of a misconception that the matter is not new, or giving the owner 'excessive control over a particular area'.[82]

## Health care

Nanotechnology could play a key role in future improvements in the detection, diagnosis and treatment of major illnesses. Some potential developments could include specialized devices for early detection of heart problems, infections and tumours; new applications for hearing and vision; and innovative tests to identify disease vulnerability and potential responses to drugs.[83] Additionally, it is expected that medical repair systems will be improved in many significant ways. For example, some scientists hope to design medical devices that will be able to clean the arteries or repair injured tissue and organs or discover and destroy tumours and cancer cells. Through nanotechnology, medicine is likely to become more effective, more personalized and more cost-efficient.[84]

Most of the nanotechnologies developed for medical purposes come in the form of very small objects that are absorbed by blood, water or 'a complex experimental concoction'.[85] Scientists are constructing special devices able to manipulate very small amounts of such liquids, known as microfluidic systems pump solutions.[86] Some microfluidic devices allow scientists to undertake complex experiments that would not otherwise be possible, such as delivering solutions to designated parts of particular cells.[87] Such bloodstream-based devices have already succeeded in curing type 1 diabetes in rats and other equally impressive achievements are close to unfolding.[88] When integrated into clothing, nano-particles can help monitor the body's function, temperature, heart rate, and so on. Such monitoring will facilitate the administration of telemedicine for people in rural areas or for elderly people who live alone but require regular monitoring by medical professionals.[89]

Meanwhile, implants based on developments in nano and biotechnologies are likely to result in replacement human organs.[90] Nanotubes and nano-particles can potentially be used to regulate glucose, $CO_2$ levels and cholesterol levels from within the body, thus providing earlier warning systems for potentially major health problems.[91]

In combination, such developments could have a wide-ranging effect on the state of global health and security. According to Piotr Grodzinski, director of the US organization the Nanotechnology Alliance for Cancer, 'Nanotechnologies could revolutionize health care in developing countries and make treatments more readily available for diseases that claim millions of lives around the world each year'.[92] For example, it might one day be possible for citizens of a country such as Bangladesh 'to place contaminated water in inexpensive transparent bottles that will disinfect the water when placed in direct sunlight, or for doctors in Mexico to give patients vaccines that can be inhaled and do not need to be refrigerated'.[93] Although these technologies are mostly still at the developmental stage, they all have great potential to save human lives and improve the quality of human life on a global scale.[94]

### Energy and the environment

Worldwide energy requirements are likely to double by 2030, and nanotechnology has the potential to generate cleaner energy resources that are not only safer than existing energy resources but renewable as well.[95] Moreover, nanotechnology may offer the creation of goods and services in a more environmentally friendly manner.

Potential energy and environment-related advances in nanotechnology are numerous and varied. As nanoscience and nanoengineering increase our understanding of nanoscale-related processes and their molecular composition, they provide opportunities to improve the condition of the environment through the development of new renewable technologies and more advanced control of greenhouse gas emissions.[96] Among the possibilities are new hydrogen storage systems based on carbon nanotubes, photovoltaic cells and organic light-emitting devices based on quantum dots, and hybrid protein-polymer biomimetic membranes.[97] It is also possible that nanosensors will be able to monitor levels of pollution in the air, identify toxic materials and leaks, and facilitate the degradation of air-pollutants, including $CO_2$.[98] Such positive potential does not come without similar potential dangers.

Nano-particles could also behave as micropollutants. Already, such particles have been dumped by pharmaceutical companies, and they are proving to be a problem for the quality of surface water in Switzerland and other places.[99] Indeed, disposal of nano-products goes far beyond the pharmaceutical industry. In order to prevent the permanent introduction of dangerous nano-particles into the environment, any dispersal of these particles into the atmosphere should be carefully controlled.[100] On a similar note, it is imperative that scientists study the interaction between nano-particles and living organisms in-depth.[101]

In the area of solar energy, nanotechnology will result in silicon-based photo voltaic cells being replaced by thin-film solar panels, which 'are flexible, unlike silicon PV cells, and can be "printed" in a process similar to printing newspapers'.[102] This would make solar power more cost-competitive with conventional energy generation. Indeed, in a similar vein, one of nanotechnology's most promising elements is the possibility of fairly quick development of inexpensive and large-scale hydrogen or solar power projects. 'Such unexpected advances would dial back some of the economic pressures coming from fast-rising energy demand and slow-rising oil investment and production, possibly make climate change a problem that countries could manage without major sacrifices, and reduce the geopolitical leverage of Russia and the Middle East'.[103] Nanocomponents are also essential to storing hydrogen that could potentially power a car. Nobel Laureate Richard Smalley believes that such developments are less than ten years away.[104]

Importantly, nanosensors could help in environmental monitoring, charting the impact of climate change, because although natural events are largely uncontrollable, many of them can be at least partially detectable.[105] In this context, 'nodes are placed in appropriate positions in the environment, such as on glaciers, water sources, or surfaces. The corresponding measurements of temperature, water flow, and water levels can be transmitted over standard (for example cellular) or ad hoc networks'.[106] With the help of nanotechnology, scientists are better able to track the indicators of global climate change, thereby providing crucial information to global decision makers.

According to the International Energy Agency, future increases in oil prices will stimulate demand for more energy efficient cars and vehicles. Nanotechnology is a promising response to this problem. In the future, the tradition of using carbon fibre as a composite filler will be replaced by using nanotubes, which are much stronger and lighter than steel.[107] These characteristics will help vehicles become more efficient without sacrificing speed, quality or durability.

Although nanotechnology has much potential in the fields of energy and the environment, it is important not to understate the magnitude of the changes and investment that would be required to bring nanotechnology to its full potential. In many cases, and particularly in places such as Latin America and India, shifting to entirely new fuel sources would require trillions of dollars in new investments.[108] Thus, in the short run, nanotechnology and related energy innovations are likely to have the biggest geostrategic impact on global efforts to reduce $CO_2$ emissions and to offset global warming. In time, the new energy sources could also be a 'body blow to the oil-producing states, including Russia and Iran, and over time reduce the geopolitical significance of the Middle East'.[109]

## Military issues

As is the case with many of the other strategic technologies addressed in this book, nanotechnology is of interest to military forces around the world – for a variety of reasons. In 2007, for example, the US military was spending over USD 417 million a year on nanotechnology-related research.[110] Technical superiority is often a key component of military strategy and perceived strength and, for this reason, the countries that invest the most in nanotechnology-related research and development are likely to gain a competitive advantage both on the battlefield and in terms of national prestige.

There are numerous potential military applications for nanotechnology. For example, miniature, mobile and autonomous sensors could be used to penetrate the remote and secure facilities of an adversary or rival government.[111] Past and current research projects include artificial insects, underwater micro-robots and micro air vehicles.[112] Such sensors would provide virtually 'total awareness' of the battle field, an obvious advantage to the army which has such technology.[113] As supercomputers become more ubiquitous, the replacement of humans with artificial systems and intelligence will increase, and new materials will lead old weapons to become obsolete.[114] Some goals for military nanotechnology researchers include battle suits that protect against bullets, new chemical and biological agents that could be used either as weapons of mass destruction or in targeted attacks, materials that compress or splinter on impact, and an exoskeleton with 'muscles' made out of artificial molecules. This latter possibility would be especially useful for soldiers carrying heavy and unwieldy loads.[115]

Nanotechnology also has the potential to enhance or alter existing versions of nuclear weapons, both in terms of improving the safety and reliability of these weapons and adding new types of explosive to the fission primary.[116]

Overall, nanotechnology is likely to change the way we envisage and plan for war, possibly giving additional advantages to terrorist groups and guerrillas who practice asymmetrical warfare.

### Privacy issues

In certain applications where nanotechnology and ICT are combined (for example in wireless sensor networks), they pose a risk to individual and national security by threatening to completely eliminate the notion of privacy. Nanosensors can potentially be placed in any environment – not just military ones – to monitor the movement of people in their home or business. For example, nanosensors could be mixed with paint used to paint a house and its walls. Such an installation would be permanent and would have the potential to dramatically reduce the privacy of the people living in that house without their knowledge or consent.[117] Another concern on this front is the question of unintended side effects. What if governments place nanosensors in something like house paint only to discover in 20 years that such paint is as deadly as we now know asbestos to be?[118]

As a concluding thought on this topic, De Micheli speculates that eventually we may reach a tipping point with nanotechnology, and public opinion will turn against it because of its potential privacy violations. De Micheli recommends the creation of policies that balance the security of communities and the security of individuals against the possible misuse of information.[119]

### An overview of key trends and developments in nanotechnology

Nanotechnology is one of the most important strategic technologies discussed in this book because it can be combined with other strategic technologies to create truly revolutionary innovations. While still in its early phases of development, the next two decades show enormous promise for this budding field, and, when evaluated in conjunction with biotechnology, ICT and cognitive science, nanotechnology's potential seems limitless. In the coming years and decades, it is predicted that nanotechnology will play an important role in manufacturing, electronics, improved health care, pharmaceuticals, chemical plants, transportation and sustainability. The scope of this potential is reflected in practical terms by the fact that nanotechnology is one of the fastest-growing industries globally. Geopolitically, nanotechnology will facilitate drug development, possibly help offset environmental pollution and, in its more advanced stages, allow us to create any kind of material essentially from scratch. The science of nanotechnology will translate into important societal implications, covering such areas as medicine, energy, ICT, privacy and the environment.

However, nanotechnology also presents a set of dangers. From a military perspective, nanotechnology may help surveillance, giving technologically advanced militaries a major advantage over their counterparts without such technology. On another note, because materials behave differently at the nano-level than they do on their normal scale, nanotechnology may result in a multitude of unforeseen consequences for the environment or human health. Such unintended outcomes and the huge diversity of nano-particles and nano-research make strong and coherent regulation of this emerging technology extremely challenging. Certain industry groups and governments have started to develop a set of industry best practices, and this should be encouraged as a good first step to more comprehensive oversight. The private sector has also shown a willingness to engage in self-regulation, and this should be fostered.

# 8
# Materials Science

## General overview

From the Bronze Age to the Iron Age, humans have framed their views of the world based on the materials available to them and the opportunities those materials afforded in terms of quality of life. Today, we live in a world with over 300,000 known materials and, thanks to the scientific and technical discoveries of the past half-century, the study and manipulation of these materials has now risen to the level of its own interdisciplinary science.[1] Materials science looks at how different materials – anything from wood to ceramics to metal to semiconductors – respond to different conditions and how the properties of these materials can be modified to enhance the materials' functions or to create new ones.[2] Using knowledge from fields such as chemistry, physics and electrical and mechanical engineering, materials science is influencing developments in nanotechnology and micro and electronic materials, among other fields.[3] Advances in materials science have encouraged multiple innovations in science and technology that are already improving our quality of life.

## The materials science industry and relevant technologies

The materials science industry is increasingly interdisciplinary, so it can be hard to quantify the total size of the industry or to highlight specific areas of research that are exclusive to materials. Some particularly interesting areas of interdisciplinary materials research are outlined below.

### Organic electronics and green plastics
Organic electronics is a field of electronics that includes plastics, conductive polymers and small molecules. This breed of electronics has

been dubbed 'organic' because the molecules making up the products are carbon based, like the molecules of living things. Organic electronics is more flexible than its traditional counterpart,[4] and it offers the potential for things such as bendable electronics and the production of semiconductors that are cheaper and more flexible than their non-organic counterparts.[5] Some electronic devices, such as a Sony Walkman launched in 2009, are available with these technologies, and many more are expected to use them in the near future.[6] Organic solar cells cut the cost of solar power[7] and could help generate power for rural developments and for military operations in remote locations.[8]

Green plastics are plastics made with the help of biotechnology from biorefineries. They make it possible for petroleum-driven polymers to be replaced with biological ones. Plant material-based sugar can be used to make polyesters and plastics, creating polylactic acid packaging materials and clothes.[9] Importantly, green plastics are biodegradable, potentially giving them an added environmental benefit over their traditional counterparts.[10]

### Carbon nanotubes, silicon nanowires, and molecular electronics

Carbon nanotubes are long, thin cylinders of carbon that are unique for their size and physical properties.[11] Essentially, they can be thought of as a sheet of graphite rolled into a cylinder. They are an important type of material because they have a broad range of electronic, thermal and structural traits, which can be modified in response to the changes in the nanotube's type, for example, to a different diameter or degree of twist.[12] Carbon nanotubes' unique electronic and mechanical properties offer enormous potential because of their notable strength and their highly elastic moduli. They have also been shown to hold up well under pressure.[13] Depending on their structure, nanotubes can be metal or semiconductors, and wires or active components in electronic devices.

Silicon nanowires are a related technology that has the potential to convert waste heat into electricity as well as potentially helping to cool computer chips, build refrigerators, or make car engines more efficient. This is possible because silicon nanowires have a lower thermal conductivity than traditional silicon. As silicon is cheaper than the commercially popular but very costly bismuth telluride, which is currently one of the most efficient thermoelectric conversion materials, silicon nanowires could help in the development of thermoelectric devices that are cheap and more readily available.[14]

Carbon nanotubes and silicon nanowires are just two elements of the broader materials science shift towards molecular electronics. Although traditionally, inorganic insulators, semiconductors and metals have been the core of the electronics industry, it is likely that over the next two decades, molecules will become the active device components in electronic circuitry.[15]

## Spintronics

Spintronics is the science of using electrons to store data.[16] Combining materials science with nanotechnology in order to exploit unique characteristics of the electron, spintronics results in smaller, more versatile versions of silicon chips and circuit elements. Whereas traditional electronic devices depend on the electrical charge of electrons in a semiconductor such as silicon, spintronics uses the spin of the electron.[17]

> All spintronic devices act according to the simple scheme: (1) information is stored (written) into spins as a particular spin orientation (up or down), (2) the spins, being attached to mobile electrons, carry the information along a wire, and (3) the information is read at a terminal. Spin orientation of conduction electrons survives for a relatively long time (nanoseconds, compared to tens of femtoseconds during which electron momentum decays), which makes spintronic devices particularly attractive for memory storage and magnetic sensors applications, and, potentially for quantum computing where electron spin would represent a bit (called qubit) of information.[18]

Potentially, this industry could be worth hundreds of billions of dollars per year.[19]

## Wide band-gap semiconductors

Wide band-gap semiconductors are semiconductor materials with electronic band gaps wider than one or two electronvolts. They can reduce the energy consumption and increase the hardware reliability of most electric devices, and they are particularly useful in very high temperature situations.[20] Gallium Nitride and Silicon Carbide are among the most promising materials for future electronic components. Compared to traditional semiconductors, they offer enormous advantages in everything from power capability to radiation insensitivity and even low noise capability.[21] They have specific strategic applications in the development of next generation systems.[22] Yet further research and development is needed.

## Smart materials

Smart materials, including piezoelectric, magnetostrictive and shape-memory materials, sense characteristics of their environment and respond to cues from those characteristics.[23] In many ways, these materials act in a similar way to biological systems.[24] Developments in these fields could lead to innovations such as a self-repairing house or an antenna that moves in response to a signal.[25]

Piezoelectric materials, such as quartz, are smart materials that are able to produce an electric current when compressed or distorted and that move in response to an applied current.[26] They are produced from cellular polymers and have the ability to conform to almost any size or shape.[27] There are numerous applications of such materials. For example, researchers have used them to convert electronic stimuli into mechanical reactions, developing artificial muscles that are stronger than natural human muscles.[28] Australian scientists are hoping that they can construct a tiny piezoelectric motor that will be able to enter the body, take photographs, deliver medicine and possibly even perform surgery.[29]

Smart textiles are essentially wearable technology. They will be able to sense how much to absorb, to measure pulse and breathing rates, to repel stain-causing agents, and heat up or cool down depending on the weather.[30] They can also interact with their environment.[31] MIT Research Labs have produced fibres that both identify and produce sounds, or, as one journalist has put it, that are able to 'hear and sing'. One possible use of this technology would be to develop clothes that function as microphones.[32] In sum, smart materials offer new possibilities for monitoring biological processes.[33]

Magnetostrictive materials are materials that undergo a change when their magnetization is changed.[34] This effect recognizes that certain materials, such as iron, nickel, cobalt, and ferrite, undergo 'slight changes in size under the application of a magnetic field'.[35] The most prominent commercially available magnetostrictive material is an alloy known as Terfenol-D. Terfenol-D exhibits large magnetostriction and relatively small applied fields at room temperature, a unique trait for a magnetostrictive alloy.[36] Often used as an alternative to piezoceramics, Terfenol-D 'is a more rugged material that can handle higher temperatures, change shape faster, and exerts more force per volume of material than piezo'.[37] Terfenol-D can also be used to make any flat surface into a speaker. Terfenol-D and other magnetostrictive materials have the potential to help detect earthquakes and be used in hearing aids.[38]

Shape-memory materials can restore 'their initial physical shape after undergoing deformations'.[39] They have been used in a variety of medical, military and robotic applications. Shape-memory devices are useful for developing less invasive surgical tools. Shape-memory alloys 'can be bent and twisted into any orientation and subsequently heated to a threshold temperature to recover their original shape'.[40]

Shape-memory alloys are at the core of several cardiovascular devices (e.g., the Simon filter). They are used 'to control flow through blood vessels to avoid pulmonary embolism'.[41] Shape-memory materials can also be used 'as an intelligent fastener system, potentially eliminating the need for screwdrivers, wrenches, and rivet guns in high-end fastener applications'.[42] Some possible future applications of shape-memory materials include aircraft and automotive manufacturing, robotics and the security field. By 2020, it is predicted that smart materials will have been incorporated into many aspects of medicine, and the energy and environmental sectors.[43]

## Pharmaceutical materials

Although materials science is most often associated with the development of structural materials such as metals and ceramics, there is much potential for materials research into soft matter such as polymers and biomaterials.[44] This is particularly true of the pharmaceutical industry, where materials science developments are helping to increase the efficiency and productivity of bringing new drugs to market.[45] Increasingly, pharmaceutical materials research is becoming its own entity. For example, a joint research venture between Cambridge University in the United Kingdom and the pharmaceutical company Pfizer looks at 'all aspects of the structure, manufacture and behaviour of solid dosage forms, such as tablets, at all relevant scales of operation and use. Research spans all length scales from modelling the processes of molecular crystallization through to achieving better powder compaction, tableting, diffusion and release'.[46] This research has yielded advancements such as controlled crystallization on surfaces, internal imaging of whole tablets, and better computational simulation of amorphous materials.[47]

## Composites and coatings

A composite is a material created by combining two or more materials to obtain a specific new set of characteristics and properties.[48] Coating is a process that generally occurs at the end of the manufacturing process in order to impart certain properties to enhance the product's desired function, or to protect that product from environmental threats. For example, coatings are often used as thermal barriers to protect sensitive materials from extreme temperatures; they can also help to strengthen a surface and prevent corrosion.[49] Advances in coatings and composites are reducing costs for the airline and space industries, allowing them to use cheaper steel with special coatings, instead of more expensive superalloys. Anti-erosion coatings can lead to as much as a 5 per cent increase in the efficiency of materials.[50] Recently, researchers have made substantial progress in the development of metals that are able to repair their own surfaces when they are damaged.[51]

## Metamaterials and invisibility cloaks

Metamaterials are composites with the ability to bend electromagnetic waves in such a way that they negatively refract light.[52] They have been described in more technical terms as a 'matrix of exceptionally tiny, sometimes nanoscale, metal wires and loops to control electromagnetic radiation in ways natural substances can't'.[53] Metamaterials can essentially make objects invisible if their structural array is smaller than the electromagnetic wavelength being used. To date, scientists have managed to make objects invisible to microwaves, and the hope is that, eventually, such metamaterials will be successfully applied to wavelengths in the visible light spectrum.[54] Developments in the field of metamaterials could possibly help in higher resolution optical imaging, nanocircuits for high-powered computers, and eventually possibly even cloaking devices that could make objects invisible to the human eye.[55]

Metamaterials are significant because they can alter the way that light normally behaves. Speaking of invisibility cloaks, the metamaterial has the potential to curve light waves completely around an object, similar to the way in which river water flows over a rock.[56] Because the light waves would reconnect on the other side of the object, the person looking at it would have the sensation of looking through it. Different variations of such technology are being developed in laboratories around the world and, while a true invisibility cloak the likes of which would make Harry Potter proud is still a long way away, invisibility is completely consistent with the laws of physics and light. For example, at the University of California in Berkeley, researchers have observed negative refraction from red light wavelengths, the first instance of bulk media bending visible light backwards.[57] This has been heralded as an important step towards invisibility cloaks. Shortly after, a team of researchers succeeded in creating the first material with this same isotropic negative index, which makes light waves run backwards.[58] Before actual invisibility cloaks can be created, however, these metamaterials are likely to help in other practical areas such as reducing antenna interference or removing the Doppler effect.[59]

Such materials would have significant geopolitical implications, especially in the context of the military where soldiers, weapons and tanks could be more effectively concealed from enemy eyes.

### Nanopiezoelectronics

As is noted above, carbon nanotubes and nanotechnology have enormous potential to revolutionize everything from electronics to health care. Nanoscale sensors, in particular, are extremely sensitive and power efficient, and they could be used across fields to detect everything from molecular signs of disease in the blood, to minute amounts of poisonous gas in the air and trace contaminants in food.[60] The main challenge to the widescale adoption of such technologies is the power source: it is notoriously difficult to make the batteries and integrated circuits that are necessary to drive these devices on a miniature scale.

However, scientists at Georgia Tech have made notable advances in miniscule power generators that take advantage of the above-mentioned piezoelectricity. These advances may allow nanosensors to power themselves. Crystalline materials that produce electrical pressure when stressed are known as piezoelectric materials, and Georgia Tech's Zhong Lin Wang was the first scientist to demonstrate this potential at the nanolevel.[61] This offers the potential to create a nanoscale power source that could be powered by things as basic as sound, the wind or even blood flow over an implanted device. Although nanopiezoelectronics are still in the early stages of development, the potential for them to create a sustainable, clean power source is promising. Wang even envisages a time when nanopiezoelectronics could be woven into fabric, allowing the rustling of a shirt to power an MP3 player or similar device. Other possible applications include a neuralprosthesis hearing aid, the affordable and effective detection of mechanical stress in an aircraft engine and medical diagnostics.[62]

## The regulation of materials science

International efforts in materials science have so far focused more on opportunities for collaboration on research and development than on formal regulatory structures. In lieu of a specific international framework for the regulation of materials science, the overall trend is for ad hoc regulation of the industry at the national or regional level. Often, these regulations are not specifically targeted at materials science. Rather, they affect the industry because of the nature of its products. For example, in the United States, the International Traffic in Arms Regulations controls the export of materials that could be used to manufacture weapons.[63] As many such products are those developed by materials scientists, these regulations affect the industry overall.

As is the case in many areas of scientific research, research and development in materials science is becoming increasingly globalized as more and more transnational companies become involved. This trend is reinforced by the ease of communication and information sharing brought about by the ICT revolution.[64]

## Geopolitical implications of materials science

### Military

From blankets that provide protection from bomb shrapnel to stronger, more durable construction materials, the innovations being made in materials science could revolutionize the way armies protect and defend themselves. As a result, materials science could influence how military leaders plan for future battles against better-defended enemies.[65]

In the United States, the navy has a special centre that specifically studies computational materials science. As outlined on the US Navy's website, the potential applications of this department's research are many. For example, by investigating the underlying physical and chemical principles that facilitate energy storage and creation, scientists are able to better evaluate the relations between structure, composition and performance, and to evaluate a material's potential as a fuel cell or a battery cathode.[66] Developments in this area could mean more fuel-efficient airplanes, ships and tanks. Less need for conventional fuels could translate to lighter vehicles capable of moving more quickly and with greater agility.[67] One of the US Department of Defense's stated goals is to be able to deliver munitions to any part of the world within one hour, something that could only be achieved with vehicles constructed with this type of weight reduction.[68] Reduced fuel needs or more efficient fuel cells and batteries could also have important environmental implications, as they could reduce the need to burn fossil fuels to meet our energy needs.

Other innovations in materials science could affect everything from military communications to manpower requirements to increased likelihood of survivability for soldiers on the battlefield. For example, using magnetorheological fluid, scientists are developing lightweight, flexible body armour that turns hard when hit by a projectile.[69] Other smart materials may be able to change at the molecular level in response to biological or chemical threats, giving soldiers new levels of protection from biochemical attacks.[70] Similarly, certain alloys and super stainless steels are being developed, and these would allow submarines or other vehicles to maintain their stealth for longer periods of time. By changing their acoustic and thermal signatures, materials could help military vehicles avoid detection.[71]

Several of materials science's military applications could also easily be applied to civilian and humanitarian uses. Smart construction materials may help buildings sense damage and cancel out strong vibrations,[72] protecting them from earthquakes or military attacks. Although the application of such developments could be decades away, their widespread use could ultimately mean fewer battle and/or natural disaster casualties. There are signs of early success in this field. For example, in 2007, a bridge spanning the Mississippi River in the US state of Minnesota collapsed without warning, killing 13 people and injuring 145.[73] That bridge has since been replaced with a smart bridge that has an embedded early-warning system made of hundreds of sensors. In the future, researchers hope to develop a cement-based sensing skin that will monitor an entire bridge. Such technology would be better than individual sensors because individual sensors may not be placed exactly where a crack occurs.[74]

Despite their potential, the challenges of applying advances in materials science to the military and civilian worlds are numerous. For example, there is currently a very long time period between when a material is created and when it is mature enough to be used in practical applications.[75] Even when cutting edge materials make it from the research and development phase to the practical application phase, questions often remain about how the material will age.[76] Whether it is electromigration, dielectric breakdown or thermal-induced stress, new materials may alter or change in a variety of ways over the medium and long term.[77] Questions like this are important for military planners to consider when they commission new weapons and vehicles that could potentially be in use for decades.

Combined, materials science developments will affect geostrategy in numerous ways. For example, a country in which the military has lightweight, fuel-efficient vehicles and lightweight, durable, highly protective armour will have an advantage on the battlefield over an army that does not have these assets. Such technologies could reinforce the military strengths of the traditional military powers (or at least those with the money to invest in materials research) and put weaker militaries with fewer resources to invest in technology at a comparative disadvantage.

The impacts of certain natural disasters could be greatly reduced if smart construction materials were employed to make buildings and structures more resistant to the elements and forces of nature.[78] This would reduce the need for emergency aid and support and would allow the military and related organizations to focus more on development in the aftermath of a disaster instead of spending all their resources on basic reconstruction. Overall, the purpose of materials science is to make structures, products and environments safer and to help us live in a better, safer world. As is the case with many other technologies, the applications of materials science in the military realms can quickly expand and be applied to issues and needs in the civilian realm. For this reason, military research and the development of materials will eventually benefit the wider population.

Although materials science has many potential applications for militaries around the world, it is important to stress that much of the current literature on these possibilities is focused on speculation, ideas and plans. Thorough research results, the presentation of the technical details of new materials, and fully developed products that are readily available and usable in conjunction with existing structures and technologies remain elusive.[79]

### Research and development

Research and development across all scientific fields is crucial to economic and technological development, and the countries that invest in these areas and develop and maintain technological superiority will, in the future, be at an advantage geopolitically. At the same time, cooperation between countries on non-defence-related research will reduce overall costs, while the sharing of information and research will speed up scientific discoveries.

The main challenge is the transition between developing materials in a laboratory and using these materials in day-to-day life. To minimize the difficulties of this transition, 'funds need to be allocated specifically to take promising materials from research through the developmental steps necessary for the materials to become reasonable candidates for service use'.[80] High levels of participation from system designers, the private sector and engineers will be a key part of identifying the most viable materials for military and private sector development.[81]

## Space

There is an interesting and important connection between materials science and space science. Space's microgravity environment allows a better understanding of different materials. NASA, the European Union Space Agency and the Russian Space Agency are using the microgravity environments at the International Space Station (ISS) to advance research in the fields of metals and alloys, electronics and photonic materials, and glass, with the aim of understanding the role of gravitational forces in crystallization, solidification and property measurement.[82]

Space is not just a research forum for materials science; new materials are regularly being developed specifically to be used in space. For example, new optical materials are especially valuable in the context of space exploration. Whether it is fibre optic sensor materials or nanostructured materials, the unique environment of space demands materials and products that we might not necessarily use in our lives on Earth.[83] Indeed, for a space-based fabrication to be effective, its weight and power requirements must be minimized. Currently, researchers are working on an efficient space heating system to satisfy these characteristics. NASA researchers are looking at a microwave approach to space-based materials processing.[84] As more and better materials are developed there will be opportunities to do new and valuable forms of research in space, and we will also be able to launch more cost-effective and safer space missions.

## Health

Smart materials are already being adapted to enhance wearable devices such as clothing and watches. With new materials, such clothes will be able to make us feel more comfortable, cooling us when we are hot or drying us when we sweat. Eventually, they will also be able to monitor bodily functions such as temperature, heart rate and breathing. Such developments will help people and their doctors better monitor health, especially chronic conditions.[85] This kind of monitoring will also facilitate telemedicine, making health care more accessible to people in remote areas or without easy access to medical professionals.[86] Sensors embedded in clothing can help monitor falls by the elderly or disabled, again facilitating their access to quick medical care.[87]

Also, developments in biological materials science may lead to breakthroughs in pharmaceuticals and drug delivery. For example, small nano molecules can be filled with a medicine that will target a very specific part of the body, at a specific time and in a specific dose. The potential for such nanomaterials to help eradicate diseases such as cancer are quite promising.[88]

Other health-related applications of materials include biodegradable and bioactive fibres that are being used in surgical implants and tissue engineering structures. Certain materials are also used to make hygienic textiles that can help sterilize medical facilities and improve general hygiene.[89]

### Nuclear energy

Nuclear power might not emit greenhouse gases, but it does produce radioactive waste as a by-product. Much of this waste comes from the water used to cool the core of pressurized water reactors, the most common type of nuclear reactor. Researchers in Germany and India have come up with a way of offsetting this by mopping up the radioactivity with a new type of polymer plastic material. This polymer binds to cobalt but ignores iron.[90] This is useful in nuclear waste clean-up because in a pressurized water reactor, hot water circulates through pipes made out of steel. This dissolves metal ions from the metal the pipes are made of. When the ions re-enter the reactor's core, they are bombarded by neutrons and sometimes become radioactive. This is especially true for cobalt, a material used to strengthen the steel pipes in reactors. The new polymer that is being developed could trap any cobalt ions, and a small amount of the polymer can swallow significant proportions of the radioactive cobalt.[91] If radioactive waste could be concentrated in these polymer beads, it would be cheaper to dispose of, especially compared to the large volumes of low-level waste now produced. As nuclear power undergoes a renaissance, polymer beads and similar materials science developments may help make nuclear power more politically feasible and more mainstream.[92]

## An overview of key trends and developments

Whether it is the development of completely new materials or the use of existing materials in new, cutting edge ways, materials science is creating both opportunities and challenges in terms of military, energy and outer space uses. This interdisciplinary science is responsible for developments ranging from superalloys that can operate at temperatures in excess of 2000°F to construction materials that are lighter weight and stronger than traditional products. Some of the most promising developments in this field include the elaboration of smart materials, which can detect and respond to changing environments; organic electronics, which can help harness solar power; and spintronics, a branch of materials science that uses electrons to store data.

The interdisciplinary nature of materials science has made it difficult to regulate, and this is posing intellectual property challenges to scientists and researchers around the world. In spite of regulatory difficulties, collaboration and joint research and development projects across borders are common in materials science, and this helps to foster broader global cooperation in the sciences. From a geopolitical perspective, materials science is producing materials that will make militaries faster, stronger and more efficient. Developments in this field promise flexible but highly protective armour, as well as lighter weight vehicles that can speed up transportation and refuelling processes, among other things. Like many other strategic technologies, materials science is also promising for health care. Many of these developments are rooted in materials science's convergence with fields such as biotechnology and nanotechnology.

Convergence is a key theme of this book, and like materials science, cognitive science and technology – the next and last emerging strategic technology this book will specifically address – will create a variety of new opportunities and innovations. These new ideas will affect us on an individual human level and on the geopolitical stage.

# 9
# Artificial Intelligence

Cognitive science is an interdisciplinary field that broadly encompasses the study of the mind and intelligence. Some of the sciences and specialties included under the umbrella of cognitive science are: artificial intelligence (AI), linguistics, anthropology, psychology, neuroscience, philosophy and education. From a strategic technology perspective, and especially when looking at strategic technologies in the context of their influence on human nature and geopolitics, the sciences of AI and neuroscience are especially relevant and significant. This chapter looks more closely at the AI dimension of cognitive science. The neuroscience component is evaluated in Part II of this book.

## Artificial intelligence: a general overview

The Singularity is near, or so argues Ray Kurzweil, inventor, futurist and author of the book by the same name. In Kurzweil's words, the Singularity is a 'future period during which the pace of technological change will be so rapid, its impact so deep, that human life will be irreversibly transformed'.[1] All the emerging strategic technologies discussed in this book will affect and be affected by the potential arrival of the Singularity, but the technology of AI, and specifically smarter-than-human AI, is among the most central technologies in this context. AI is 'simply the application of artificial or non-naturally occurring systems that use the knowledge-level to achieve goals. A more practical definition that has been used for AI is attempting to build artificial systems that will perform better on tasks that humans currently do better'.[2] AI essentially involves the creation of a computer with thinking ability that outstrips that of a human, potentially by a factor of trillions.[3] These machines, created by humans, will eventually be able to operate and communicate much more effectively than humans ever will because their communication interfaces will not be strictly limited to language, as ours are.[4] Moreover, these machines will be interconnected and will be able to share information much faster and easier than we as humans can.

The holy grail of AI is to create computers with intelligence that equals or surpasses that of humans.[5] This is the grand goal of AI, but many scientists have abandoned such lofty projects, preferring to focus their programming on more precise, manageable goals.[6] That said, today, robots can already drive cars, play sports and independently find information on the Internet.[7] As scientists build on these milestones, it is not unreasonable to at least consider the possibility that the creation of a super-intelligent computer is possible. Computers are already able to perform calculations and data analysis much faster and more efficiently than human beings can, but there are clear limits to computers' current intelligence. For example, it is extremely difficult to program so-called common sense knowledge on to a computer, and tasks that would be overwhelmingly easy for even a two year old (e.g., telling the difference between a tomato and an apple) are almost impossibly difficult even for highly sophisticated computer systems.[8]

Should a smarter-than-human machine be created, it will have the capability to produce a machine or computer program even smarter than itself. With the help of nanotechnology and other emerging technologies, such a second generation machine will, in turn, be able to devise, create and construct an even smarter machine. Each subsequent generation of machines will presumably be smarter and more capable than the last. Suddenly, challenges that are simply beyond the abilities of the human brain to process and solve will become solvable. Problems such as climate change and curing genetic and viral diseases as well as other formidable obstacles that we currently face could potentially be resolved at lightening speed once the Singularity has arrived. As is explained on the website of the Singularity Institute for Artificial Intelligence:

> Human intelligence is the foundation of human technology; all technology is ultimately the product of intelligence. If technology can turn around and enhance intelligence, this closes the loop, creating a positive feedback effect. Smarter minds will be more effective at building still smarter minds. This loop appears most clearly in the example of an Artificial Intelligence improving its own source code.[9]

Without doubt, the post-Singularity world would be dramatically different from the world we live in today. Vernor Vinge is credited with coining the phrase 'Singularity', and he has argued that having a true understanding of our lives in the post-Singularity world is like trying to apply our model of physics to the singularity of a black hole – our understanding of the way things are just breaks down in this context.[10] It is simply impossible for us to truly grasp what the Singularity would look like:

> We're trying to guess what it is to be a better-than-human guesser. Could a gathering of apes have predicted the rise of human intelligence, or understood it if it were explained? For that matter, could the 15th century have predicted the 20th century, let alone the 21st? Nothing has changed in the human brain since the 15th century; if the people of the 15th century could not predict five centuries ahead across constant minds, what makes us think we can outguess genuinely smarter-than-human intelligence?[11]

In spite of this admittedly massive shortcoming in our ability to predict the future, we can still establish pre-Singularity frameworks and policies that push science and AI in positive directions for as long as we have control over them.

## AI industry

The AI industry involves contributions from a number of scientific disciplines, including computer science, robotics and mechanical and electrical engineering. To quantify the size or value of the AI industry per se would be a daunting task because of its interdisciplinary nature, but it is safe to say that AI has applications in everything from banking systems that detect credit card fraud to speech-recognition telephone systems to computer chess programmes.[12] In fact, every email sent and cell phone call made is routed through the help of AI.[13] In 2007, the Business Communications Company estimated that the overall AI market was worth roughly USD 21 billion and projected annual growth at 12.2 per cent.[14]

The origins of the AI industry date back to the 1950s. In its earliest phases, AI was primarily focused on using cognitive and biological models to simulate and explain human information processing skills, and on developing robots capable of perceiving and interacting with their environments.[15] It was not until later in the century that the science and technology of AI really took off with the development of so-called expert systems,[16] or systems in which a 'computer applies heuristics and rules in a knowledge-specific domain [in order] to render advice or make recommendations, much like a human expert would'[17]

Today, AI can integrate learning, vision, navigation, manipulation, and even reasoning and speech.[18] Increasingly, AI is able to perform in complex real-world situations, and its pace of progress and development, after a seemingly slow period of development in the 1980s and 1990s, is now increasing quickly. AI has proved itself especially valuable in day-to-day life not just for its own applications but also because it often results in the creation or development of spin-off technologies. Such inventions include the laser printer, the personal computer and high-level symbolic programming languages, to name a few.[19]

## Relevant technologies: AI

Artificial Intelligence is a broad concept. Often, it entails a significant robotic component, which acts as the physical manifestation of the intelligence, but, in many instances, it can include many other types of technology and innovation. Some of the most promising contemporary AI technologies include neural networks that stimulate the working of neurons in the brain; computers that can understand, translate and communicate in human languages; computer programs that can solve and discover new mathematical problems and theories; simulations of how humans reason based on past experiences; and vision-related tasks such as face recognition.[20]

AI and intelligence-driven computers and machines have numerous advantages over their human counterparts. For example, machines are able to share information and learning experiences much more easily than humans thanks to worldwide grids of computing resources that can come together to form massive supercomputers.[21] Moreover, machines and AI consistently perform at peak levels, synthesize and process all available information, and have exacting memories. Unlike humans, AI machines and program do not have 'bad days'. 'For these reasons, once a computer is able to match the subtlety and range of human intelligence, it will necessarily soar past it and then continue its double-exponential ascent'.[22]

In broad terms, AI can be classified into three types: symbolic, connectionist and evolutionary.[23]

### Symbolic AI

Symbolic AI 'is the branch of artificial intelligence research that concerns itself with attempting to explicitly represent human knowledge in a declarative form (e.g., facts and rules)'.[24] Based on logic, symbolic AI uses sequences of ordered rules to instruct the computer what to do next.[25] It requires that procedural characteristics of human thoughts, actions and reactions be converted into symbols and rules that can be understood and manipulated by computers.[26] The technological challenges of symbolic AI are many. For example, symbolic AI depends on reactions and decisions being very specifically programmed after natural human ones. On the surface this may seem like a straightforward task, but what researchers and scientists have underestimated is the significance and complexity of common sense knowledge, or 'the vast amount of implicit knowledge we all share about the world and ourselves'.[27] Programmers can create symbols and rules so that a computer knows, for example, that Queen Elizabeth is in England. For humans, common sense knowledge would dictate that if Queen Elizabeth were in England, her left foot would also be in England. Although this is obvious to our human brains, it would require a whole separate line of programming and equations for a symbolic AI computer to specifically recognize this. As humans, we take this sort of logic for granted, but when you consider the infinite number of common sense judgements that we as humans make every day, the prospect of creating algorithmic rules for each possible thought seems virtually impossible. Thus, programming this type of knowledge on to a machine is a daunting task and one that researchers are still grappling with.

Another challenge for symbolic AI is that many human decisions that AI scientists try to replicate depend on procedural or implicit knowledge. Again, this is a difficult task to address in the framework of Symbolic AI.[28]

## Connectionist AI

Based on some of the most fundamental concepts in neuroscience, connectionist AI works off a series of artificial neural networks. These 'neurons' are each connected to their neighbours by links that 'can raise or lower the likelihood that the neighbor will fire',[29] and, collectively, these neural networks can be trained to recognize simple patterns.[30] Ray Kurzweil describes neural nets as follows:

> Each point of a given input [...] is randomly connected to the inputs of the first layer of simulated neurons. Every connection has an associated synaptic strength, which represents its importance and which is set at a random value. Each neuron adds up the signals coming into it. If the combined signal exceeds a particular threshold, the neuron

fires and sends a signal to its output connection; if the combined input signal does not exceed the threshold, the neuron does not fire, and its output is zero. The output of each neuron is randomly connected to the inputs of the neurons in the next layer. There are multiple layers (generally three or more), and the layers may be organized in a variety of configurations. For example, one layer may feed back to an earlier layer. At the top layer, the output of one or more neurons, also randomly selected, provides the answer.[31]

A powerful, well-taught neural net could mimic a broad range of human capabilities, including pattern-recognition faculties. Increasingly, researchers are using more complex and realistic models of actual biological systems in order to structure their neural nets.[32]

Importantly, these networks are able to learn by doing and can respond to things based on past experience. For this reason, connectionist AI is more dynamic than symbolic AI; even in the absence of full recognition or programming for each and every contingency, connectionist AI may still provide the right answer simply based on previous trial and error. That said, connectionist AI has shortcomings in the areas in which symbolic AI is the strongest, namely logic and following directions.[33]

Visual perception, language processing, and financial analysis procedures such as loan risk assessment and real estate valuation all depend on different variations of connectionist AI.[34] More recently, neural networks have been replaced by support vector machines (SVMs) and other high-level statistical analysis tools.[35] These machines take 'kernels' of information and using pattern analysis algorithms, they find and study various types of relations between data. These relations can include everything from rankings to correlations to clusters to principle components. SVMs transform these data into a set of points in Euclidean space, allowing for unique analysis of relationships between individual data points and for machines to learn from previous experiences. SVMs have applications in industry, such as geostatistics, bioinformatics, text categorization and handwriting analysis.[36] SVMs and other AI-based statistical techniques have been responsible for the rise of machine learning and have made data mining into an increasingly profitable business.[37]

## Evolutionary AI

Evolutionary AI involves the creation of multiple programs that then compete against each other for survival.[38] In the process of this 'evolution', the programs make random changes to their own rules and, through a series of algorithms, choose the optimal alteration to pass on to the next generation.[39] This type of AI is based on evolutionary patterns, sexual reproduction processes and genetic mutations that have been observed in biological systems.

To cite one example of the successful implementation of evolutionary AI, in the United States, scientists at NASA have successfully used evolutionary AI to create antennae for communicating between Earth and satellites. Thanks to evolutionary AI, researchers were able to sample and eliminate millions of potential designs before settling on the optimal model that met NASA's very precise specifications. Not only did evolutionary AI help to optimize the antenna design, it did so in a remarkably short time frame.[40]

Indeed, one of the greatest advantages of evolutionary and other forms of AI, and one of the reasons that these technologies are likely to accelerate the arrival of the Singularity, is that they can effectively solve problems much faster than human beings could do on their own.[41] In fact, 'the key to [evolutionary AI] is that human designers don't directly program a solution; rather, they let one emerge through an iterative process of simulated competition and improvement'.[42]

**Other key AI terms and technologies**

AI, which has an intellectual ability on a par with that of a normal human being, is called 'strong AI'. It is essentially the holy grail of the AI industry.[43] Strong AI would be able to think, reason, imagine and so on – all the things we currently associate with the capabilities of the human brain.[44] In contrast to 'weak AI', which is a simulation of a cognitive process but not in itself a cognitive process, 'strong AI' is a digital computer that is programmed to actually *be* a mind, 'to be intelligent, to understand, perceive, have beliefs, and exhibit other cognitive states normally ascribed to human beings'.[45] Although futurists speak of 'strong AI' as a fairly imminent discovery, it is more likely that such goals will remain elusive for at least 50–100 years.[46]

At the more practical level, some scientists are working to develop AI that although admittedly more conservative in its aims, nonetheless makes tangible contributions to daily life. One such example is Siri, a virtual personal-assistant software program that would help users complete basic, everyday tasks more quickly and efficiently. Siri has been described as a 'do-engine' – an evolution of the more commonly known search engines.[47] Siri depends on AI to process what a user is looking for and then to streamline Internet-related tasks that would normally require a person to visit several websites. A rather mundane example of Siri's possible applications deals with restaurant reservations.[48] Siri would allow a user to speak casual commands and would then decipher the user's intentions based on learning and context. If a user asked for a mid-priced French restaurant near work, Siri would be able to find and present options online, taking that user's past choices into account. Once the user made his or her decision, Siri could make the reservation. While strong AI is still somewhat elusive, programs like Siri provide practical and tangible examples of how AI can be used to handle menial and basic tasks.[49]

In 2009, IBM's 'Watson', the world's most advanced question-answering computer, beat humans in the TV game show 'Jeopardy'. The difficulty of the game is that it challenges players with wordplays and linguistic finesses, and that answers have to be provided in a very short space of time. Previously, this had been considered an impossible task for a computer.[50]

In the same year, an AI-driven robot went beyond menial tasks and identified the role of around 12 genes in a yeast cell. Although it was not a major scientific discovery in its own right, it was remarkable that a robotic system was able to make a new discovery with virtually no human input.[51] The robot's inventor, Ross King, argues that this robot (nicknamed Adam) and others have almost limitless scientific potential and will one day discover a concept akin to Einstein's theory of relativity.[52]

## Regulatory structures

Thinkers like Kurzweil believe that AI will eventually result in the merger of biology with technology. Even more radically, some suggest that if machines become so much more intelligent than our human selves, there is the possibility that we will no longer need our human bodies and could exist entirely as a machine. Given these revolutionary visions of what the future of AI could mean for humanity and society, it is somewhat remarkable that there are few, if any, government regulations that deal with the specific oversight of AI. In fact, many are even reluctant to discuss what such a regulatory framework could look like.

For example, at the 2006 Singularity Summit, not one speaker was invited to speak about the regulatory challenges facing AI. Any suggestions of regulation were merely made in passing or as a side note to the central issues of the conference. As Bill Hibbard points out, this is a major oversight for several reasons. Most significantly, AI potentially poses many risks, from intelligent weapons to machines capable of destroying the human race. Despite these concerns, many thinkers in the field of AI seem sceptical about either the potential for successful regulation of AI, or even for the need for such regulation.[53] Hibbard takes a different perspective, arguing that such regulation is both possible and necessary. As he explains, 'to be effective, regulation should be linked to a widespread public movement like the environmental and consumer safety movements. Intelligent weapons could be regulated by treaties similar to those for nuclear, chemical and biological weapons'.[54]

James Hughes is a futurist who has spoken about regulating AI with more specificity than most in the field. His suggestions include governments and cyber security firms developing detectors for and countermeasures against self-willed machine intelligence. Additionally, Hughes argues that humans should 'pursue cognitive enhancement and cyber-augmentation in order to give them a competitive chance against machine minds, economically and in the event of conflict'.[55]

Bill Joy, the co-founder of Sun Microsystems, argues the other side of the issue. While he echoes Hughes' fears that the current pace of technological change could threaten the human race, he maintains that humanity's best possible response to this scenario is to halt dangerous research until the ethical guidelines for such research have been clearly and firmly defined. Joy concludes that in the case of smarter-than-human technology, Murphy's Law will rule, and anything that can go wrong will.[56] In his words, 'I think it is no exaggeration to say we are on the cusp of the further perfection of extreme evil, an evil whose possibility spreads well beyond that which weapons of mass destruction bequeathed to the nation-states, on to a surprising and terrible empowerment of extreme individuals'.[57]

Joy is by no means a Luddite, and his conclusions are quite notable. He says, 'I know, knowledge is good, as is the search for new truths [...] But despite the strong historical precedents, if open access to and unlimited development of knowledge henceforth puts us all in clear danger of extinction, then common sense demands that we re-examine even these basic, long-held beliefs'.[58]

One of the challenges with regulating AI is that many of its potential risks are very abstract, and yet we concretely benefit from AI-related technologies every day. Additionally, much of AI is a matter of computer programming, something done every single day by countless people around the world. Defining boundaries of what can and cannot be programmed or how these things should be overseen would be akin to regulating creative writing. It would be impossible to effectively regulate what is, in many senses, the product of someone's imagination. In sum, AI is too broad a concept to regulate under one umbrella. The most dangerous aspects of AI need to be identified and a more surgical approach to regulation needs to be taken. Regulatory efforts are most likely to be successful in the areas of assigning responsibility for decisions made by machines. Although it will be difficult to set up a universally agreeable framework on this front, deciding the legal status of intelligent machines and the degree to which the machine itself as opposed to the machine's programmer, for example, is responsible for its actions, is urgent and crucial.

This is especially important in the fields of health care and the military, where intelligent machines may be responsible for making decisions that result in the death of a human being.[59] Questions of particular concern include whether a machine can be held to a contract, at what standard would a machine be measured, and to what extent can a machine's creator be held accountable for the machine's actions?[60] Mistakes are a part of life, and this extends to the battlefield, often with painful poignancy. Soldiers and military commanders inadvertently kill innocent civilians or sometimes even members of their own unit by accident. In some ways, unmanned aerial vehicles working in conjunction with AI may be better able to predict the consequences of a suggested attack and provide computer generated suggestions of how to approach an attack with minimal collateral damage.[61] However, when mistakes are made, the structure for placing blame or assigning responsibility is ambiguous at best. 'Sometimes the blame is placed on the humans behind the machines [...] Other times, the data itself is bad [...] The blowback from such mistakes is worsened'. For example, 'America's technology is viewed as almost magical, but much of the world also sees the United States through jaundiced eyes. When something goes wrong, the immediate assumption is not that a mistake occurred but that it was planned malice'.[62] Once again, technological innovations have outpaced the evolution of institutions. There are currently no rules of international law for the use of remote control robots. In other words, robots going to war are not subject to regulation.[63] Without clear, transparent frameworks for assigning responsibility to the actions of robots and AI, the chances for misunderstandings and the escalation of violence or global tensions are high. As machines gain more and more autonomy and increasing levels of intelligence, these legal issues will only become more problematic.

Such issues are not limited to the military sphere. In fact, the question of general legal rights for intelligent robots or machines is one that lawmakers have already begun to consider. In the United Kingdom, for example, a panel tasked with assessing potential major developments in the next 20 years began to consider the question of the legal status of intelligent robots.[64] The panel's report essentially argued that intelligent robots could be worthy of having many of the same rights and responsibilities as humans: 'If intelligent beings, even artificial ones [...] were expected to serve in the military, in turn society would have its own responsibilities towards its "new digital citizens"'. Countries would be obliged to provide 'full social benefits to them including income support, housing, and possible robo-health care'.[65]

## Geopolitical implications of artificial intelligence

Depending on how AI develops, it could have a variety of both positive and negative effects on geopolitics. Among its most salient potentially positive applications are its military, health and space-related applications. AI also has some potentially more menacing applications, particularly in the case of 'runaway AI'.

### Military and space applications

*Military applications*

AI is profoundly and durably altering the way war is conducted.[66] Ever since its earliest phases, AI has been very closely linked to the military. In fact, in its very beginnings, AI research was done almost exclusively on the military's behalf.[67] Today, AI is widely used in the military context, and the number of robots used on the battlefield has dramatically increased. For example, the US military uses pattern-recognition software to guide autonomous weapons to highly precise locations. Importantly, AI takes into account changing weather and wind conditions, unexpected ground terrain, and so on. This ensures successful weapon delivery even in the absence of a human-directed flight plan.[68] An unmanned air vehicle can already be flown into battle, controlled by a pilot sitting in an office 10,000 miles away.[69] Thanks to the capacity of computers to 'observe, pinpoint and then attack', humans controlling the weapons are no longer exposed to risks.[70] According to US Colonel Michael Leahy, Unmanned Combat Air Vehicles (UCAV) go out in packs of four. If one UCAV is damaged, the remaining three are programmed with enough independence that they can instantly adapt and figure out how to guide the damaged vehicles' weapons.[71] Current weaknesses in UCAVs, such as difficulties processing large amounts of information on their own, are being quickly overcome. For instance, the US Airforce is working on designing algorithms that would allow drones to independently avoid collisions on landing or take-off.[72] AI is also aiding militaries with planning and logistics tasks;[73] it can help to create and manage complex supply systems and to develop sophisticated military training exercises and simulations.[74]

To quantify the US military's dependence on AI-related systems and robots, the number of unmanned systems on the ground in Iraq has gone from zero in 2003 to over 12,000 today.[75] From a financial perspective, UCAVs are often cheaper than their manned counterparts as they are smaller, need less fuel, carry less weight and have fewer safety and precautionary measures that need to be taken in order to make them mission-ready.[76] There are less tangible benefits to using AI in battle as well. For example, computers have the advantage of processing and communicating information much faster than human beings, and they are not affected by the emotional toll of battle and warfare.[77] They need no sleep or clothing, and AI robots can be designed to be expendable.

AI is also helping government intelligence analysts to improve war games and other defence simulations. A team of researchers at the University of Maryland believe that AI will help create a virtual world in which terrorist behaviour can be mimicked and predicted based on analysis of a variety of factors, ranging from social to political to religious.[78] The researchers hope that this virtual world program will exploit advances in AI to develop a digital mock-up of regions such as the Middle East. AI will help estimate how different actions, such as building schools or burning drug crops, might affect complex, real-life interactions. While these programs do not currently take into account the complexities of human ethical and religious beliefs or the histories of various civilizations, it is one tool that can be used to predict the possible outcomes of different situations and the course of action in high-profile military and ethnic conflicts.[79]

Some envisage a day when all military battles and wars will be fought entirely by AI-driven robots. In the words of Will Warner, 'If robot weapons will afford some protection from the slaughter of modern warfare, and can be produced, they will and should be deployed'.[80] As weapons become increasingly intelligent, warfare and attacks will become more and more precise and, in all likelihood, the number of casualties and injuries as a result of warfare will decrease thanks to AI technology.[81] The possible outcomes of this situation are numerous and contradictory. The growing numbers of successful UCAVs will continue to reduce the need to put human troops in harm's way while still allowing mission objectives to be met.[82] Robots can take risks that humans would not or could not take and run less risk of making lethal mistakes in the heat of the moment or in the face of uncertainty.[83] Thus, the barriers to war will have been lowered, and countries might be more willing to engage in military operations. In other words, AI-driven robots may make war more likely, as the human costs of war will decrease and it will be politically easier for governments to sell their populations on the idea of offensive war. Unmanned weapons also have the potential benefit of demoralizing the enemy because, while an opponent might risk heavy odds if it means the chance of killing his attacker, death by remote-controlled machine is seemingly less noble.[84] Such logic depends implicitly on the assumption that an army with sophisticated AI robots would be warring with a more traditional army. One of the leaders in robotic sciences, Dr Robert Finkelstein, argues that AI-controlled robots will be important in warfare that becomes increasingly centred on non-state actors such as terrorists, guerrillas and tribal groups.[85] Interestingly, Finkelstein also stresses that in order for AI to be adopted by much wider levels in the military, it is not necessary for AI to have human-level intelligence. Instead, the most important thing is that the robot's level of intelligence is appropriate to its missions and tasks.

However, if AI robots become the de facto method for waging war, war could actually become obsolete. Consider that historically, the foundation of traditional warfare is based on a country's willingness to sacrifice its men and resources for a greater cause. If humans no longer need to participate in battle, however, those countries that choose to wage war against others may no longer have the leverage of mounting human casualties to use as a bargaining chip to end the war and pursue peace. Instead, warfare will become a matter of who has the most sophisticated technology, whose robots are tougher, and who can afford to produce this technology on a large-scale. In short, AI and other emerging strategic military technologies offer the possibility of changing the entire nature of traditional warfare, potentially allowing for the rise of new superpowers with might rooted specifically in technological sophistication rather than manpower or political power. From an ethical perspective, 'the implications of dealing with an enemy that does not value its own survival are deeply troublesome and have led to controversy that will only intensify as the stakes continue to escalate',[86] but what if neither the attacker nor the attackee have survival at stake?

If robotic war evolved fully, this is the kind of question that would need to be addressed.

Long before we get to the question of robot-on-robot warfare, there are more urgent ethical and moral issues to address. For example, authorizing robots to kill also has significant moral implications. As the Massachusetts Institute of Technology's John Hansman has said, 'A human must always be in the loop to authorize weapons release [...] In warfare, you may want to give the planes the authority to return fire automatically', but, he adds, 'if they can return fire automatically, then you've made a huge moral shift in warfare. The responsibility for deaths that occur is much harder to assign. Is it the operator's fault? Or the computer programmer's? Either way, the person responsible is very far removed from the situation'.[87] The idea of a robot killing a human being, even if that human being is a militant enemy, is disconcerting from moral and human perspectives. AI will only make these issues and questions more pressing.

Another issue to consider when it comes to AI and the future of military battle is the increasing significance that technological asymmetry will have or, to put it slightly differently, the fact that technological superiority will become increasingly unrelated to the likelihood of victory on the battlefield. This can present multiple problems. On the one hand, commanders using new, cutting edge technologies may not be familiar enough with the technologies to fully take advantage of them, or they might use tactics and strategies that worked when employing older technologies but that are less relevant with new AI-robots.[88] Another challenge stems from the changing face of war and the growing prevalence of transnational and non-state actors using military tactics to achieve their goals. For example, despite huge technological superiority, the United States is still struggling to combat insurgent forces and their improvised explosive devises in Iraq. As Chinese military planners have reportedly noted, 'On the battlefields of the future, the digitized forces may very possibly be like a great cook who is good at cooking lobster sprinkled with butter. When faced with guerrillas who resolutely gnaw corncobs, they can only sigh in despair'.[89] These military challenges and scenarios go far beyond the realm of AI, but they are worth addressing here as AI's potential to increasingly remove humans from the battlefield raises many questions over the future of military conflict and the role technology will play in this evolution.

Finally, it should be noted that even if humans develop AI capable of running military missions on its own, a human element may prevent that from happening.

> Even once computers can make air-combat-related judgments reliably (or at least as reliably as humans do), many politicians and military officers nonetheless will resist surrendering decision making authority to machines. Most leaders value their authority and are reluctant to surrender it, while the notion that machines could make tactical decisions without human input is unsettling to many people.[90]

Just because we as humans can do something does not mean we necessarily will, especially if it undermines our own perceptions of our place in life and society. This is certainly another important consideration to take into account when evaluating the geopolitical implications of technology.

*Space applications*

Similar technologies are also being applied to space technology, and spacecraft with missions to far destinations such as Mars and Jupiter are being equipped with the software and technology to enable the vehicles to make their own tactical decisions mid-mission. This offsets the challenges posed by long communications lags between spacecraft in mid-mission with their ground support teams.[91] NASA is also researching new land-based robotic telescopes that can make their own judgements on where in the sky to look in order to find the desired phenomena. These telescopes organize and optimize data and search parameters in order to dramatically increase the likelihood of finding the target in the sky.[92]

Other AI-related NASA projects include the 'Autonomous Sciencecraft Experiment', which uses on-board science analysis to increase science returns through the enabling of intelligent downlink selection and autonomous retargeting. Distributed spacecraft use new technologies to control groups of spacecraft with collective missions, as opposed to having separate command sequences for each individual spacecraft.[93] Above all, AI-related innovations will make studying and exploring outer space more efficient.

### Global health applications

Health care is another area in which AI has potentially enormous geopolitical implications. Currently, virtually every major drug company is using AI to help in intelligent data mining for the development of new drug therapies.[94] AI also has applications in diagnostics, in understanding the function of genes and their roles in disease, and in identifying inconsistencies in a treatment plan, among other things.[95] Indeed, the role AI can play in medicine is manifold. AI 'provides a laboratory for the examination, organization, representation and cataloguing of medical knowledge; produces new tools to support medical decision-making, training and research; integrates activities in medical, computer, cognitive and other sciences; and offers a content-rich discipline for future scientific medical specialty'.[96] Robotic surgery is another example of AI in health care. While such robots lack human empathy, they are able to drill and cut with much greater precision than a human doctor, reducing the chance of human error.[97]

Cyber care is another frequently discussed application of AI to health care. Automated assistance to delay the moment when people have to move to a nursing home,[98] robots serving as companions for hospitalized people or machines keeping track of dietary requirements are only some of the ways in which AI is being used.[99]

AI can facilitate all facets of medicine, from diagnosis to treatment. In California, for example, researchers and doctors at Cedars-Sinai Medical Center are developing:

> Software to allow computers to process and analyze three-dimensional images of the heart in much the same way an experienced human operator would. The program applies artificial intelligence techniques to the measurement of parameters critical to understanding the state and behavior of the human heart. This automated approach allows information to be obtained very quickly that is quantitatively accurate and does not suffer from intra-observer or inter-observer variability.[100]

An innovation to be expected on the market soon is what has been dubbed 'smart pills'. About the same cost and size as traditional medicine, smart pills contain sensors with the ability to track the effects of the pills on the body, thereby monitoring both body reactions to the drug and compliance of the patient with drug-taking instructions.[101]

In developing countries, lack of trained physicians often means a higher mortality rate and leaving treatable diseases untreated due to simple lack of resources. Advances in AI could ameliorate this challenge by assisting doctors in making decisions and providing diagnosis, thereby reducing waiting times of patients.[102] Overall, these advances in health care will directly and indirectly help to improve global health.

Although the idea of distributing health-related knowledge through AI-based computer programs has tremendous potential, there are significant legal hurdles that need to be overcome before the technology can be adopted in a widespread fashion.[103] Even in cases where machines have been shown to perform better than humans, legal questions over who is responsible for the decisions enacted by expert systems prevent the technology from being fully embraced by hospitals and the greater medical community. Issues of liability and insurance must be resolved before AI's potential in the health care industry can be maximized.[104]

### International finance

AI is being increasingly used across a variety of industries and businesses to help in the design of new products or the optimization of certain processes and procedures. Banks, brokerage firms and insurance companies have been relying on AI for decades to help with fraud detection, credit checks, and so on.[105] AI also plays a crucial role in the storage, organization and analysis of the massive amounts of data compiled by these financial services firms.

AI has applications in financial data mining, arbitrage, hedging and trading strategies, supply chain management and fraud detection.[106] For example, to successfully take advantage of arbitrage opportunities, financial markets require the use of thousands of transactions of various securities over holding periods that can be mere seconds long. These quick actions and such data processing are outside the realm of human capabilities and AI is required in order to benefit from arbitrage.[107] Evolutionary AI-based algorithms are also important in finding optimal solutions to problems, especially in the discontinuous functions that often appear in market modelling and asset allocation.[108]

Moreover, in instances where data are too large to be analysed manually or by humans, AI-based computers can help to process financial data with great speed and accuracy. Because of their ability to process complex algorithms, these computers can also assemble data and predict patterns in ways that the human mind cannot.[109] Indeed, one of the greatest contributions AI has made to the financial world is the ability to make important, complex calculations in a timeframe appropriate to the sometimes lightning-quick pace of international financial markets.

Although it is tempting to try to draw causal links between the rise of AI in the international finance industry and market volatilities, such links do not hold up under analysis. In fact, AI has helped to reduce the dangers and volatilities in international markets through advanced fraud and abuse detection. AI's ability to detect credit card fraud is one of the most banal examples of this. Expert systems analyse each credit card transaction and compare it with the customers' previous purchases. Based on algorithmic analysis, AI systems can flag unusual behaviour, which can then be assessed by a human being to determine whether the irregularity merits follow-up with the consumer. The same concepts apply to other financial transactions such as information security and e-business. Ultimately, 'AI could formalize legal common-sense principles to detect fraud and abuse autonomously and thus allow lawmakers to deal with the challenges posed by the information society'.[110]

### Artificial intelligence and online search engines: questions of privacy

Online search engines such as Google depend, in varying degrees, on AI to function. For its part, Google hopes that advances in AI will make Internet searches more intuitive and less technical. It uses AI-based statistical learning methods to determine how to rank its listings, and such innovations have revolutionized Internet searches.[111] However, there is still much room for improvement, and AI's application to online searches is admittedly imperfect. Even with the help of AI, search engines still cannot understand the context of words, and thus a search for 'American chips' is just as likely to turn up information on computer chips as potato chips.[112] As Google co-founder Larry Page has explained, 'The ultimate search engine would understand everything in the world. It would understand everything that you asked it and give you back the exact right thing instantly'.[113] The search engine Clusty uses AI to group search results into different categories, thereby helping people to quickly navigate through sometimes endless search results more rapidly than they can when search results are merely organized by popularity.[114] Of course, in this context, it is important to keep in mind that search engines keep records of different search terms and can link search parameters to specific computer network IP addresses.[115] Given AI's ability to collate, organize and analyse data, consumers must consider the types of personal information their searches reveal and how such information could potentially be turned against them. For example, for people who get their news via a search engine, the news stories they click on could be correlated to search terms, thus making it possible to potentially draw conclusions about their political leanings, or former and future actions.[116] This is an especially weighty concern because there are very few regulatory structures to govern how search engines and the private companies that run them use the information they gather. So far, it has been up to each individual company to set its own privacy policy, but few consumers really read or understand such policies and, thus, the opportunity for abuse of the information is high.

## The dangers of AI

AI also has potentially deadly applications. As is touched on above, there is concern over 'runaway AI' or AI that develops so rapidly as to find ways around the limits we as humans have tried to impose on its capabilities. Although Kurzweil maintains that such rapid, sophisticated advances are a key component of arriving at the Singularity, uncontrollable AI could easily turn against humans. With the help of nanotechnology, runaway AI could rapidly recreate and reproduce itself and update versions of itself, and it is possible to envisage a scenario in which AI and AI-driven robots turn against humanity and begin killing off our race. Thus, given that self-improving and strong AI cannot be recalled, scientists must get the design of human friendly features and checks right the first time. As Eliezer Yudkowsky puts it, self-improving strong AI's initial design must have 'zero nonrecoverable errors'.[117] Renowned physicist Stephen Hawking has recommended that we enhance our own bodies and brains so that we are better prepared to respond to and potentially fight off AI that becomes unfriendly towards humans.[118] The themes of human enhancement and human destiny are explored in greater depth in Part II.

Most importantly, we should not be fooled into thinking that there is any perfect protection from strong AI. Kurzweil calls for a rather open way of regulating these issues, in order to avoid the illicit and illegal implications of secretive regulatory programmes:

> Although the argument is subtle I believe that maintaining an open free-market system for incremental scientific and technological progress, in which each step is subject to market acceptance, will provide the most constructive environment for technology to embody widespread human values [...] [Strong AI] will reflect our values because it will be us. Attempts to control these technologies via secretive government programs, along with inevitable underground development, would only foster an unstable environment in which the dangerous applications would be likely to become dominant.[119]

Even before we reach this dramatic point of no return, we run the risk of AI machines making an ever widening array of human jobs obsolete, or criminals gaining control of AI to facilitate fraudulent activities.[120]

## An overview of key trends and developments in artificial intelligence

Cognitive science encompasses many different fields, but one of the most significant from a geopolitical perspective is the science of AI. The emerging strategic technologies in these fields have the potential to dramatically accelerate the rate of scientific and human progress to levels we cannot even imagine. The arrival of the so-called Singularity will be greatly facilitated by AI and smarter-than-human computers, which will presumably be able to solve problems faster and more efficiently than humans can. The three different areas of AI research – symbolic, connectionist and evolutionary – will be important factors in bringing about the Singularity, if and when it happens. In addition to the Singularity, AI will also have geopolitical implications in the way militaries plan for and fight battles and in the global provision and quality of health care. International regulation is almost non-existent and can certainly not keep pace with the rapid developments in technology. More menacingly, out-of-control AI could ultimately destroy the human race.

The implications of this science, especially in conjunction with the other emerging strategic technologies outlined in this book, are so significant that they are addressed in Part II of this book.

# 10
## Conclusions of Part I

Part I examines emerging strategic technologies at the global level and through a lens that has been trained primarily on issues of geopolitics and geostrategy. Starting with information and communications technology (ICT) and moving on to energy, the environment, health care, biotechnology, genomics, nanotechnology, materials science and cognitive science, Part I has looked at emerging strategic technologies on what could be described as the macro or global level. In a sense, Part I has provided a bird's eye view of technologies and their impact. Discussing the social and political movements being fuelled by ICT, the environmental benefits and dangers of nanotechnology and nano-particles, the possibility of improving food safety and security through biotechnology and genomics, and other potential implications of emerging strategic technologies, the first part of this book has approached technology from the perspective of states, societies, cultures and socio-economic groups. It has argued that in a globalized, interconnected world, technology pervades virtually every facet of life; moreover, its role in shaping global dynamics will only become deeper and more profound in the future.

By contrast, Part II of this book looks at these technologies from a more micro perspective. Specifically, it addresses the potential for emerging strategic technologies to be used in individual human enhancement – a concept that covers everything from cosmetic surgery to brain implants and beyond.

Emphasizing the theme of convergence, Part II breaks away from the format used in Part I, looking at different technologies as standalone entities, and instead considers how all the technologies described individually in Part I are combining and merging in order to improve, enhance and generally alter our individual human bodies and minds.

This is a subject that has attracted quite a bit of attention in certain, specialized circles in recent years, but it has yet to be intensively debated at the global or even societal level.

At its core, this book is about the intersection of strategic technologies and global politics, security and strategy. Considering that mission, it may seem strange to delve so deeply into issues of human enhancement – a process that many of its advocates strenuously argue should be strictly a matter of individual choice. Such a highly personalized and individual issue seems to belong to a completely different discourse from issues of global and transnational security. However, this is a false impression. When taken collectively, individual human enhancements have tremendous potential to change human nature, thereby gradually altering geopolitics. *Even more significantly, human enhancements, evaluated collectively and over the long run, may have a vast potential to change the overall destiny of the human race.*

Such a seemingly dramatic statement is not made lightly. Part II builds on the scientific details and technological explanations presented in Part I to highlight the different technologies that can be used for human enhancement. In presenting the case for and against enhancement, I touch on some of the major advantages and the potential dangers of such uses of technology. Technologies used for human enhancement may make us faster, smarter and stronger, but they also carry the enormous risk of unintended consequences. Especially when it comes to things like germ-line genetic engineering or brain modifications and implants, it is likely that what starts out as an 'enhancing' technology could inadvertently result in deformities or reduced human capacities. These issues are all addressed in Part II.

The central argument of Part II is that when human enhancements are adapted on a large scale, something that will happen inevitably for the reasons I outline below, they will gradually alter the fundamental traits and characteristics that make us human beings. Collectively and over time, such changes will mean that humans will change to a point where they can no longer be considered humans. In other words, they will have moved beyond the human phase of evolutionary development. This is at the heart of how emerging strategic technologies will affect our human destiny.

Indeed, when considered in this context, individual human enhancement suddenly takes on a much more urgent tone, and its relevance to global security and justice becomes much more evident. Using my *'multi-sum security principle'*, I outline how emerging strategic technologies and their human enhancement applications apply to five substrates of global security: human, environmental, national, transnational and transcultural. Whether it is from having more and more human beings living longer and longer lives and thus straining the Earth's limited resources, or the notion that governments could use genetic engineering for eugenics and human enhancement is not merely a matter of individual choice; it is a call to action for global society.

Unfortunately, policymakers have been loath to try to regulate technology's applications to human enhancement. This is true for a number of reasons. On the one hand, human enhancement technologies are extraordinarily diverse and, in many cases, it is unclear whether a technology is being used to treat a legitimate medical problem or to simply improve someone who already has 'normal' physical and mental capabilities. Thus, any regulatory framework to oversee human enhancement would have to be extremely nuanced. Perhaps even more significantly, many of the potential dangers of human enhancement are ones that will really come into play only in the distant future. Policymakers, sensitive to short term election cycles and the most recent poll numbers, are reluctant to take the political risk of regulating and overseeing politically and religiously loaded issues such as human enhancement and reproductive rights.

This book concludes by forcefully arguing that a lack of strong, clear regulations on emerging strategic technologies and human enhancement would lead to extremely dire consequences for the human race. Whether it is issues of justice and equality and the inherent unfairness of an enhanced individual competing for the same jobs or in the same sports as an unenhanced individual, or a race of post-humans that seeks to destroy unenhanced humans, our application of technologies to ourselves has the potential to get out of control and cause grave problems. We must establish a truly global framework for dealing with these issues now. By the time it is politically expedient to address these issues, it will be too late. Not acting now is akin to implicit acceptance of the end of the human race as we know it.

Part II addresses all these issues in some depth. Although traditional approaches to international relations often emphasize the primacy of states, militaries and economies, the key theme of this book is that technology and individual choice are proving to be the potent and resilient forces of twenty-first-century geopolitics and geostrategy.

# Part II

# Emerging Strategic Technologies: Human Enhancement, Dignity and Destiny Implications

Part II

Emerging Strategic Technologies,
Human Enhancement, Dignity
and Destiny Implications

# 11
# Introduction: Definitions, Terms and Concepts

## General overview

In the world we live in, the issues surrounding science and technology often seem black and white. Scientific innovations are governed by the laws of physics, rules of reason, or by any number of seemingly strict guidelines and givens. Science and technology, we tend to believe, are realms where logic and order are king.

However, some of the emerging strategic technologies detailed in this book have enormous potential to upend many of the apparent 'laws' of science and nature that we often take for granted. Whether it is fundamentally altering a material at the atomic level or radically extending our natural life span by hundreds or thousands of years, emerging technologies may force us to question some of the things we have previously accepted as the fundamental rules of life and nature. Needless to say, this sort of paradigm shift could have enormous implications for everything from human dignity to equality to geopolitics.

Part II specifically examines the human-related applications of emerging strategic technologies and outlines some of the major ways in which the ongoing scientific developments outlined in Part I might fundamentally alter our human nature, our concepts of human dignity and our understanding of the role that science plays in our lives. Specifically, it looks at the power and potential of emerging strategic technologies to alter or enhance the natural mental and physical capacities of human beings, and what this may mean for the future of the world and, indeed, for the future of humanity.

## Definitions

### Human enhancement

Before addressing weighty issues of human destiny, it is instructive to clarify my perspectives on and definitions of a few key terms frequently used in the dialogue on technology, science and human nature. I begin with the concept of human augmentation or, as it is more frequently called, human enhancement. I define human enhancement as the use of innovative technologies to augment or enhance human functions and abilities beyond the replacement of dysfunctional cellular groups and organs. In other words, human enhancement includes anything that goes above and beyond restoring normal human physiology and functions. This covers anything from the seemingly mundane use of reproductive technologies such as in-vitro fertilization (IVF) and the increasingly popular cosmetic surgery to more radical enhancements such as germ-line genetic engineering, deep brain stimulation, brain-computer interfaces or even uploading one's entire brain on to a computer. As far-fetched as some of these prospects, especially the latter ones, seem at first glance, they will in all likelihood become our reality in the not too distant future.

### Human evolution and the course of human destiny

Both the excitement over and the fear of enhancement can be traced back to the fact that enhancement will fundamentally alter our status as human beings and the overall destiny of the human race. Indeed, as enhancements become more widespread and mainstream, we will evolve into a new species. The possible steps of this transition are outlined below.

#### *Transhumans and transhumanism*

Transhumans are an intermediary form of humans, somewhere between humans and post-humans on the evolutionary path that we as emotional, amoral and egoistic humans will ultimately choose.[1] In short, a transhuman is 'a human in transition'.[2] More specifically, 'We are transhuman to the extent that we seek to become posthuman and take action to prepare for a posthuman future. This involves learning about and making use of new technologies that can increase our capacities and life expectancy, questioning common assumptions, and transforming ourselves ready for the future, rising above outmoded human beliefs and behaviors'.[3] As transhumans, we will still be recognizable as humans and still have much in common with unenhanced humans.

Prostheses, plastic surgery and intensive use of telecommunications are some (but by no means the only) types of enhancement and behaviours indicative of transhumans.[4] As developments in nanotechnology make it possible to build and reconstruct matter from the atomic level up, this technology is likely to be applied to the human body, reconstructing damaged cells and systems. These reconstructed bodies and systems will be a central component of transhumans.

The definition of transhuman is relatively straightforward, but it becomes slightly more complex when taken in the context of transhumanism. The term 'transhumanism' is distinct from transhuman and refers to an 'intellectual and cultural movement that affirms the possibility and desirability of fundamentally improving the human condition through applied reason, especially by developing and making widely available technologies to eliminate aging and greatly enhance human intellectual, physical, and psychological capacities'.[5] In other words, we can use technological tools to alter and improve our human organisms. Therefore, individual choice and the centrality of technology are key components of the transhumanist philosophy.

Transhumanism has also been described as 'a philosophy that humanity can, and should, strive to higher levels, both physically, mentally and socially. It encourages research into such areas as life extension, cryonics, nanotechnology, physical and mental enhancements, uploading human consciousness into computers and megascale engineering'.[6] While transhumans are humans in a certain physical and mental state of being, transhumanism is more of a philosophical movement, embracing the characteristics of transhumans but also focusing on the ontological and eschatological elements of humans overcoming their natural limitations.[7]

### Post-humans

A post-human belongs to a race of beings so fundamentally and categorically different from our own human race that it can no longer be considered human, even if its evolutionary roots were in humanity.[8] This transition can occur physically, mentally or in both but, overall, posthumanism will be defined by 'sweeping modifications to our inherited genetics, physiology, neurophysiology and neurochemistry'.[9] A posthuman would have overcome the biological, neurological and physical limits imposed on humans by the evolutionary process.[10] 'Post-humans would have a far greater ability to reconfigure and sculpt their physical form and function; they would have an expanded range of refined emotional responses, and would possess intellectual and perceptual abilities enhanced beyond the purely human range.'[11] Post-humans are not necessarily good or bad, evil or moral. The key point is that they have moved beyond the accepted definitions and characteristics of a contemporary human.

Ironically, it is our own human nature that has huge potential to drive us towards enhancements that may completely alter not only the characteristics of our species but also our core nature. Our human biology and physiology are actually quite inefficient compared to their potential, and post-humanism is about capitalizing on the opportunity to improve these basic systems. As Daniel Ust explains:

> To start, there is nothing special about the current form of human-ity. This includes not only social systems and cultures, but also the organs of the body. This does NOT mean these various organs and systems are unnecessary, but that they leave room for change and improvement. For example, there is no reason why our average life-span should be around 70 years and not 200 or more. There is no compelling reason to accept things as they are. There is no reason to accept 20/20 vision as the final goal of all corrections of vision.[12]

If we can improve ourselves, the desire for gratification that is embed-ded in human nature will inevitably pull us in that direction.

Post-humans are related to transhumans in the sense that being a transhuman is a step on the evolutionary path towards becoming a post-human.[13] Perhaps post-humanist Max More best captures the sen-timents of post-humanism on his website when he says, 'Let us blast out of our old forms, our ignorance, our weakness, and our mortality. The future belongs to posthumanity'.[14]

## Eugenics versus enhancement

As we advance this discussion of human enhancement, it is important to clarify and emphasize that human enhancement, modification and augmentation are not synonymous with eugenics, a term coloured by strong negative connotations. Eugenics aims to 'improve the genetic constitution of the human species by selective breeding'.[15] In other words, eugenics involves determining who is 'allowed' to reproduce, based on characteristics that are perceived to be desirable or advanta-geous.[16] In its traditional definition, eugenics was intended to mean the science of 'improving stock [...] which takes cognizance of all influ-ences that tend, in however remote a degree, to give the more suitable races or strains of blood a better chance of prevailing speedily over the less suitable'.[17] When the term was first conceived, it was considered from both social and scientific perspectives, a fact that ultimately led to widely varying historical interpretations and understanding of eugenics as a concept.[18] In fact, 'by tracing the historical development of the concept through the twentieth century, one encounters a larger and more complex semantic framework involving historically specific understandings of the relation between the individual and society, science and politics and responsible and irresponsible reproductive behavior'.[19]

Today, the term eugenics often evokes thoughts of Nazi Germany and the systematic extermination of the Jewish race – a form of negative eugenics. However, at one time forms of eugenics were so accepted as to be legally mandated in certain parts of the United States. In Indiana in the early twentieth century, for example, mentally ill and physically handicapped people underwent forced sterilization, incarceration and sometimes euthanasia.[20]

Although human enhancement can involve genetic modifications, the term encompasses a much wider array of alterations and augmentations. Hearing aids and glasses can be considered forms of human enhancement, as can pacemakers, steroids and mood-modulating drugs. Genetic enhancements have the potential to limit the transfer of genetic diseases and, one day, they may be able to increase human intellectual capacities, among other things. So while the term eugenics implies the selective breeding of desirable traits and the methodical elimination of undesirable ones, human enhancement is more focused on improving human beings' overall quality and duration of life. Its purpose is to benefit the individual, not to create or eliminate a specific type of person. While some raise the valid concern that genetic enhancement may lead us down the slippery slope of eugenics, the phrase 'human enhancement' and the debate over its future must be understood in much broader terms.[21] Eugenics clearly violates principles of human dignity. The potential ramifications of human enhancement are more nuanced.

One of the key arguments made by supporters of human enhancement is that choices about whether to enhance or not, and which enhancements to make, should be left up to individuals. This is an important point to make in assessing the differences between enhancement and eugenics, as eugenics often connotes strong government intervention and government mandates about which characteristics are to be prioritized. Enhancement, its advocates argue, should be a freely made personal choice. Parents, for example, should be able to decide on their own whether to do embryonic screening to test for genetic diseases in their unborn child, or whether to make intrinsic improvements, such as increased intelligence, to their child's genetics.[22] If parents opt not to do genetic testing or alterations, then that is their choice to make. Similarly, no one should be forced to take mood-altering drugs or other related psychopharmaceuticals, but if an individual wants to be able to improve his or her mood through a mood-modulator or to have a differently shaped nose through cosmetic surgery, those choices should be his or hers to make. Although some overall regulation of human enhancement is strongly and urgently needed, it is important that individuals take responsibility for their own actions and choices.

## Foucault's bio-power

Philosophically, the French thinker Michel Foucault has detailed some concepts, which are useful and interesting in understanding contemporary discussions over enhancement. In the first volume of his *History of Sexuality,* Foucault introduces the idea of bio-power, which he argues emerged when governments began exercising their highest powers not by putting people to death, as had been the case throughout much of history, but by ensuring, sustaining and multiplying life and then providing order to this life.[23] Bio-power refers to a way of managing people as a group and is a fundamental component of the emergence of the modern state and the capitalist economy.[24] The state's future depended not only on having bodies to run the machinery of production, but also on the cultivation of these human resources, thereby bringing about greater state investment in universities and workshops and greater state concern over questions of health care and migration.[25] Notably, Foucault argues that such issues were important to the state as means of control over populations; they were not motivated by benevolent feelings.

Because life has become so important to the power basis for sovereignty, states have an easier time justifying the eradication of an entire group if that group is perceived to be a threat to the existence of the life the state seeks to protect.[26] This is one of the reasons genocide has become so prevalent in recent centuries. Massacres or the opportunity to eradicate threatening populations can be justified in the name of preserving the life of the main population.[27] As Foucault puts it in the context of the atomic bomb, 'the power to expose a whole population to death is the underside of the power to guarantee an individual's continued existence'.[28]

By extension, some of Foucault's ideas seem to advocate eugenics and genetic engineering in order to preserve the life the state seeks to protect and control as a basis for its power; in fact, for Foucault, eugenics seems to be the best way to achieve and maintain control over a population.[29] The key issue to keep in mind is that Foucault's concept of bio-power shows the strong interest states and governments have in allowing or preventing the enhancement of their citizens according to whether it suits the state's needs. It is an important method for preserving the existence of an entire population. The centrality of life to the concept of power results in the importance for governments of seeking access to and control of the body.[30]

This is certainly clear in the area of reproduction and the politicization of sex. For governments, 'sex was a means of access both to the life of the body and the life of the species'.[31] Given the power of sex and sexual reproduction, the control of human reproduction becomes an important tool for governments looking to perfect their species or for preserving 'prestigious' blood over the blood of the lower classes.[32] These themes are important to keep in mind when discussing how states should regulate and address human enhancement technologies.

## Human nature

Human nature will play a key role in determining how the evolution of the human race unfolds and in how emerging strategic technologies are created and used. As I have outlined in my previous book *'Emotional Amoral Egoism': A Neurophilosophical Theory of Human Nature and its Universal Security Implications,* human nature is rooted in our human biology. In summary, I argue that humans are motivated by survival instincts that are based partially in emotions and that are partially predisposed through our genetic make-up. Certain variations in instincts and emotions can be accounted for because of different genetic make-up among individuals, heterogeneous genetic variations within individuals, and the personality traits, which are unique to an individual. These emotions and genetic predispositions are mediated through neurochemistry as well as by more individual-specific issues such as upbringing, education, society and culture. Infrequently, morality can also influence human survival interests, although such morals will probably be rooted in concerns over an individual's self-interest.[33] In fact, I argue that humans are not inherently moral beings, nor are they immoral, although we may have some moral sensitivities. We are actually *amoral* beings, capable of acting morally or immorally as it suits our survival needs. Because this argument is the crux of the arguments made in the rest of this book, I will spend time reviewing the emotional and genetic elements of human nature in greater detail.

Our self-preservation instincts manifest themselves through emotions, including, but not limited to, reward, fear, sadness and grief, anger, jealousy, pride, shame, embarrassment, surprise, happiness, disgust and contempt. Some of these emotions are primal, some have more complex, multifaceted origins but, importantly, the most dominant emotions are *negative* responses.[34] In other words, the most visceral, primal human emotions are designed to drive us to *avoid* something, such as injury or harm.

Not all of our emotions are based on instincts or the need for self-preservation. Some, like shame and embarrassment, are connected to our conscious selves, an indication that they may be provoked as much by thoughts and self-reflection as by instincts. Similarly, positive emotions such as happiness are also important but more as a means of ensuring overall well-being once our basic survival needs are met. Regardless of whether our emotions are positive or negative, instinctive or not, they are all rooted in our neurochemistry, which means they come about because of changes in our brain. From a neurochemical perspective, emotions are primarily located on the right side of the brain and, more specifically, in the dorsal-lateral frontal cortex.[35] Interestingly, positive and negative emotions have been shown to come from different sections of the brain, underscoring the idea that these variations in emotions have different evolutionary contexts.[36]

We are genetically and evolutionarily programmed to focus on survival, and our survival instincts are manifested through emotions. These emotions are shaped and driven by numerous factors – primarily by our individual genetic profiles but also by environmental factors such as our culture, our upbringing and our education. Contrary to many other philosophers, I maintain that humans are only occasionally driven to act based purely on reason, and that they act in response to conscious moral judgments even less frequently. Ultimately, human nature is neither a matter of free will nor decided entirely by biology and genetics. Rather, 'human nature is a predisposed tabula rasa'.[37] As humans, we have 'no innate ideas, but we do have predilections that are coded by genetics and influenced by the environment' and that are mediated through neurochemistry.[38]

Thus, in this framework for human nature, genetics assumes a role of central and fundamental importance. Genetics are a function of the way we as a human species have evolved and changed over time. Outwardly, genes and genetic predispositions are expressed in our phenotype or our physical appearance, but, less visibly, genes also affect our personality traits. Genes are rooted in the structure of our DNA, something that is unique to each and every individual. They may incline us towards aggressive reactions to events or increase our predisposition to things such as depression. Genes influence us to act, look or behave in certain ways, but they are not the only factor in determining who we are or the entirety of our human nature. Equally important is the role of environment and the variations different people experience in terms of geography, education, culture, upbringing, and so on. Admittedly, even some of these traits can be tied to genetics. For example, Aleutians from Alaska are genetically predisposed to hoard fat cells as a way of insulating their bodies against harsh winter temperatures. This is an example of a community with similar genetic and evolutionary roots that are characteristic of an entire population and not limited to a single individual. Other genetic traits are more heterogeneous. They are unique to an individual and come from the unique 'editing' process that genes go through during the course of sexual and evolutionary selection.

It must be acknowledged that other factors defining human nature are truly independent of genetics. For example, whether someone has access to educational resources is more a question of culture and finance than genetic characteristics. Education is a good example of how environmental traits can shape and influence human nature as, at its best, it can foster transcultural synergy, reduce stereotypes and shape our responses to day-to-day life. Environmental factors, such as education, family, the state of society, and culture, all build on our genetic make-up to define our individual psyches and behaviours. These driving forces of human nature are less susceptible, but not immune to, alterations from emerging strategic technologies. A drug that improves a person's focus or mental capacity would touch on the educational force that influences that individual's response to their survival instinct. This is perhaps a less obvious but no less real impact of emerging strategic technologies on human nature.

Having explained the *'emotional amoral egoism'* framework for understanding human nature, the next question is: what is the connection between this concept of human nature and human enhancement? The link is fundamental, for human survival instincts and emotions *can be modified* by any number of emerging strategic technologies, including, but not limited to, psychotherapy, medicines, genetic engineering, neural prostheses and similar neuro-technological developments. As is mentioned above, human nature is connected to human neurochemistry and, above all, to the visceral and emotional need of humans to protect and promote their own self-interest.[39] This neurochemistry is susceptible to being changed by technology, thus altering the survival instinct that our species has developed over millennia of human evolution. This reality means that our entire human destiny is at stake in the arena of emerging strategic technologies and human enhancement.

As an extension of our human nature, we as humans are genetically and neurochemically programmed to be motivated by a number of factors, chief among them power, profit and pleasure. This fact is intimately connected to how we find meaningfulness in life, something outlined in my theory of *'sustainable neurochemical gratification (SNG)'.* In short, 'what gives meaning to our lives are those things that serve our neurochemically based emotional self-interest in a sustainable way'.[40] Such gratification is highly individualistic and can serve a variety of forms, some benign, some not.[41]

Because our actions and nature are primarily guided by emotional self-interest and the need for neurochemical gratification, it is a natural conclusion that we will *inevitably* go down the path of human enhancement. If a technology is available that increases our pleasure and reduces our pain, our neurochemically mediated emotions and survival instincts will intuitively push us in that direction. The brain is pre-programmed to feel good, and technologies have the potential to activate reward mechanisms in the brain and simulate feelings of extreme pleasure.[42] Keeping in mind the key aspects of *'SNG'*, it is clear that such technologies will be irresistible as we look to create meaning in our lives. Indeed, most human behaviour is a form of addiction rooted in an individual's neurochemistry. Addiction has negative connotations, often relating to drugs and alcohol, but, in this context, addiction can also refer to ongoing gratification from more positive forces such as family life, scholarship or faith in God. As emerging strategic technologies become more advanced, we run the risk of becoming addicted to technologies and what they do for our neurochemistry rather than to things that are truly beneficial to us as individuals and members of society.

It can be argued that the greatest danger to the human race will come when we begin depending entirely on technologies for *'SNG'*, as it is unlikely that any neurochemical gratification from technology could be truly sustainable and, thus, it is likely to undermine the sustainability of the human race and human civilizational history in the long term. These themes are revisited below where I define my previously published philosophy of *'sustainable history'*. For now, it suffices to say that technology has a growing ability to shape who we are and how our race will evolve and change.

Overall, it is my contention that *radical human enhancement through emerging strategic technologies is not a question of 'if' but of 'when' and 'how'*. How these inevitable alterations will change our human nature, our societal morals and our fundamental dignity is the next natural topic to address.

## Human dignity

The United Nations recognizes human dignity as an inalienable right in the opening sentence of the preamble to the Universal Declaration of Human Rights: 'recognition of the inherent dignity and of the equal and inalienable rights of all members of the human family is the foundation of freedom, justice and peace in the world'.[43] Dignity, the declaration maintains, is a birthright that belongs to each and every human being.[44] It is an easy idea to agree on, but creating a concrete definition of human dignity and identifying the specific criteria that need to be met in order to ensure universal human dignity has proved a more difficult task.

I set out my perspectives on the minimal criteria for human dignity in my book *Sustainable History and the Dignity of Man*, but some of the central points are worth reiterating here as background for this discussion on emerging strategic technologies and human destiny. In the simplest terms, human dignity is a concept that is fundamental to human well-being. Preservation of this dignity depends on coordinated and collaborative efforts between individuals, communities, nation states and the international community as a whole. We live in an era where, largely thanks to technological innovations, we have the tools and wealth to offset or eliminate so many of the causes of human suffering and affronts to dignity. Given our global resources and knowledge, there is no excuse for our global society to allow human suffering to continue.[45] The core challenge is one of fostering a more inclusive and compassionate global society. Valuing human dignity is a key part of this challenge.[46]

Many philosophers throughout history, and especially in the aftermath of World War II, have attempted to define human dignity. For the Romans, dignity had moral, political, social and legal roots and was not necessarily an intrinsic or inherent property of human beings.[47] Hobbes also viewed dignity as a characteristic that was attributed, something based on the worth others ascribed to one.[48] Others base dignity not on attribution but on other accounts, including ostensive, rational, and social.[49] Ostensively, the intelligibility of dignity is rooted in 'the intersubjective recognition that each human being is connected to an eternal being'.[50] In other words, dignity is connected to the existence of a higher being, a concept which is not universally accepted and which varies dramatically across cultures. Thus, this ostensive definition of dignity makes it difficult to apply inclusively and comprehensively.

Rationality, a core part of Kantian philosophy and that concept of human dignity, connects dignity to autonomy. To me, human dignity is universal. In other words, it applies to all human beings and therefore using criteria such as autonomy to judge whether someone is deserving of dignity is not comprehensive enough; the link between autonomy and dignity is a decidedly Western concept, and therefore it is again not universal or inclusive enough to apply to global policymaking.[51] To me, the best definition of human dignity is one that considers human dignity an intrinsic quality. In short, human dignity is 'the status of human beings that warrants respect [...]. Someone has dignity, simply because he or she is human – because he or she is *someone*'.[52] Although some philosophers relate human dignity to humans' capability for rationality, my explanation of human nature as emotionally driven leads me to conclude that a comprehensive explanation of human dignity has to go beyond rationality, especially since reason is a trait humans do not necessarily maximize or even employ. As is outlined above, humans are motivated by their emotional repertoire, and this displays itself in a need for attachment, physical security and a sense of personal and collective identity.[53] Reason is one of several aspects of human motivation, but is by no means the most central one. Although some may find it surprising, the reality is that we use reason only occasionally – most of our actions are rooted in habit or instinct.[54] This reality has important implications for approaching human dignity. That said, 'a life governed by reason is likely to be more dignified than one shaped by dogma and unbridled emotions'.[55]

Rather than focusing on the philosophical definition of human dignity, I prefer to approach the question from the perspective of what requirements must be met in order to ensure human dignity. I believe that human dignity is contingent on nine core factors: reason, security, human rights, accountability, transparency, justice, opportunity, innovation and inclusiveness (See Figure 2). I consider each of these components to be essential to human dignity. More importantly, at this point in human history, given the access we have to education, wealth, and technology, there is no longer an excuse to allow or disregard human suffering. Universal human dignity is obtainable and just, and it is our moral obligation to work towards this goal.

The steps towards achieving universal human dignity are clear, and they are similar when evaluated at both the national and the global levels.

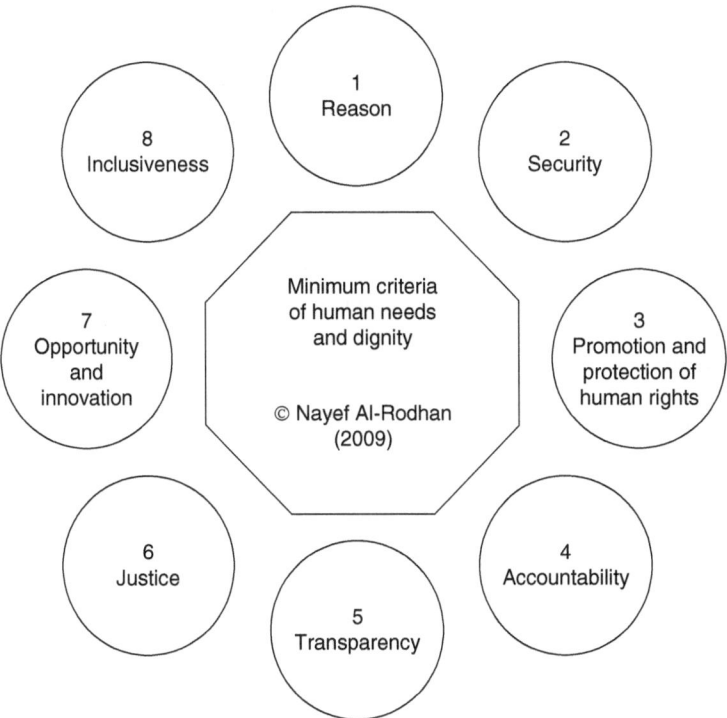

*Figure 2*  Minimum criteria of human needs and dignity

Source: N.R.F. Al-Rodhan (2009) *Sustainable History and the Dignity of Man: A Philosophy of History and Civilisational Triumph* (Berlin: LIT), p. 210. Reproduced with Permission from LIT.

First, governments must focus on meeting the basic needs of their citizens. This includes food, housing, clothing, health care, education and security. Basic welfare provision and security should be staples of any and all societies and states.[56] Similarly, allowing for universal participation and inclusion in society is another way to promote human dignity, as dignity is contingent on feeling valued both as an individual and by the community at large.[57] Ways to promote greater inclusiveness range from freedom of expression to participation in decision-making processes.[58]

Socio-economic justice is equally important in the pursuit of human dignity. Equal distribution of resources within a community, especially in the context of population growth, is a growing challenge and one that must be addressed in order to avoid reducing human dignity.

This list goes on. Gender equality deserves attention from states, especially but not exclusively in the context of domestic violence and ensuring the protection of women who are vulnerable to human trafficking or to falling into industries such as prostitution. On a broader level, ensuring the protection and promotion of human rights, whether they are civil, political, social or cultural, should be a top priority in the pursuit of universal human dignity.[59] Finally, the protection of the environment and maintaining an ecological balance affects quality of life and the future of the planet for ourselves and future generations. Thus, environmental protection, the avoidance of discrimination in aspects such as waste disposal, and fair treatment with regard to environmental issues are all key parts of human dignity as well.[60] These objectives must be met at both the national and the global level. Global objectives build on national criteria and also include measures such as the avoidance of conflict.[61]

Although it should come as no surprise that there are many challenges to reaching these goals, they are still valuable as a benchmark for measuring and ensuring our progress towards universal human dignity.

### 'Sustainable history' and human destiny

The issues of human enhancement, human evolution, human dignity and human nature are important and significant for any number of reasons but, collectively, they are crucial to creating a *'sustainable history'* for mankind and ensuring that we do not lose control of our human destiny. In my book, *Sustainable History and the Dignity of Man*, I define *'sustainable history'* as 'a durable progressive trajectory in which the quality of life on this planet or any other planet is premised on the guarantee of human dignity for all at all times and under all circumstances'.[62] Ensuring *'sustainable history'* is a core element of the destiny of the human race.

According to my *'sustainable history'* approach, history is 'propelled by good governance paradigms that balance the ever-present tension between human nature attributes (emotionality, amorality and egoism), on the one hand, and human dignity needs (reason, security, human rights, accountability, transparency, justice, opportunity, innovation and inclusiveness), on the other'[63] (See Figure 3). *'Sustainable history'* does not necessarily mean an automatic universal acceptance of Western liberal democracy. Instead, it depends on the right governance structures and the progression and promotion of human dignity. The exact form of governance is not as important as the government's

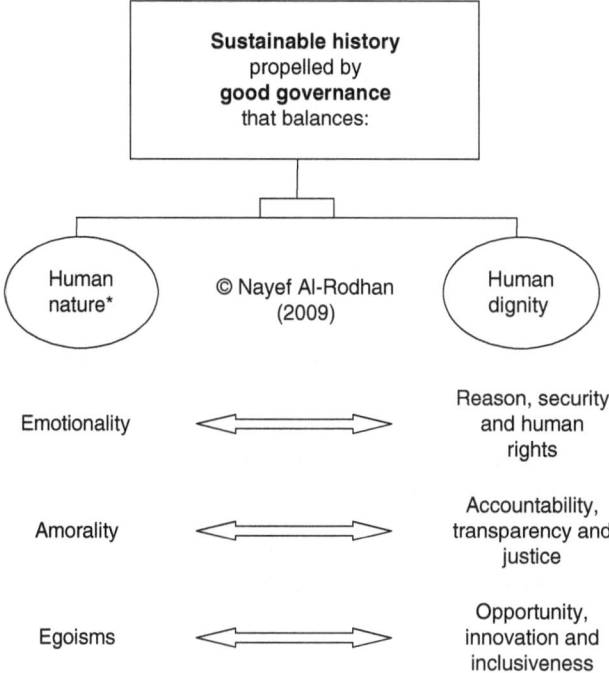

*Figure 3* '*Sustainable history*'

*See Nayef R.F. Al-Rodhan (2008) '*Emotional Amoral Egoism*': A Neurophilosophical Theory of Human Nature and its Universal Security Implications (Berlin: LIT).

Source: N.R.F. Al-Rodhan (2009) *Sustainable History and the Dignity of Man: A Philosophy of History and Civilisational Triumph* (Berlin: LIT, 2009), p. 15. Reproduced with Permission from LIT.

ability to meet a minimum criteria of governance and its appropriateness to the specific cultural domain(s) over which it rules.

Although there is not one proper or better form of governance, there is still a common global standard that must be reached if we want to have the highest levels of political and moral cooperation. We must stop looking at our world as a place in which one society or country will succeed over the other. As I note in *Sustainable History*, 'A sustainable progressive trajectory also depends on our collective triumph. For this to occur, transcultural synergy is essential. This is because the success of any one geo-cultural domain is likely to be dependent on that of another: a geo-cultural domain cannot excel in isolation from others'.[64]

Criteria for good governance include limiting the excesses and encouraging the moral and rational aspects of human nature and ensuring an atmosphere of happiness and productivity. The latter part of this can and should be achieved by promoting reason and dignity.[65] In other words, individuals should exercise reason as opposed to just accepting dogma in order to lead a more dignified life. By doing this, we will be able to identify the institutions and arrangements that will best contribute to and foster a *'sustainable history'*.[66]

Human enhancement threatens to irrevocably change or possibly even destroy our race. One of the biggest differences between our human race today and our ancestors is the access we have to increasingly advanced technologies.[67] These technologies could elevate our species. Certain technologies promise to increase lifespan, quality of life, cognitive abilities in old age, and so on, but they could destroy us as well. For example, smarter-than-human artificial intelligence could turn anti-human and in conjunction with other forms of technology, it could turn against humans and ultimately replace us. Nor can we limit our analysis to life as we currently know it today. Some thinkers argue that we are on the road to planetary civilization. Nikolai Kardashev, for example, proposes classification of technologically advanced civilizations by the efficiency of their energy consumption for interstellar communication.[68] At the most basic level, a civilization would have the energy for planetary-wide communication. The most sophisticated civilization would have the energy to control an entire universe.[69] At the crux of the Kardashev scale is the idea that more advanced civilizations consume more energy than less advanced ones.[70] Other civilizational scales such as the Torino scale base civilizational sophistication on ability to withstand disasters; some maintain civilizational advances will eventually flatten out or that we will evolve to a post-human future.[71] Whichever scenario you prefer, the fact remains that future civilizations may be very distinct from our human society of today. We must consider how technology fits into a variety of civilizational scenarios.

Equally important to civilizational development are the ethical issues human enhancement technologies pose. Fairness and safety are key concerns in this context. Overall, the only way to ensure the survival of the human race is to prioritize the preservation and advancement of human dignity above all else. The *'ocean model of human civilization'* is a useful one in this context because it recognizes 'civilization as a single collective civilization, comprised of a number of different geo-cultural domains'.[72] Instead of 'thinking in terms of *multiple civilizations*, we need to think in terms of *one fluid human story* with internal characteristics linked to the time and the place in which it manifests itself. A philosophy of history needs to encompass a span of human time that captures human nature and its mastery of the environment'.[73] In everyday terms, we are all facing the challenges and opportunities of human enhancement technologies together as a global civilization.

## Conclusion

This chapter outlines some key theories and definitions that are used throughout the debate on human enhancement. The rest of Part II builds on my theories of human nature, human dignity and *'sustainable history'*. I have therefore also spent some time introducing these concepts. At the heart of my argument is the fact that human nature and our desire for *'sustainable neurochemical gratification (SNG)'*[74] will inevitably lead us down a path of enhancement towards trans- and post-human futures. In the face of this inevitability, we must take steps to preserve and promote the minimum criteria of human dignity. This is the only way we can ensure a *'sustainable history'* for all civilizations, regardless of our stage of human evolution.

# 12
# Human Enhancement: The Nature of the Debate

The debate over the morality of human enhancement, and what constitute acceptable enhancements and what fundamentally alters our essence as human beings is, not surprisingly, a heated one. Broadly, the debate can be broken down into two sides: the bioconservatives, including thinkers such as Francis Fukuyama, Leon Kass, Jürgen Habermas and Charles Krauthammer; and the transhumanists led by theorists such as Nick Bostrom, Ray Kurzweil and John Harris. Naturally, there are also some more intermediate, nuanced positions, but the two extremes serve as a good starting point for introducing the debate.

After outlining these two central poles, this chapter explains my assessment of human enhancement and my proposed theory of 'inevitable transhumanism'. Related to the transhumanist philosophy, 'inevitable transhumanism' is especially focused on the intersection of human nature and enhancement technologies, and how that interaction will affect our destiny.

## The case against enhancement

Those who argue against human enhancement come from both sides of the political spectrum, but the general nature of their argument is the same. Overall, these thinkers believe that human nature is sacred and delicate and that it should not be tinkered with, particularly since the consequences of modifying the human form are unpredictable. While such alterations or enhancements may not seem worthy of concern when carried out on an individual level, at the societal level they undermine something special, valuable and intrinsic about the human race.[1]

Bioconservativism, as it is sometimes called, 'is a social, political, and moral stance that urges regulation and relinquishment of biotechnologies regarded by bioconservatives as dangerous, dehumanizing or immoral'.[2] Among the technologies bioconservatives frequently deride are genetic modification of both plants and animals, pre-implantation genetic diagnosis, all forms of cloning and radical life extension.[3] Although bioconservatives share strong ideas on the topic of human enhancement, they often agree on little else politically. For example, bioconservatives are just as likely to be religious conservatives as liberal environmentalists.[4] Bioconservatives on the right side of the political spectrum are most likely to object to enhancement on religious grounds.[5] In many cases, religious values prohibit reproductive rights due to needs for population reduction or because enhancement seemingly tampers with God-given traits.[6] These groups are wary of using technology to 'play God'. Such ideas have been criticized by the more liberal elements of the human enhancement debate, who compare this religious reasoning to 'human racism' or the idea that humanness is the sole criteria for personhood.[7]

By contrast, left-leaning bioconservatives are more concerned by the capitalist and patriarchal undertones that they associate with the development of new technologies.[8] Left-leaning bioconservatives are also concerned about the potential impacts that enhanced human beings might have on the environment.[9] The critics of leftist bioconservatives accuse the group of believing that the 'yuck factor' of human enhancement 'trumps individual liberty for germinal choice and biotechnology, but [ironically] not in matters of sexuality or abortion'.[10]

Francis Fukuyama is one of the most outspoken and eloquent critics of human enhancement, and his concerns are manifold. On one hand, he is concerned about the possibility of creating a superior breed of humans, thereby undermining the worth or dignity of normal, unmodified humans.[11] Additionally, Fukuyama worries about the inherent inequalities of the enhancement process. Enhancements such as cosmetic surgery or embryo screening are often expensive, meaning that the rich are better poised to take advantage of the available technologies. Human enhancement thus has the potential to increase the gaps between the rich and poor and to more firmly solidify differences and inequalities across generations and cultures.[12]

The US President's Council on Bioethics echoes and expands on Fukuyama's concerns about fairness and raises a more fundamental concern. According to the Council, 'some people see in biotechnical intervention a way to compensate for the "unfairness" of *natural* inequalities, say, in size, strength, drive or native talent'.[13] While this may seem like a strong argument in favour of enhancement technologies, the fact remains that 'even if everyone had equal access to genetic improvement of muscle strength or mind-enhancing drugs, or even if these gifts of technology would be used only to rectify the inequalities produced by the unequal gifts of nature, an additional disquiet would still perhaps remain: the disquiet of using such new powers in the first place or at all, even were they fairly distributed'.[14]

Widening inequalities may threaten both social stability and institutions. As the gap between enhanced and non enhanced humans widens, the worse-off may revolt against the continuous disadvantage. Moreover, diverse societal institutions such as sports, retirement programmes and even job searches would have to be rethought in order to cope with these inequalities.[15]

A more subtle argument against human enhancement deals with questions of self-worth, the inherent abilities and characteristics of human beings, and the nature and intrinsic value of human achievements. This is a point expanded on by Michael J. Sandel and Jürgen Habermas. Sandel writes, 'It is one thing to hit seventy home runs as the result of disciplined training and effort, and something else, something less, to hit them with the help of steroids or genetically enhanced muscles [...]. As the role of enhancement increases, our admiration for the achievement fades – or, rather, our admiration for the achievement shifts from player to pharmacist'.[16] Sandel goes on to argue that this reaction 'suggests that our moral response to enhancement is a response to the diminished agency of the person whose achievement is enhanced'.[17] For this reason human enhancement should be approached with caution.

Jürgen Habermas, a vehement opponent of human enhancement and any activity implying the modification of human genomes or the destruction of human embryos, argues that human enhancement at birth implies a loss of self-determination of the individual. This also has implications for equality. A whole generation affected by 'eugenic decisions', as he calls it, may therefore not be considered equal and free.[18]

Philip Brey frames his concerns over enhancement in terms of personal identity, arguing that new, enhanced abilities may alter the way people look and behave, thus changing the way they experience the world and themselves.[19] Although Brey does not perceive human enhancement as categorically wrong or immoral, he does think that enhancement has the potential to undermine self-esteem and that it therefore presents a potential moral danger.[20] Although Brey's final position on enhancement is somewhat ambiguous, his concerns over enhancement are reflective of some of the worry people have over the slippery slope of enhancement.

Another argument against human enhancement could be described as the *Brave New World* scenario. In this classic book by Aldous Huxley, biotechnologies are used and abused in the name of creating social stability.[21] In this society, almost genetically identical embryos are raised and hatched in incubators and assigned to castes based on their predetermined levels of intelligence and physical strength. Huxley's world is one free of war, pain, hatred and poverty, but it is also a world lacking in diversity, where emotions are controlled through drugs and where moral complexities are ignored. Although Huxley published this book in 1932, many of the characteristics of Huxley's world are reflective of the concerns contemporary bioethicists have with regard to the dangers of modern, converging human enhancement technologies. Indeed, the evening out of human emotions, abilities and personalities through technologies is a major concern for those worried about the dangerous ramifications of human enhancement. As C.A.J. Coady notes, the *Brave New World* scenario presents 'an intrinsic moral objection to the very idea of changing human nature'.[22]

A final argument that appears throughout the literature against human enhancement is that of hubris versus humility. The President's Council frames this argument against enhancement in terms of the fact that man should not try to interfere with what God or nature has produced naturally. As the Council argues, there is inherent value in the way we have been created or evolved, and to tamper with that is to brazenly defy the profound forces of nature and the cycle of life.[23]

In a similar vein, Fukuyama argues that human enhancement is dangerous because it has the potential to destroy something he dubs 'Factor X'. According to Fukuyama, Factor X is the essential human quality that is worthy of a certain minimal level of respect. Factor X, Fukuyama says, is what we find when 'we strip all of a person's contingent and accidental characteristics away'.[24] Human enhancement and its unintended consequences, Fukuyama argues, have the potential to fundamentally and irrevocably alter or damage Factor X, thus destroying our humanity. What makes the prospect especially dangerous is the fact that there is no line in the sand that shows when Factor X has been damaged or irretrievably lost. The boundaries are grey and fuzzy, but once the tipping point has been crossed, there is no going back. For this reason, human enhancement should be avoided and discouraged. Fukuyama has even gone so far as to label human enhancement and transhumanism as the 'world's most dangerous idea'.[25]

## The case for enhancement

Biotechnology and other emerging strategic technologies can cure diseases, dramatically improve human lifespan, increase human mental and physical capacities, and improve people's overall quality of life. For all these reasons, advocates of human enhancement technologies believe that the biotech revolution and the convergence in the fields of nanotechnology, biotechnology, information and communications technology (ICT) and cognitive science should be widely and almost universally embraced. Enhancement technologies can be justified morally, medically and humanely.

The leading thinker in the advocacy of human enhancement is Nick Bostrom. Bostrom and his fellow pro-enhancement colleagues generally belong to a school of thought known as 'transhumanism', and they believe first and foremost that enhancement technologies should be made widely available to society.[26] For transhumanists and other supporters of human enhancement, it is a common refrain that it is the right of an individual to plan his or her own life and make his or her own decisions regarding whether to enhance. Bostrom points out that 'many methods for enhancing cognition are of quite a mundane nature, and some have been practiced for thousands of years. The prime example is education and training, where the goal is often not only to impart specific skills or information but also to improve general mental faculties such as concentration, memory and critical thinking'.[27] It is important to keep in mind, Bostrom notes, that the 'most dramatic advances in our effective cognitive performance have been achieved through non-biomedical means'.[28] Thus, the use of technology to improve our bodies and brains should not be viewed as such a radical idea; many of the objections over these technologies seem to come from the fact that they are novel and experimental more than from the fact that they are bad. What is considered actual enhancement, and if it is morally desirable, is seen as context-dependent.[29] Thus, transhumanists maintain that enhancement should be assessed on a case-by-case basis. Not all enhancements are necessarily good but, in many instances, it can be argued that it would be unethical *not* to enhance ourselves.

Indeed, transhumanists believe that we have not only a right but a moral obligation to improve ourselves with available technologies.[30] This moral obligation is rooted in the fact that technologies and enhancements offer us the unique opportunity to alleviate human suffering. The possible scenario presented by Ron Bailey, author of the book *Liberation Biology,* is reflective of the potential transhumanists see in technology:

> By 2100 the typical American may attend a family reunion in which five generations are playing together. The great-great-great-grandma is 150 years old, and she will be as vital as she was when she was 30 and as vital as her 30-year-old great-great-grandson, with whom she's playing touch football [...] No one in her extended family will have ever caught a cold. They will be immune from birth to the shocks that human flesh has long been heir to: diabetes, cancer, and Alzheimer's disease. Her granddaughter, who recently suffered an unfortunate transport accident, will be sporting new versions of the arm and lung that got damaged in the wreck, and she'll be playing in that game of touch football with the same skill and energy as anyone else in the family. Infectious diseases that terrified us at the beginning of the 21st century, such as HIV-AIDS and the avian flu, will be horrific historical curiosities for the family to chat about over their plates of super-fat farm-raised salmon, which will be as tasty and nutritious as any fish any human has ever eaten [...] Not only will this family enjoy all these benefits, but nearly everyone they work with, socialize with, and meet with will enjoy them as well. It will be a remarkably peaceful and pleasant world. Beyond their health and their wealth, they'll be able to control things such as anti-social tendencies and crippling depression. And they'll manage these problems by individual choice, through new biotech pharmaceuticals and personalized genetic treatments.[31]

Although this scenario may seem naively utopian to some, transhumanists and their close cousins the technoprogressives fully believe that such developments are well within the realm of possibility, especially given the convergence of nano, bio, cognitive and ICT technologies. Transhumanists see technology as merely an extension of existing, natural enhancement processes, and they stress that in many ways human enhancement is already taken for granted as a moral good. When we educate ourselves or engage in healthy diet and exercise, for example, we are essentially improving or enhancing ourselves in fundamental, albeit non-technological, ways. Transhumanists argue that using technology to facilitate human improvement should not be considered any differently from non-technological enhancements.[32] As human beings, we are only partially evolved, and we should look to technology to help facilitate our natural transition to a more advanced species. Moreover, it can be argued that as human beings, we have fundamental flaws, and that the state of being human, as it exists, is not perfect. Cancer, dementia, and disease are all shortcomings of the human condition, and if the technology exists to fix these flaws, transhumanists argue that we should take advantage of it.[33] Here, again, the rational individual argument takes a central role. As long as rational, free individuals choose to participate in technological improvements, there should be no real basis for objection to human enhancement technologies.[34]

In his 2005 book *More Than Human*, Ramez Naam succinctly offers four specific arguments in favour of human enhancement, and in doing so, he sums up much of the transhumanist school of thought. His list includes the pragmatic benefits of enhancement, the inevitable failure of any attempts to regulate enhancement, respect for individual autonomy and freedom of choice, and the fact that the desire to enhance is an inherent part of human nature that cannot be suppressed.[35]

Yet another debate concerns the question of whether human enhancement will actually lead to more happiness.[36] Supporters of human enhancement argue that it offers an avalanche of promising prospects. Technology has the potential to offer happier, longer, more prosperous lives. Certain technologies may reduce disease and increase lifespan while others may reduce or eliminate disability or improve the productivity of the individual and society. Others, however, fear that human enhancement will not be the 'holy grail of happiness we might believe it to be, that is, we will still be dissatisfied with ourselves, no matter how much we enhance ourselves'.[37] In a similar vein, it has been argued that in human enhancement, since it will make enhanced humans very different from non-enhanced humans, there will be a loss in commonality. In other words, since the needs of enhanced humans will change, neither side will be certain what makes the other side happy anymore.[38]

It is also worth noting that transhumanists do not accept all enhancement as valid. Indeed, there are certain technologies that cross ethical lines and undermine human dignity, and these technologies should not be allowed even in the name of individual choice. For example, Bostrom cites ethical concerns over genetic manipulation of embryos and over the unintended consequences of such actions. Although he stops short of saying that such practices should be universally banned, he does acknowledge that 'it may be wise to avoid using this technology until it is advanced enough for us to be sure that the expected benefits outweigh the risks'.[39]

To counter some of the key concerns bioconservatives have over human enhancement, transhumanists such as Bostrom emphasize the difference between enhancements with only positive externalities and those with negative ones. Positive enhancements are those that have intrinsic benefits for society and enable an individual to better contribute to his or her community. An enhancement that improves a person's health or eliminates a disease-carrying gene would fall under this category. By contrast, an enhancement with a negative externality is one that has no advantage for society as a whole and offers only a competitive advantage to the enhanced individual (as opposed to an intrinsic advantage). Height, often thought to be a desirable trait but not one that contributes to or improves society, is a classic example of this type of enhancement.[40]

Although transhumanists acknowledge that human enhancement does have the very real potential to undermine human dignity, they nonetheless maintain that human enhancement offers more distance to rise than to fall in terms of dignity and, for this reason, that human enhancement should be embraced and approached with a libertarian attitude.[41]

Because so many of the bioconservatives' arguments against human enhancement focus on the risk to the intrinsic value and importance of human nature, it is worth looking at how transhumanists and those in favour of human enhancement respond to these arguments. Norman Daniels of the Harvard Public School of Health argues that human nature has been misrepresented as a concept, and that this shortcoming has coloured the debate over human enhancement in a negative way. For Daniels, human nature is not a concept that can be evaluated at the individual level. Instead, it is a 'population concept'.[42] In other words, one cannot look at a single human individual and draw conclusions about the nature of the entire human race. Human nature is an aggregate concept that must take into account a wide spectrum of diversity. The fact that a person such as Mother Theresa is uniquely selfless cannot be extrapolated to mean that the entire human race is driven by benevolence and self-sacrifice.[43] Indeed, that would be a grave misconception.

Because human nature is an aggregate concept, Daniels argues that enhancements taken at the individual level are practically incapable of fundamentally changing our collective human essence: 'We can modify human nature, but it takes a very tall tale. We must affect the (or at least a) whole population of humans, and we must do so with a trait central to that nature'.[44] While still acknowledging bioconservatives' fears that human nature is vulnerable to human enhancement, Daniels clearly demonstrates that human enhancement, especially when considered at the level of the individual, is not an imminent threat to the nature of our species.

Arthur Caplan is also somewhat dismissive of bioconservative fears of human enhancement, saying that while some of their concerns carry 'emotive force,' they are not a sound basis for rejecting choices that individuals or parents might make to improve or optimize their children.[45] Caplan takes particular issue with bioconservative arguments that happiness or satisfaction achieved through engineering will result in a deformation of our character and that improvement through engineering is not morally commendable. As Caplan says, 'laying the blame for vice, sloth, or the willingness to settle for cheap thrills at the feet of enhancement ignores the inconvenient fact that the desire for quick returns, easy money, and instant gratification have nothing at all to do with whether some or all of us choose to use biotechnology to become enhanced beings with different experiences'.[46]

Transhumanists and those in favour of enhancement are not oblivious to the risks of such choices. That said, they believe human enhancement should still be made widely available and that the individual can best decide what technologies are most necessary or appropriate for him or herself. Peter Hagoort puts it this way, 'I find it a precarious thought that some people may change themselves under social pressure. Freedom *not* to use a technology should be warranted'.[47]

Perhaps the most damning counterargument to the bioconservative school comes from Rebecca Roache who aptly highlights the lack of concrete arguments made by bioconservatives. She writes that the debate over human enhancement 'has reached an impasse, largely because bioconservatives hold that we should honor intuitions about the special value of being human, even if we cannot identify reasons to ground those intuitions'.[48] Roache correctly identifies the neurological basis of human intuition and its potential fallibilities. As she says:

> Intuitions – including intuitions about enhancement – are subject to various cognitive biases rendering them unreliable in some circumstances. We argue that many bioconservative intuitions about enhancement are examples of such unreliable intuitions. Given this, it is unrealistic of bioconservatives to expect others to rely on their unexamined intuitions. Furthermore, refusing to engage in debates

about the reasons and values that underpin their intuitions about enhancement will have the effect of making bioconservative voices less relevant in policy debates about enhancement than they would otherwise be.[49]

## Technodemocrats

A more middle ground between bioconservatives and transhumanists would be what John Hughes has dubbed a 'technodemocrat'. Technodemocrats, also referred to as technoprogressives, are in favour of technological advancement, often including enhancement technology, but they insist on strong regulatory frameworks. Such regulation should ensure the safety of all enhancement technologies and guarantee that the technologies are widely available in order to avoid issues of discrimination and inequality.[50] Other considerations come into play too. For example, consumer groups and technodemocrats worry about the potential for governments or third parties to abuse or collect information from those who have had certain types of brain enhancement.[51] Again, these concerns underscore the need for very strong regulation. A gradual and incremental approach to human enhancement is one core component of the technodemocrats' approach.

Dale Carrico's explanation of techno-progressivism captures the movement's essence perfectly:

> At its heart technoprogressivism is simply the insistence that whenever we talk about 'progress' we must always keep equally in mind and in hand both its scientific/instrumental dimensions but also its political/moral ones. From a technoprogressive perspective, then, technological progress without progress toward a more just distribution of the costs, risks, and benefits of that technological development will not be regarded as true 'progress' at all. And at the same time, for most technoprogressive critics and advocates progress toward better democracy, greater fairness, less violence, a wider rights culture, and such are all desirable but inadequate in themselves to confront the now inescapable technoconstituted quandaries of contemporary life unless they are accompanied by progress in science and technology to support and implement these values.[52]

My vision of the future of human enhancement is closely aligned with the technodemocrats, and those are opinions I expand on in the section below.

## My view: the inevitability of human enhancement

My perspective on the validity of human enhancement is somewhat distinct from the three perspectives outlined above, falling somewhere between that of the bio-conservatives and that of the technodemocrats. For me, the central question over human enhancement is not a matter of if (as the bio-conservatives suggest it might be) but when. Human nature is an emotionally driven, neurochemically mediated phenomenon that will inevitably drive us towards increasingly ambitious forms of human enhancement. I therefore believe in *'inevitable transhumanism'*. Whether the ultimate implications of these enhancements are for better or worse remains to be seen. However, in the long run, I believe that our human impulses towards *power, permanency, profit and pleasure*, and our emotional, amoral, egoistic human nature will more than likely take us down a dangerous path in which we will ultimately be replaced by beings more intelligent than us – ironically, products of our creation. In the long term, it is highly probable that we as humans will cease to exist as we know ourselves today.

Others share my view that human enhancement is an inevitability in spite of the potential dangers it poses. For example, Francoise Baylis and Jason Scott Robert argue that the inevitability of such technologies will result 'from a particular guiding worldview of humans as masters of the evolutionary future'.[53] For these scholars, we cannot control whether enhancement occurs and, instead, we should focus our attentions on which types of enhancement we as a society decide to allow.

The inevitable human drive towards widespread, dramatic enhancement – and possibly towards ultimate extinction – will not come from a flaw of human nature or from any misunderstanding of enhancement, per se. Rather, we will pursue enhancement because our fundamental, neurochemical composition does not leave us any other option. In my book *Emotional Amoral Egoism*,[54] I outline the major driving forces of human motivation. Some of the central factors are self-interest, the core foundation of human nature; fear, grief and pain; ego, pride and reputation; pleasure; greed; specific individual inclinations; occasional reason; and, less frequently, reflection and morality. Certain of the more dominant factors will push humans towards enhancement. Take, for example, pleasure. Human enhancement technologies such as neuropharmacology offer huge opportunities to control emotions, to stimulate pleasure and to reduce pain. On a visceral level, we as a race will seek such technologies out. Pleasure benefits us and makes us feel good, therefore, our nature will lead us down any path that offers the chance to increase our pleasure. Weighty considerations about human dignity and the future of human destiny will be easily pushed aside for immediate individual gratification.

Similarly, our survival instincts and our fundamental, primordial predispositions to recreate and ensure the continuation of our species will lead us down a path of reproductive enhancement. Although not all people are intent on reproducing, those who are often have problems due to infertility, age or illness. Thanks to technologies such as artificial insemination, in-vitro fertilization, and fertility drugs, the opportunities for reproductive enhancement abound. Since ultimately the survival of our species is dependent on such reproduction, it naturally follows that our emotions and instincts will inevitably draw us towards the technologies that facilitate this process or that make a process once thought to be impossible for a particular couple possible again. Interestingly, it is inevitable that we will pursue these technologies, and the root of our motivation will be survival instincts, ensuring the overall survival of our species and the manifestation of human destiny. However, the same technologies that we will inevitably use to preserve and enhance our species and to ensure the fulfilment of our destiny are extremely likely to destroy us.

In other words, we will inevitably seek enhancement technologies because they can easily and effectively fill needs that our neurochemically mediated survival instincts demand. In the barest terms, we will inevitably pursue emerging strategic enhancement technologies as a way to survive, but, these same technologies, especially when evaluated over time and when taken in the aggregate, will more than likely undermine our survival – at least in terms of the human race as we know it today. This is because as we use technologies to modify our emotions, our bodies and our neurochemical balances, we will undermine and alter the instincts that have developed over millions of years in the process of human evolution. With technology, we will – in a very short period on the evolutionary timescale – fundamentally change, enhance or suppress the very things that make us human. What we initially seek out for survival will be the same things that ruin our species by transforming us into something different, inevitably into transhumans and, possibly, ultimately into post-humans. Whether this will be positive or negative for human destiny and human dignity is an open question but, for the sake of this discussion, the central point is that this shift is *inevitable*. Thus, our policy debate should focus on how best to regulate emerging strategic technologies and their application to human enhancement now, before we lose control of our humanity.

## 'Inevitable transhumanism' and a framework for the human enhancement debate

The Institute for Ethics and Emerging Technologies (IEET) has framed the debate over human enhancement by creating a scale of biopolitics.[55] This spectrum, which ranges from the most liberal category of libertarian transhumanists to the most conservative BioLuddites, is a useful tool for explaining the variety of perspectives in this debate. Accordingly, it is worthwhile spending some time evaluating where my position falls on the scale. From far left to far right, the IEET scale is as follows: libertarian transhumanists, technoprogressives, left-wing bioconservatives and right-wing bioconservatives – also known as Bio-Luddites.[56]

As described by Hughes, libertarian transhumanists are on the far left of the biopolitical scale. These transhumanists 'embrace technologies that extend and enhance regardless of their effect on "natural" life spans'.[57] Almost universally, transhumanists are libertarian, advocating freedom of individual choice and extremely limited government involvement in the lives of citizens. Like the similarly oriented technoprogressives, libertarian transhumanists believe 'human beings are free to determine their own future, guided by prudent reason. There are no obvious natural or divine limits on human aspiration'.[58] Bio-Luddites, on the other hand, call for a universal ban on all technologies that threaten the 'natural'.[59]

In between these two extremes are the more moderate positions. IEET defines the middle ground as consisting of 'technoprogressives' and left-wing bioconservatives. My perspective is what I call *'inevitable transhumanism'*. This concept falls somewhere in the area of the biopolitical scale between technoprogressives and left-wing bioconservatives. For example, on the issue of human choice in the matter of enhancement, I believe that when left to our own devices, our human nature will drive us to pursue enhancement. However, I also believe that specifically because of this nature, strict government regulation of enhancement and clear ethical and moral guidelines are necessary. On this issue, my biopolitics are more right-leaning than left. While libertarian transhumanists argue that 'technology is uncontrollable'[60] and that 'government intervention always has bad unforeseen consequences',[61] *'inevitable transhumanism'* recognizes that even if regulations are imperfect, there absolutely must be some system of checks and balances to ensure that the impulses and attributes of our own human nature do not lead us down a path of self-destruction. The fact that government regulations often have unforeseen consequences only underscores the universal value and upmost importance of *good* governance.

With regard to topics such as the accessibility of enhancement technologies, I am much more left-leaning, believing that, like the IEET website explains, 'democracies should work toward social equality and provide universal access to enhancement technologies'.[62] Once governments and civil society have set the limits of human enhancement and decided what is medically, ethically and personally acceptable, I believe such technologies should be made universally available, regardless of economic, social or geographical differences. Such a concept is a key component of global justice.

In terms of procreative liberty, *'inevitable transhumanism'* again takes the middle ground. While Bio-Luddites reject any reproductive enhancements based on religious reasons and often on religiously motivated population control, they are certainly the extreme. *'Inevitable transhumanism'* embraces the idea of the limited procreative liberties advocated by right-leaning bioconservatives, specifically that reproductive rights and technologies should be allowed, but they must stop short of germline alterations.[63]

Overall, *'inevitable transhumanism'* rejects the idea of libertarian transhumanists that the market will take care of inequalities, the environmental concerns of human enhancement, and the advancement of positive enhancements but not the negative ones. As I explained in detail above, in the absence of regulation, human nature and our neurochemically mediated survival instincts will draw us towards enhancements whether they are ultimately good for us or not. Although the market is a powerful force, it simply cannot compete with our most visceral survival needs, and that is why it is so necessary to have a strong regulatory structure in place. My recommendations for regulatory guidelines on human enhancement are outlined in greater detail below. For now, it is enough to say that the preservation of human dignity is the central objective. That concept is the core of *'inevitable transhumanism'*.

# 13
# The Science and Technology of Human Enhancement

Having outlined the vocabulary of human enhancement and the debate over the issue in general terms, it is time to address specific enhancements and the potential consequences of these enhancements in greater depth.

## Physical enhancements

Physical enhancements cover everything from everyday items such as hearing aids, contact lenses, cosmetic surgery and pacemakers to more futuristic scenarios such as artificial brain synapses/neurons, prostheses run by artificial intelligence, and the development of super human muscular strength. With regard to the more basic technologies, physical human enhancements are most notable for their ubiquity. Indeed, many physical human enhancements are such an integral part of our daily life that it seems almost absurd that there could be an ongoing moral debate over their legitimacy or value.

Physical enhancements, while important to the debate over human dignity, human nature and human morality, are not nearly as significant or as potentially dramatic to the debate as mental enhancements are. However, physical enhancements are noteworthy for the simple fact that they have already normalized certain forms of human alteration and enhancement. Indeed, is there anyone who thinks twice about wearing glasses or contact lenses to improve their vision? How could we really fault an accident or attack victim for seeking out plastic surgery that will reduce the appearance of scars or other similar damage? At least on one level, human enhancement is already mainstream. The pervasiveness of these types of human enhancement therefore opens the door for easier, wider acceptance of other, more serious enhancements.

That is not to say that all types of physical enhancement are accepted without question or controversy. One particularly divisive type of human enhancement involves athletic enhancement or the use of steroids and similar drugs to boost physical capabilities. Especially in the world of competitive sports, drugs such as anabolic steroids, Insulin Growth Factor and even Viagra are often used to improve endurance and stamina among professional and amateur athletes alike.[1] For professional athletes, the financial incentives to dope and enhance can be enormous, and taking performance-enhancing drugs or steroids can have a dramatic effect on an athlete's abilities and his finances.[2] However, such enhancements lead to fundamental questions over the nature and value of athletic achievement. Do we value athletic achievement under any circumstances or only when it reflects an individual's ability to discipline himself and his body through diet and exercise? Of course, that presents the follow-up question: if diet and exercise are valid, accepted forms of human enhancement, why not take that idea one step further to accept enhancing drugs? The question is a complex one, and the fact that there is no clear boundary between the two illustrates the slippery slope of the debate over enhancement. Some prominent types of physical enhancement are outlined below.

### Performance enhancing drugs

Perhaps the most high profile form of human enhancement – performance-enhancing drugs such as steroids, diuretics or creatine – is widely used and somewhat controversial. Their high profile stems largely from their use by professional athletes looking to improve their speed, strength and endurance ahead of an athletic competition. They can also be used by people who are looking to speed up their recovery after an injury or people who are averse to physical exercise but would still like to improve their overall fitness.[3] For example, anabolic-androgenic steroids such as testosterone can help athletes improve their strength and muscle mass; they also allow for quicker recovery and can promote aggression – a characteristic which can be an advantage in certain sports.[4] Diuretics can help athletes drop weight quickly or can dilute their urine so that other enhancing drugs are undetectable in tests.[5] Creatine is a naturally occurring compound in our bodies, and athletes often take creatine supplements to benefit from short bursts of power.[6] When speaking about human enhancement, creatine is an interesting example, as it is something our bodies produce naturally. Thus, taking a creatine supplement seems like as much of an enhancement as taking a vitamin pill (an imperfect analogy, but nonetheless thought-provoking). This underscores the fuzzy line in the human enhancement debate over what should and should not be acceptable. Indeed, whether such enhancements undermine the value of athletic achievement or give doping athletes an unfair advantage over their competitors is a hot issue for debate, and even if enhancements undermine professional sports they may still have positive impacts in everyday life.

## Cosmetic surgery

Globally, cosmetic surgery is a multi-billion dollar business, and its prevalence is so wide that it is easy to overlook it as a form of human enhancement. Each year, the number of people seeking to improve their eyesight through laser surgery, increase their breast size through implants, reduce wrinkles through Botox injections, have their noses or chins reshaped, as well as any number of other procedures, increases.[7] In some cases, cosmetic surgery can be used to correct a physical disability or abnormality, such as poor eyesight or a cleft palate. In other instances, cosmetic surgery is done purely for aesthetic reasons. This tension between therapy and enhancement can be seen throughout the debate over the value and legitimacy of human enhancement technologies. The question of whether cosmetic surgery changes how we see ourselves or how others see us is a central ethical and philosophical question that colours the debate over the validity of this branch of human enhancement technology. Pressure to enhance or to seek out cosmetic surgery strictly to comply with perceived societal norms is another danger in this field. It has even been argued that, since being physically attractive can have important social advantages, cosmetic plastic surgery can be considered a 'fair game for survival'.[8]

## Anti-aging technology

For many transhumanists, longer lives and even immortality are viewed as desirable goals. In this context, anti-aging technologies are of particular interest because if lifespans are to be increased, we have to improve our bodies' physiologies. At some point in each individual's life, the body's cells deteriorate and have been damaged to the point where death becomes inevitable. Reversing or stopping the aging process is thus quite appealing,[9] making anti-aging and rejuvenation medicine a key sector of enhancement technology. Some things are topical. Any number of lotions and creams available at pharmacies around the world promise to reduce laugh-lines, age-spots and wrinkles, but there are other more sophisticated advances on the horizon. For example, in the future, developments in nanotechnology will help us to rebuild dying cells and organs in our bodies and ensure that our appearances and our cognitive and physical abilities function better and longer.[10]

Aging is a multifaceted and complex process that involves all systems in our bodies and is affected by environment, diet and genetics. As we develop a more sophisticated understanding of the human genome, it is likely that we will be able to target certain genes responsible for aging and alter or eliminate them. Such steps, when made in conjunction with others such as a healthy diet, will help slow and change the aging process as we currently understand it.[11] Moreover, some studies have shown that regular injections of the Human Growth Hormone can slow the aging process. Although the evidence on this front is conflicting and such procedures have noticeable side effects, the studies in favour of this finding are nonetheless indicative of the potential technology has to influence the aging process.[12]

### Artificial retinas

Artificial retinas are a fascinating technology currently being developed by the US Department of Energy. These devices, surgically implanted into the eyes of someone who has been blinded by a condition such as retinitis pigmentosa, use an array of microelectrodes to help recreate light patterns and provide a sense of vision. The device works with the help of a miniature camera that is mounted in a pair of eyeglasses. This camera captures images and wirelessly sends those images to a microprocessor, which is typically worn on a belt. This processor converts the data into an electronic signal that is then relayed to a receiver in the eye.[13] These signals bypass defunct photoreceptor cells in the eye and go directly to the remaining viable cells. After that, the pulses travel to the optic nerve, allowing the person wearing the device to distinguish light and shadow.[14] Although these artificial retinas are still far from restoring full vision capabilities, they are a significant step forward for people blinded by often hereditary diseases, and further improvements can be expected soon.[15] In addition, types of electronic contact lens can provide enhanced capabilities and additional information. For instance, they could allow the wearer to see by night or to receive and send information, or could provide telescopic vision.[16]

### The future of physical enhancement

Beyond these somewhat mainstream physical enhancements, there is much speculation about more radical physical enhancements that would truly alter our core human essence. Although much of this is still hypothesis, some of the potential developments in physical enhancement that have been floated include: altering skin pigmentation in order to reduce the likelihood of sunburn and skin cancer; adding artificial ligaments and tendons to different parts of the body (such as knees) in order to make those parts stronger and less prone to injury or deterioration; and hormonal modifications that would slow the aging process.[17]

For those people who will not live long enough to see such developments, there is the option of cryonics, or the science of freezing an entire human body after death in the hopes of later restoring life to it when technology permits.[18] With cryonics, a patient is 'infused with a substance to prevent ice formation, cooled to a temperature where physical decay essentially stops, and is then maintained indefinitely in cryostasis (for example stored liquid nitrogen)'.[19] If and when future technology allows it, 'patients hope to be healed, rejuvenated, revived, and awakened to a greatly extended life in youthful good health, free from disease or the aging process'.[20] An alternative possibility is 'neuro-cryopreservation', whereby only the head of a dead person is removed and cryopreserved.[21] Although the science of cyronics has not been proven, it has groundbreaking potential in the field of human enhancement. Nanotechnology and its ability to manipulate individual atoms or molecules and to eventually build and repair virtually any physical object are at the heart of cyronics' promise, because nanotechnology could help to rebuild and repair the damaged organs and tissues that led to the human death in the first place.[22]

Brain-machine interfaces, although still in the early stages of development, present another main field of research for future enhancements. Projects have shown that by implanting micro-electrodes into the brains of rats, the experimenter can steer the rats to take new paths. As with all emerging technologies, brain-machine interfaces have huge potential benefits, for instance, in helping paraplegic patients, but ethical issues arise should they be applied to healthy persons to enhance performance.[23]

## Reproductive enhancements

As science advances, we are gaining more and more control over the reproductive process. Whether it is through fertility drugs, artificial insemination, in-vitro fertilization or the screening of embryos for diseases, technology is giving us increasing opportunities to create and manipulate life in unconventional ways. For couples who are unable to conceive a child naturally, the emotional and physical toll of infertility can be extraordinarily high. This fact has pushed many Western societies, as a general rule, to be open to and accepting of at least the premise of reproductive technologies. Indeed, the rapidity with which certain reproductive technologies have gained legitimacy and acceptance is remarkable. In the United States alone, the number of Assisted Reproductive Technique (ART) cycles has almost doubled since 1998. In 2007, for instance, there were more than 142,000 cycles of assisted reproduction resulting in 43,412 live-birth deliveries and 57,569 infants.[24]

The fact that procedures such as artificial insemination and in-vitro fertilization have been more or less normalized in certain societies does not mean that the moral debate over reproductive enhancement has ended. As is demonstrated throughout this book, technology is advancing at a rapid rate. Soon, we will have more morally complex decisions to make over which reproductive enhancements and technologies should be allowed.[25] New discoveries in genetics and genetic engineering make so-called designer babies an imminent reality. Already, sperm banks allow potential parents the opportunity to select sperm based on the physical, academic/intellectual and social traits of the donor. Similarly, those seeking to conceive are willing to pay top prices – in some instances, as much as USD 50,000 – for the eggs of intelligent, athletic women whose physical traits match the traits of one of the prospective parents.[26] With new technologies, babies are less and less tied to having an appealing or compatible donor. Genetic engineering and other advanced reproductive technologies will gradually make sex selection, preference for certain physical traits and the elimination of genetic diseases possible. While even the most carefully picked eggs and sperm still leave much to chance in the genetic lottery, new genetic enhancement technologies and germ-line genetic engineering will enable the control of children's genes with much higher levels of scientific precision. With increasing levels of surrogacy, egg and sperm donation, artificial insemination and direct genetic engineering, we are gradually blurring the lines of what it means to be a parent and what nature's genetic lottery may have originally intended.

When discussing reproductive technologies and human enhancement, it is important to reemphasize that, for better or worse, the pursuit of these technologies is inevitable. Human nature and parents' desire to seek out the best for themselves and their children will ultimately lead to widespread adoption of these technologies. What this means for society at large is explored in more detail below. Some of the most prominent reproductive enhancement technologies are outlined in this section.

## Artificial insemination

Artificial insemination is a method for treating infertility in both men and women by inserting sperm directly into a woman's cervix, fallopian tubes or uterus.[27] Artificial insemination is an enhancement in the sense that it helps to correct a perceived abnormality in the reproductive process – low sperm count or endometriosis in women, to cite two examples.[28] Often, artificial insemination is coupled with the prescription of fertility drugs in order to increase the likelihood of conception. Again, this can be considered another form of enhancement. Additionally, sperm is 'washed' before being introduced into the uterus, removing certain chemicals that make the sperm less appealing to the female body. Then, a chemical is added to sort out the 'best' sperm.[29] Each of these steps is designed to make fertilization of the woman's egg more likely. Because of the technological expertise required at each step, artificial insemination is a key meeting point between reproductive enhancement and emerging strategic technologies.

## In-vitro fertilization

In-vitro fertilization involves a man's sperm fertilizing a woman's egg in a laboratory dish. The resulting embryo is then implanted into the woman's uterus.[30] The treatment is especially helpful to women with blocked or damaged fallopian tubes, men with low sperm count, or women who suffer from endometriosis. Like artificial insemination, it is often coupled with fertility drugs, and the enhancement component of the technology is further reinforced by the fact that there are certain steps that can be taken to enhance the embryo's viability. For example, assisted hatching can improve an embryo's chance of taking hold in the uterus, and pre-implantation genetic diagnosis can help identify the healthiest, strongest embryos.[31] The use of medical and pharmaceutical technologies helps to offset perceived physical and hormonal abnormalities.

## Germ-line genetic engineering

Germ-line genetic engineering is perhaps one of the most radical forms of reproductive enhancement – for now, more in theory than in practice. It not only alters the genes of an embryo, but alters them in such a way that any changes will be passed down to the next generation.[32] This is a notable difference from the alteration of somatic cells. Germ-line genetic engineering offers parents the opportunity to select genes for their children, potentially reducing their risk of developing certain genetic diseases, increasing their learning ability or improving their immune systems, among other things.[33] Germ-line genetic engineering also has some specific advantages from a scientific perspective.

That said, our understanding of the human genome is far from perfect, and it is highly likely that in seeking to enhance a child's future health, a gene modification has unintended consequences. Thus, instead of improving that child's health, the genetic modification could result in an inadvertent deformity. Because this deformity would have the potential to be passed down to the next generation, germ-line genetic engineering is a form of reproductive enhancement that raises serious ethical concerns. Although it is not being applied to human reproduction yet, we must begin considering the challenges of this technology now.[34]

### Selection and designer children

In some senses, genetic selection begins when one partner chooses a mate, some of whose genetic features will be passed on to the child. Some of this is conscious and based on individual tastes and attractions. Other elements of this basic selection are more instinctive, such as an intuitive sense that one man will be a better provider. Thus, at its most basic level, selection can be considered a fundamental part of the human mating and reproductive processes and one that is unrelated to technology.[35]

Technologies enable the selection process to be taken even further, and it is here that ethical issues and affronts to human dignity become more prominent. Genetic pre-screening essentially allows parents to select the sex and a large range of attributes of their baby. Eventually, technology will move beyond just identifying undesirable traits and will allow parents to select the genes they want in their children, creating a phenomenon that has been dubbed 'designer children'. Even if such technologies are rooted in health issues (e.g., the desire to avoid having a baby with a genetic inclination to a gender-specific disease), the technologies are quickly taking on a life of their own in order to cater to parents who simply prefer a child of one sex or the other. This will have global implications, especially in countries such as India and China where boys are much more desired than girls. Another possibility offered by selection through emerging strategic technologies is the enhancement of otherwise healthy children through genetic engineering. For example, beyond just choosing the sex of a child, a parent might seek to increase the baby's genetic disposition towards intelligence and athleticism. This presents a number of moral and ethical dilemmas to society, not least of which is equality and the danger that parents feel pressure to create a 'perfect' child.[36] Habermas, for instance, warns that a certain set of attributes chosen by parents for their child before birth may in fact not correspond to the child's preferences when it grows up. In other words, the future of the child has been determined by others, and the enhanced individual is not the sole author of its life.[37]

Some researchers predict that by 2020, scientists will have a better understanding of many more genetic variations that increase the risk of common problems such as heart disease and diabetes. While scientists originally thought that the DNA variations leading to certain common diseases would not be amenable to embryo screening, new developments in the field are forcing re-evaluations of this initial assessment. As knowledge improves on this front, so too will the ability to detect these genetic variations in embryos. Embryo screening will become dramatically more common and more popular.[38] It is practically inevitable that such embryonic diagnoses will become widespread. The key issue will not be whether to do embryonic screening but where to draw the line. As Dr David Goldstein of Duke University has said, 'We should think about an appropriate dividing line. Most people are in favour of allowing this when a disease is severe but are more uncomfortable with marginal disease risks. It's something we are going to have to think hard about'.[39]

## Mental and cognitive enhancements

Some forms of human mental enhancement are independent of technology and as old as time. Education and training, for example, could be considered mental enhancements, as could drinking a caffeinated beverage to help stay awake and focused at work or school.[40] These kinds of non-technological enhancements have genuine physiological effects on the brain, a point that is important to remember as we move on to discuss technologies, drugs and their roles in mental enhancement.

In the broadest terms, mental enhancements include cognitive, emotional and behavioural changes to humans, and they are significant in terms of altering human nature and potentially undermining human dignity.

Because human nature is so intimately connected to neurochemistry, there is huge potential for human instincts and impulses – the foundations of human nature – to be moderated or altered by a variety of ESTs, including psychopharmacotics, medicines and genetic engineering. Emerging strategic technologies, and neuroscience and neuropharmacology in particular, are opening up a floodgate of potential methods for improving the mental functioning and capacity of human beings.

Already, due to drugs such as Ritalin (used to treat conditions such as Attention Deficit Disorder) and Prozac (a mood-modulator), we are using strategic technologies to alter and improve our mental functioning and capacities. They, however, merely scratch the surface of the future potential for strategic technologies and mental enhancements. Future drugs may enhance intelligence, creativity and personal productivity,[41] and significant advancements are already being made with regard to memory enhancing drugs. While the initial basis for the development of these drugs is therapeutic (e.g., to help with age-related cognitive decline), the transition of these drugs to pure enhancement applications is not too difficult to imagine.

Touching on similar themes as physical performance enhancing drugs, the Dutch Ministry of Health, Welfare and Sport has asked the Dutch doping authority to consider creating a list of doping drugs to be banned from so-called mind-sports, such as chess.[42] This would include cognition-enhancing substances and possibly psychoactive and psychopharmacological substances. As is noted above, such products already exist for the treatment of medical conditions such as age-related cognitive decline or head injury, but new moral questions arise when they are applied to gaming or competition.[43] Since everyday and widely accepted substances such as caffeine also fall under this umbrella of cognition-enhancers, regulatory questions and ethical frameworks become that much more difficult. Interestingly, the Dutch Ministry has concluded, at least for the time-being, that 'there is a principal difference between doping in physical versus mental sports, for example that in mental sport it is, at least at present, impossible to push maximum intellectual capacity using drugs'.[44] It will be interesting to see how this issue unfolds in the future.

More dramatic mental enhancements include brain mapping and the possibility of uploading one's brain, memories, feeling, and neurological reactions on to a computer. Cybernetic implants could be placed in the brain to repair malfunctioning or damaged parts of the cerebrum. Although such technologies now seem like the stuff of science fiction, we are much closer to such radical innovations than many people realize.[45] The neuroscientist Henry Markam of the Ecole Polytechnique Fédérale de Lausanne stated in 2009 that it is possible to build a human brain, and that this may be done within the next ten years.[46] As more and more of the human brain is mapped and understood, the opportunities for us to treat mental illnesses and to improve brain function are enormous. But by the same token, such technologies and research open the door to more menacing prospects as well. Brain control and mind manipulation are some of the potential dangers of such innovations.[47]

Looking at the long term, some of the most realistic mental enhancement innovations include brain-computer interfaces (BCI) and cognitive prostheses.[48] BCIs offer a direct communications channel between the brain and an external device such as a computer. With these devices, neuronal signals can be translated into motor commands.[49] These machines are already helping to restore mobility and motor control to disabled patients.

Another type of technologically enabled mental enhancement is personal sensory device interfaces. This type of technology, a fusion of nanotechnology, ICT and others, has the potential to dramatically improve the ways in which humans interact and communicate with each other, as well as the ways in which they perceive the world around them. As the World Technology Evaluation Center explains in a recent report:

> Research can develop high bandwidth interfaces between devices and the human nervous system, sensory substitution techniques that transform one type of input (visual, aural, tactile) into another, effective means for storing memory external to the brain, knowledge-based information architectures that facilitate exploration and understanding, and new kinds of sensors that can provide people with valuable data about their social and physical environments.[50]

Concerns over mental enhancements are twofold. First, basic safety is a key concern for medical service providers, patients and society alike. A drug that improves your memory for one day is not worth much if you wake up dead the next. Less straightforward are concerns about ethics and how neuropharmaceuticals and other mental enhancements may alter individual identity and human nature.[51] On the surface, perhaps a memory-enhancing drug is not such a great leap of imagination or such a huge risk for human nature. Indeed, humans can already improve their memories without drugs through training and basic memory exercises.[52] However, what if a drug was developed and had the capability to make people forget? If memories could be erased or forgotten, would we be undermining the core of our human experience? Would we be eliminating the patchwork of experiences that make us who we are?

By the same token, would our society be happier and healthier if war veterans could get rid of the battlefield images, the memories of which often provoke serious psychological disorders such as post-traumatic stress syndrome? Unintended consequences also come into play with these types of technology. For example, on a more practical, immediate level, some studies have suggested that invasive brain technologies such as BCIs may slow brain waves, ultimately shortening attention span. These are just some of the serious ethical, medical and personal questions raised by human mental enhancement.

## Neuroscience and human enhancement

Although neuroscience is a cutting edge field that is making rapid advances through modern technology, its history goes back thousands of years. In fact, it was the ancient Egyptians who were responsible for the first-known written account of the brain. (Interestingly, the Egyptians did not regard the brain very highly, choosing not to preserve the organ after death.)[53]

Today, neuroscience involves the study of many facets of the brain, from behaviour issues to studies of neurological systems (e.g., auditory and visual) to the membrane and genetic levels of neuronal functions.[54] In short, it is 'the study of the nervous system – including the brain, the spinal cord, and networks of sensory nerve cells, or neurons, throughout the body'.[55] Using everything from computers to special dyes, neuroscientists seek to better understand the normal development of the human nervous system and then to determine why neurological disorders sometimes develop.[56] Contemporary neuroscience puts a large emphasis on the replacement of localism and holism with the concept of 'connectionism', or the idea that the lower level functions of the brain are highly localized but that higher-level functions such as memory are the result of interconnections between areas of the brain.[57] The connections between the mind, the brain and the soul, and whether these entities are distinct or one and the same are also ongoing debates in neuroscience.[58]

Cognitive neuroscience is one specific branch of the broader neuroscience field. Scientists in this specialty 'study functions such as perception and memory in animals by using behavioural methods and other neuroscience techniques. In humans, they use non-invasive brain scans – such as positron emission tomography and magnetic resonance imaging – to uncover routes of neural processing that occur during language, problem solving and other tasks'.[59]

Interestingly, neuroscience has only recently been recognized as a field in its own right. There are many significant advances and discoveries taking place in this area, and many of them have significant consequences for our global community.

Today, drugs are increasingly being used to improve cognitive function and these have raised a number of serious ethical questions. Often, these medical advances are rooted in treating illnesses, and the bigger ethical questions arise when medicines designed to treat specific ailments are adapted and used to improve rather than treat the cognitive function of otherwise healthy individuals.

## Neuropharmacology

Neuropharmacology deals with drugs that change or alter the functioning of cells in the nervous system. In the vast majority of cases, these drugs target synaptic activity or physiological processes that are directly related to synaptic activities. A related science, neuropsychopharmacology, looks at the connection between neuroscience and how drugs affect the mind. Mood modulators and attention-enhancing drugs such as Ritalin and Adderall are examples of drug technologies that have been developed in these fields. Through neuropharmacology and neuropsychopharmacology, doctors and scientists are increasing their ability to alter the chemistry of the brain, something that may eventually lead to a fundamental alteration of our human nature.

## Addiction advances

Drugs such as marijuana, heroin and methamphetamines act on specific receptors and brain areas, potentially to create a physical dependence. Increasingly, addictions to drugs, alcohol, cigarettes and the like are being understood not simply as a problem of individual weakness or a lack of moral character but as the result of genetics and a real chemical imbalance in the brain. To this end, a number of drugs and medical treatments are being developed to help a patient kick addiction through physiological control of the brain. For example, naltrexone is a drug originally developed to help curb opiate addiction but that is now more widely used in the treatment of alcoholism. It blocks the euphoric sensation experienced by users of heroin and other opiates. In doing so, naltrexone has a direct impact on neurobiological processes related to the addiction cycle.[60]

Drugs such as naltrexone can be reinforced with other neuropharmacological technologies. Antidepressants, anti-anxiety agents, mood stabilizers and antipsychotic medication can work in tandem with other drug therapies to help patients overcome both addiction and the character traits that lead to such addictions in the first place.[61]

## Neural engineering

Neural engineering is an emerging area of biomedical engineering. It is an interdisciplinary field that uses an engineering approach to examine the function and manipulate the behaviour of the nervous system. To achieve this end, neural engineering relies on computational biology, electrical engineering, signal processing, chemistry and chemical engineering, as well as other separate disciplines.[62] Neural engineering can model and simulate neural systems from the level of a single neuron to a complex network of neurons. Applications include medical diagnosis and treatment of nervous system disorders, including the replacement of damaged or dysfunctional brain tissue and spinal cords.[63]

One particularly important area of neural engineering is neuroprosthetics. Using neural engineering practices, neuroprosthetics can help restore the functions lost as a result of neuron damage. Neuroprosthetics are in varying stages of development but, in general, they have been shown to help with functions as diverse as bladder and bowel control to the restoration of mobility and respiration to paralysed individuals.[64] Brain-computer interfaces, technology that can translate imagined movements into digital commands, are a type of neuroprostheses that allow increased communication, function and mobility for patients paralysed by conditions such as locked-in syndrome and amyotrophic lateral sclerosis.[65]

## Brain mapping

Brain mapping is a procedure that tries to relate the brain's structure to its function. It attempts to identify and display the billions of neuron connections in the brain, essentially the basis of neuroscience. In an effort to determine which part of our brain enables us to perform certain activities, brain mapping is intended to provide a picture of the structure of the brain. Whether it is through watching the brain perform various tasks and identifying which part of the brain is engaged at what point, or by looking at how the environment changes our brain's structure, brain mapping attempts to provide a bird's eye view of the brain and its function.[66]

Thanks to brain mapping, scientists can get a better idea of how disease and disability affect the body; and they can better target certain parts of the brain to administer various treatments.

## Brain uploading

Brain uploading is perhaps one of the most radical and seemingly far-fetched scenarios that could come about as a result of human enhancement. To explain an idea that seems to come from a science fiction film, the Institute for Emerging Ethics and Technologies Marshall Brain has compared brain uploading to the development of aircraft. This is a useful analogy, and I borrow heavily from it for the purposes of this explanation.

In 1903, the Wright Brothers flew a contraption of wood and fabric for a grand total of 200 feet. This was the world's first introduction to the aircraft. Although impressive in its own time, we would hardly pay any attention to such an achievement today. Similarly, if you had told the Wright Brothers in 1903 that aircraft would, in the space of 50 years, evolve into an aluminium body, three football fields long, with capable of travelling faster than the speed of sound, the Wrights would have expressed blatant disbelief. Nonetheless, within 50 years of 1903, such aircraft existed and had been used throughout World War II. Aircraft were not only advancing, they were changing the world. Technology evolved at a rapid rate in the first half of the twentieth century. The rate of technological change is becoming even faster now – and we are possibly approaching a point of Singularity.[67]

Today, our human bodies can be compared with the first aircraft that the Wright Brothers flew in 1903. According to Brain's analogy, we all have bodies but 'we all want out of our bodies. We would like to discard these vehicles that we currently use for transportation and we would like to replace them with something better'.[68] Brain then suggests that 'something better' might look like a video game. Video games, which are still in an early stage of development, could develop realistic simulations of everything from a football game to being in the Coliseum of ancient Rome. As a result, we will be driven out of our bodies; 'either we will install hardware that will let us emulate or connect to virtual environments and control and feel them, or we'll realize we don't need our bodies anymore'.[69]

This connection to virtual environments is the first step in brain uploading. It is the first step in separating our brains and consciousness from our physiological bodies. However, even if we can connect our brains to a virtual environment, our brain's neurons will still die. This is the reason why we may move towards brain-uploading, which will essentially allow our brain to be mapped atom by atom with the map then being uploaded on to a computer where we would be able to experience our thoughts, memories and reactions virtually. We would preserve our consciousness, but we would avoid the risk of the physiological decline of our brains and bodies, essentially allowing us to avoid death and live forever in computer form.

Having explained how we might move towards brain-uploading, it is now worth looking at what brain uploading would entail. Once again, I continue to lean heavily on the language of Brain who has summed up these issues very clearly and concisely.

If we are going to take the information and patterns that are currently stored in the intricate reactions between our brains' neurons and synapses, we would have to take all the brain's physiology, put it into a computer medium, and then figure out how to execute it all. To do this, scientists have proposed either slicing the brain very gently and scanning it or injecting some kind of nanotechnology product that can assess how to emulate each neuron and then replace it, thus making up a brain of nanotechnology.[70] The most in-depth procedure would involve looking at every single atom in the brain and then storing the type of atom, its location, its bonding to neighbouring atoms, and cork spin and then taking atomic images of this in order to have a fully comprehensive, atom-by-atom look at the brain.

Although these suggestions seem far-fetched and speculative, 'these things are probably possible, perhaps within 40 years [...] Going back, the Wright Brothers' airplane to the B-52 was hard to imagine in 50 years'.

Having put brain uploading in its general context of technological change, it is important to note that brain uploading raises many philosophical and ethical issues, and there are many questions that relate to brain uploading, the end of our physical bodies, and issues of human rights and human dignity, all of which are better dealt with now while the technology is still in its speculative phases.[71]

## Conclusion

Emerging strategic technologies and their application to human enhancement are extremely wide-ranging and diverse. In some instances, such as glasses and hearing aids, things that could be considered enhancements are so ubiquitous that we hardly give them a second thought. Other technologies, such as germ-line genetic engineering or neural engineering, pose more ethical challenges and, for this reason, they are most likely to have the biggest impact on geopolitics, geostrategy, human destiny and our *'sustainable history'*. The remaining chapters in Part II deal with these issues directly.

# 14
# The Geopolitics of Human Enhancement: Applying the *'Multi-Sum Security Principle'*

## The *'multi-sum security principle'*

In my book, *The Five Dimensions of Global Security: Proposal for a Multi-Sum Security Principle*,[1] I outline a new, multifaceted approach to global security. It is my contention that global security can no longer be considered merely a collection of different nations acting strictly on behalf of their own national security interests. In the globalized world we live in, security concerns have moved dramatically beyond national borders and must now be evaluated from all levels, starting with the individual and the nation state and working up to environmental, transnational and transcultural issues.

My model for international security involves five key security elements: human, environmental, national, transnational and transcultural. These five elements are interdependent and deeply intertwined. The security of one state, culture or individual simply cannot be guaranteed as a stand-alone entity. In other words, a person cannot enjoy security on an individual level unless he also has security on all other levels.

More and more, our world is characterized by high levels of connectivity and the strong interdependence of nations. Take, for example, the case of Afghanistan. As recently as 100 years ago, the fate of an isolated, least-developed Central Asian nation would hardly interest or affect the policies and politics of other nations on the same continent, much less half a world away. Today, however, Afghanistan is a central item on the agenda of the United States, the world's hegemon. Because of transnational issues such as terrorism and the drugs trade, and because of more traditional geopolitical issues such as Afghanistan's proximity to unstable states such as Pakistan or potential power threats such as Russia, the United States must take an interest in Afghanistan. This is just one of countless examples of a modern-day multi-sum security scenario in action.

Simply put, the state can no longer be viewed as being capable of complete self-sufficiency when it comes to maintaining security. The traditional realist concept of international relations must be expanded to a more comprehensive and collaborative approach. As an international community, we face numerous problems and challenges that simply cannot be improved or resolved within the realist framework. In today's world, there are more intangible factors at play, and the significance of soft power is growing. To achieve true global security, it is essential that we adopt a *'multi-sum security principle'* (see Figure 4), the core goal of which must be global justice. I define this *'multi-sum security principle'* as follows:

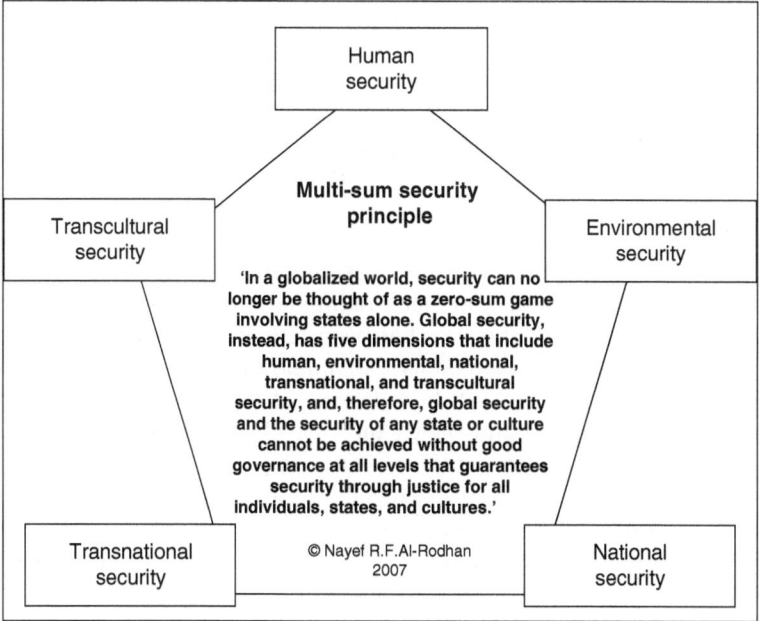

*Figure 4*   The *'multi-sum security principle'*

Source: N.R.F. Al-Rodhan (2009) *The Five Dimensions of Global Security: Proposal for a Multi-sum Security Principle* (Berlin: LIT), p. 31. Reproduced with Permission from LIT.

In a globalized world, security can no longer be thought of as a zero-sum game involving states alone. Global security, instead, has five dimensions that include human, environmental, national, transnational and transcultural security, and, therefore, global security and the security of any state or culture cannot be achieved without good governance at all levels that guarantees security through *justice* for *all* individuals, states and cultures.[2]

It cannot be emphasized enough that the only way to achieve this is through good governance at all levels. I briefly summarize each of the five substrates of my '*multi-sum security principle*' below before examining the connections between each of the five elements and human enhancement one by one.

### Human security

The concept of human security was introduced in the UNDP's 1994 Human Development Report. Security, it argues, should be associated 'with people rather than territories, with development rather than arms'.[3] It deals with the security of individuals and encompasses issues such as freedom from violence, hunger and oppression and represents a move away from looking at the nation state as the primary referent of international security. By focusing on the needs and interests of the individual, the human security principle argues that a state exists to serve its citizens and not vice versa.

Threats to human security include hunger, disease, forced migration, human trafficking and labour-market instabilities. Civil war, genocide and the displacement of human populations are among the most serious violations of human security.[4] It is important to stress the relation between human security and more traditional concepts of state-centred security. Traditionally, the state has been responsible for protecting its citizens from a foreign attack. This is an important component of human security but by no means is it the only one.[5] More basic individual needs such as food, clothing and shelter must also be met. Additionally, the human security paradigm emphasizes the importance of the state not violating the rights of its own citizens through torture, violations of basic freedoms and rights, genocide, and so on. This is a concept that is essentially unacknowledged by the realist framework of state-centred security.

### Environmental security

Environmental security is the second facet of my '*multi-sum security principle*', and it is significant because it acknowledges that issues such as global climate change, pollution and resource distribution are not peripheral, but security challenges that have a direct influence on global peace and security. The threats to environmental security are abundant and growing. They include deforestation and the subsequent decline of biodiversity, water scarcity, climate change and the impacts of pollution and resource depletion on human health and ozone layer depletion.

Environmental challenges can also cause numerous secondary problems. For example, rising sea levels caused by melting glaciers can cause the forced migration of coastal populations. Especially in poverty- or conflict-stricken areas, such climactic and environmental changes can greatly exacerbate tensions and existing problems.[6] Similarly, resource shortages may provoke or increase conflict in regions vulnerable to climactic fluctuations.[7] For these reasons, environmental security must be evaluated as a key component of the multi-sum security paradigm.

### National security

After environmental security, the next component of multi-sum security is national security. Long considered the central (and sometimes the sole) dimension of international security, in this new security framework national security cedes some of its previous importance to other facets of security. That said, national security is still an indispensable component of global security. Our international system is still composed of a Westphalian group of states which, by definition, have sovereignty over the territory and peoples within their borders. Traditionally, national security has a strong military and police component, but more contemporary assessments of national security encompass political, economic and social dynamics as well.

From a military perspective, a nation state is traditionally defined by the exclusive right to self-governance over a specific territory and population. The state's legitimacy is closely tied to its ability to effectively use force and to keep its citizens safe from foreign attack. Politically, a government must maintain some sort of legitimacy of rule. Otherwise, the state's stability and security will be undermined and it is likely to be more vulnerable to an outside threat. Economics is increasingly being seen as within the purview of the state and as an essential part of a state's overall security. States have assumed responsibility for ensuring healthy levels of growth and low levels of unemployment. Finally, related to political security, national security also demands some level of social security. In other words, the protection of a state's population regardless of whether that community shares a strong common identity.

Overall, when speaking of national security, it is important to note that a state is responsible for ensuring levels of security for its citizens in the military, economic, political and social arenas. Only when national security is truly comprehensive can genuine national security be said to have been achieved.

## Transnational security

The next component of my multi-sum security paradigm is transnational security. Transnational security refers to any cross-border threats. Terrorism, organized crime, drug trafficking and cyber crime are some of the most prominent examples. This is an idea that has emerged primarily since the end of the Cold War and the end of the bipolar security structure dominated by the United States and Europe. Threats in this substrate are rarely confined to traditional state borders; in fact, they transcend borders, threatening the social, political and economic integrity of a nation or group of people.[8] As I explain in my book *The Five Dimensions of Global Security*, 'Transnational threats are also challenging because they emerge over indefinite, often long periods of time, which means that they may not receive adequate attention from governments, which are primarily concerned with short-term problems, largely due to concerns about being re-elected'.[9] Another transnational challenge is logistical problems such as the management and deployment of crisis response capabilities to distant parts of the world.[10] In an increasingly transnational world, states and their militaries can no longer be seen as the exclusive provider of security. Greater regional cooperation across borders and intercultural understanding are becoming more and more imperative.

## Transcultural security

The final substrate of my *'multi-sum security principle'*, transcultural security, addresses the importance and value of diverse cultures and civilizational forms. Although culture has often been marginalized or overlooked in traditional security frameworks, and especially by realism, I maintain that it is an essential component of conflict and security, and, ultimately, of justice. I believe that a plurality of cultures can have a broad and sweeping effect on global security and one that is larger than if individual cultures act individually. For this reason, transcultural security is a crucial part of achieving the objective of transcultural synergy.[11] The greatest threats to transcultural security occur when diverse communities are not positively integrated into broader society, especially when this society is perceived as being a homogenous entity.

Thus, synergy between cultures – not just coexistence – is the crucial component of transcultural security.[12]

## Human enhancement in the framework of the *'multi-sum security principle'*

Having outlined the fundamental concept of the five dimensions of global security and the components of my *'multi-sum security principle'*, I now look at the topic of human enhancement and the implications it may have for each of these five components. The central argument is that strategic technologies and human enhancement have significant ramifications for human dignity and our conception of human rights, and they also affect geopolitical issues and global security in very real, often complex, ways.

### Human enhancement and human security

The most salient implications of human enhancement to individual and human security are linked to issues of equality and justice. Although the biotechnology revolution has enormous potential to improve the health of populations around the world and to eradicate certain illnesses, it is also, like so many aspects, a double-edged sword. As more and more people opt for voluntary enhancements or enhancements that give them a competitive edge over other individuals (as opposed to enhancements that have purely intrinsic benefits), human enhancement poses the very real risk of fostering a dangerous divide between enhanced and unenhanced individuals.

Specifically, this concern deals with the idea that an intellectually, emotionally and/or physically enhanced trans- or post-human would be likely to view an ordinary, unenhanced human as inferior and therefore possibly fit for exploitation.[13] As enhancements become increasingly widespread and increasingly complex, humanity runs the risk of breaking into two species: the enhanced humans and the naturals (a variation of the concepts of trans- and post-humans). An unenhanced individual is likely to be at a disadvantage when compared to an enhanced person, assuming that the two individuals started at approximately the same level of strength, intelligence, and so on.

Making matters potentially more dangerous and disruptive from a security perspective, these inequalities and differences would only be reinforced along the pre-existing lines of social classes and nationality. Enhancement technologies are, and promise to continue to be, very expensive. It will therefore be easier for the rich to afford enhancements, for both themselves and their children. Poor people will be less able to afford expensive enhancements and the gap in opportunities, abilities and possibilities between rich people and poor people and developed and developing countries will be further reinforced. Already, our world suffers from dramatic differences in terms of educational opportunities, technology access, and the meeting of basic human needs. Enhancement will only exacerbate these differences and will possibly make the distinctions even more rigid, especially when applied across multiple generations.

The existing inequalities posed by enhancement are already evident. Existing research into technological enhancements often focuses on the needs or desires of those living in rich nations with the money for purely cosmetic enhancements. Human enhancement of intelligence and physical strength, and for the elimination of disease-carrying genes would only reinforce these differences, possibly provoking war or violence between unenhanced and enhanced individuals.

At the societal level, unregulated enhancement (and even regulated enhancement) offers the potential to undermine the current 'genetic lottery' in favour of pre-selected genetics. As Michael Sandel points out, if this were to be the case, we might find ourselves in a situation in which individuals were blamed for their *lack* of enhancements and discriminated against accordingly. As things stand now, genetics are random. A teenage boy who does not make the basketball team cannot be faulted for his lack of height. In an age of designer children, however, such a failure would become the responsibility of the parents or the individual himself, depending on the specific nature of the enhancement in question. This would open new avenues for discrimination, and it would raise key questions over human rights. Currently, all individuals have a set of basic, equal, fundamental human rights (at least as defined by the United Nations Declaration of Human Rights). How would an enhanced individual fit into this scheme? Would an enhanced, improved individual merit more rights? What would become of the unenhanced individuals? Would their individual worth suddenly become relatively less? All these issues are fundamental components of geopolitics and they must be considered as the international community looks towards creating a comprehensive regulatory framework on human enhancement and human augmentation.

While enhancements pose huge moral problems on issues of rights and equality, proponents of human enhancement maintain that the benefits of enhancement outweigh the dangers, and, as such, we have a moral obligation to use the available technologies to enhance ourselves.

In other words, withholding a benefit from a person is equivalent to doing harm to that person. According to this argument, not taking advantage of existing enhancement technologies could be a violation of human rights, or if not of human rights than at least of some of the facets of human dignity.

When discussing how to ensure and improve human security in the face of human enhancement, the most essential component of the discussion must focus on equality of access. Although there are numerous controversies and debates over human enhancement, many of the issues regarding human security would be alleviated or ameliorated by simple virtue of equality of access to enhancement technologies. This is a daunting task to contemplate, as parts of the world still lack access to basics such as clean drinking water or anti-malaria pills in malaria-ridden areas. For the time being, however, we are speaking on a moral and theoretical level and, in this context, equality of access is the key. Human enhancement should be an issue of individual choice not individual budget or opportunity. Especially with regard to technologies that could eliminate disease-carrying genes, human security will be best improved and secured if access is level.

Many thinkers in the field of human enhancement also emphasize that there are possible scenarios under which fears over inequality and uneven distribution of enhancement technologies would essentially be moot. For example, Ray Kurzweil suggests that as we approach the Singularity, the rate of technological innovation and distribution will accelerate and the costs of technology will fall so quickly that any disparities will be relatively short-lived.[14] Ramez Naam argues that the wealthy will be on 'the bleeding edge, but not get as much of an enhancement as the majority do when they receive the latter, refined, massified and less expensive versions of the enhancement. In effect, the wealthy are subsidizing enhancement innovation for the many without receiving compensatory advantages'.[15]

Other arguments that minimize the significance of human enhancement on equality issues include the idea that even unevenly distributed cognitive enhancements may still yield society-wide benefits. It is possible that enhanced moral or intellectual characters would help in the innovation of a more egalitarian-minded society or the development of more just forms of governance.[16]

Moving on to the questions of dignity and equality, the argument cuts two ways with regard to human enhancement. On the one hand, enhancements and the process of becoming post-human might be degrading and undermine human dignity, a scenario clearly played out in *Brave New World*.[17] Cloning is a key example of the potential to undermine individual human dignity. A cloned individual would run the risk of constantly being overshadowed by his or her original. If the clone were inferior to the original in some way, the clone's sense of worth and value would be undermined. Failure to meet the expectations set by the original would thus mean a violation of dignity.[18]

However, enhancements beyond cloning are just as likely to offer the potential to improve or facilitate components of human dignity. For example, enhancements improving intelligence, memory and mental capacity could directly influence the levels of opportunity and innovation in society. As is outlined above, opportunity and innovation are among the key minimum criteria necessary for ensuring human dignity.

### Human enhancement and environmental security

Human enhancement and its effects could take the concept of international environmental security down a variety of paths, some of them positive and some of them not. For their part, transhumanists are quick to defend enhancement technologies: 'The technologies necessary for realizing the transhumanist vision can be environmentally sound. Information technology and medical procedures, for example, tend to be relatively clean'.[19] Ultimately, the impact of human enhancements and related strategic technologies on the environment will depend on which enhancements become most prevalent, how the enhancements affect population size, whether enhancements alter basic human needs and how enhancements may help us respond to changing environments and ongoing climate change. The timeframe on which these enhancements come into play is another key consideration.

Certain human enhancements that reduce aging processes at the cellular level or that reduce genetic predispositions to disease and illness have the potential to dramatically increase our lifespans, thereby increasing the size of the world's population. Consider that in prehistoric times, the average lifespan was 20–25 years,[20] and that even in the 1800s the average life expectancy at birth was 50 years. Today, average worldwide life expectancy at birth is 66 years.[21] With the help of enhancement technologies, improved medical care and genetic alterations that reduce the likelihood of disease, cancer and so on, it is not difficult to imagine that average human lifespan will continue to increase, perhaps quite dramatically. As lifespan increases, so too will the world's population. The natural extension of this phenomenon would be a growing strain on natural resources, from water to food to land for agriculture, and residential and commercial development. Moreover, as Richard Lipsey notes, when the average lifespan increases dramatically in the course of a very short period (as short as half a century), it requires an enormous adjustment in all aspects of the facilitating and policy structures of a society.[22]

In some more radical assessments of human enhancement and our post-human future, scientists and scholars speculate that as our understanding of the human brain increases and as our technology for mapping the brain improves, we will eventually arrive at the point where our entire brains can be fully uploaded on to a computer system. If and when this technology becomes fully developed, it is possible to imagine a situation in which an individual's memories, attitudes, values and emotional dispositions could all exist on a silicon chip.[23] Without going into too much of detail about the specific technologies and moral valuations of human uploading, there is a conceivable scenario under which an upload of your brain could be considered you. Your essence would be established on a computer, and your personality and person could develop, function and grow on this computer, participating in virtual reality and interacting with other individuals through an integrated computer network. Again, in a deliberate over-simplification of the ethical, moral and physiological components of the issue, uploading could essentially provide immortality and eliminate the need for a human body while still preserving the neurological traits that define you as you.[24] If machines can truly be organic, self-modifying and intelligent (as AI researchers and scientists hope and anticipate), we will be able to merge technology into ourselves and ourselves into technology.[25] In this scenario, we could eliminate the need for a body and presumably we could live forever.

Why speak of uploading and post-corporeal post-humanism in the context of the environment? The answer, although not necessarily obvious, is simple. If we could upload our brains, essence and existence on to a computer and eliminate the need for a body, we would reduce or eliminate our demand for food, water and other natural resources. Although the world's population and subsequent demand for natural resources may grow dramatically thanks to human enhancements, we also have to consider the scenario in which human enhancement and the prospect of a post-human future actually eliminate our need for the Earth's resources. Our needs and realities may become purely virtual, and, from this perspective, human enhancement, post-humanism and environmental security will go hand in hand.

Moving back to the realm of more immediate and dangerous environmental concerns over human enhancement, some environmental concerns arise from the fact that not all enhancements, especially physical ones, are biodegradable. Bodies may decompose after death but what about electronic and other technologies attached to the body that may damage the environment or contaminate natural resources? Polythene and other materials used in hospitals, technologies and daily life can have dramatic impacts on water supplies, ground contamination and the atmosphere.[26] Certain nanomaterials and medicines may be similarly detrimental to the environment, especially radioactive treatments or products that damage plants and animals and even entire ecosystems. Certain viral vectors used to enact different genetic enhancements or alterations may also have unforeseen environmental consequences.

## Human enhancement and national security

In the context of national security, human enhancement is a way for richer and more scientifically advanced states to potentially gain a competitive advantage over other states (especially, but not exclusively, in a military context), to improve the enhancing country's overall prestige and standing in the world, and to offer economic advantages and stimulus. Naturally, human enhancement also poses dangers to national security.

According to traditional concepts of national security, a state is defined by its exclusive right to the legitimate use of force in a given territory. As such, military power and control are fundamental concepts of national security, and human enhancement technologies have the potential to play a large role in the future of military endeavours. With the rise of terrorism and other transnational threats, traditional standing armies are becoming increasingly obsolete, and states are showing a preference for smaller groups of fast-moving soldiers and special forces.[27] In this context, consider human enhancements that improve stamina or physical strength. For example, researchers at Stanford University are developing a glove capable of cooling the blood in the body, thereby improving the endurance, mental capacity and physical resilience of troops.[28] Technologies like this combined with other enhancements that improve physical strength and stamina (anabolic steroids or even genetic engineering) could create a type of transhuman uniquely suited to the battlefield with distinct advantages over non-enhanced members of rival armies. With stronger, more capable armies, enhancing countries would be extremely well-positioned militarily and would have an easier time gaining and exerting control over rival countries or countries with valuable resources. Such advantages would primarily favour rich countries, as they would be the only ones who would be able to afford large-scale enhancement of their armies and populations.

Similarly, developments in nanotechnology and health medicine could make it easier and faster to heal wounded soldiers on the battlefield. Again, the advantage this would offer enhanced soldiers over their non-enhanced comrades or enemies would be notable. A state with broad access to human enhancement technologies could easily gain a leg up over its rivals and could possibly win military control over states with no access to enhancements.

Mental enhancements of intelligence, creativity and ambition could lead to a type of transhuman with an abnormal desire to conquer and control. If such technologies became within the purview of one specific state, the enhanced population of that state might be more inclined to provoke wars, whether for reasons of national interest or simple glory. Such enhancements could improve military capabilities and drive, but they could also backfire to create an overly aggressive transhuman with a loss of human compassion and dignity.

These issues are not just hypothetical. The United States, the world's largest military power, is actively working on contingency plans to respond to the threats posed by human modification technologies. For example, one of the Pentagon's most respected advisory councils, the JASONs, has recommended that the US military 'push ahead with its own performance-enhancement research – and monitor foreign studies – to make sure that the United States' enemies don't suddenly become smarter, faster, or better able to endure the harsh realities of war than American troops'.[29] The JASONs has also expressed particular interest in neuro-plasticity technologies that would allow troops to function on less sleep, among other things.[30]

Less tangible but no less important implications of human enhancement on national security might include issues of prestige. In much the same way that nuclear weapons programmes today are regarded by certain states as a way to gain international respect and negotiating leverage, the ability of a state to offer enhancements to its populations could in future be regarded as a symbol of progress, scientific prowess and superiority. Not only would these enhancements presumably give certain populations an advantage over non-enhanced ones, but they could lead to migration from states where no enhancement technologies are available to those where they are accessible. Whether this particular facet of human enhancement and national security is positive or negative would depend on the circumstances, but it is nonetheless worth considering. Already, there is a trend for people from rich, developed countries to seek out medical enhancements (e.g., gender reassignment surgery or rhinoplasty) in countries where such procedures are cheaper than that which is available at home (questions of legality also drive this burgeoning industry). Such 'enhancement tourism' could be a major economic boost for certain countries and could improve their national economic security. As long as there is a market, technologies or procedures that are not available in one country will quickly become easily accessible in another. This fact underscores the need for international guidelines on the ethics and regulatory frameworks of human enhancement technologies. Without such rules, we run the risk of certain pro-enhancement countries gaining significant advantages over their non-enhancing neighbours. These advantages could be economic, military, social or a combination of all three. The nature or extent of these changes is speculative, but they are nonetheless real challenges for the state of national security.

Thus far, this section has looked at the implications of human enhancement for national security based on the presumption that human enhancements will offer enhancing nations what could broadly be defined as inherent advantages over their neighbours or rivals, but this is not necessarily the case. Certain scenarios can be envisaged in which enhancement technologies might actually undermine national security. For example, while some enhancements might improve the strength and stamina of military forces, other pleasure-inducing mental enhancements might dissuade a nation's citizens from joining the army. Who would voluntarily risk pain, injury and possible death when pleasure and happiness are obtainable through simple neuropharmacology? Additionally, enhanced military personnel could easily turn against the interests of the state, perhaps exploiting non-enhanced members of society to do their bidding. Such issues of equality and potential exploitation by trans- and post-humans are recurring concerns in the debate over human enhancement. The existence of enhanced humans on the battlefield will ultimately result in asymmetric warfare, tipping the balance of power in favour of those states with the technology and funding available to develop and utilize such new technologies.[31]

Also on the topic of national security, the issue of eugenics must be addressed. Some of the more dangerous human enhancement scenarios foreshadow a situation in which imbalances in enhancement technology could result in a revival of eugenics and the pursuit of a so-called ideal race of peoples. Indeed, I outlined above some of the key differences between enhancement and eugenics and emphasized that one of the most central distinctions is that enhancement is based on individual choice, whereas eugenics is traditionally state-mandated. If a state begins to dictate certain enhancements in the name of national security (to improve the military, to gain prestige vis-à-vis other states, etc.), the state would have opened the door to eugenics and selective breeding. Such issues have even greater implications in the context of transcultural security but are nonetheless worthy of acknowledgement in this section.

The coming Singularity is the final issue to address on the topic of human enhancement and national security. As is explained above, one of the defining characteristics of a contemporary nation state is that the state has the exclusive right to use force in its given territory. However, when the Singularity arrives, it is likely to profoundly change this definition of national security. Instead of a state's power being defined by its exclusive control of the use of force, it is likely that states will be more defined by their ability to access and gain control over artificial intelligence (AI) and enhancement technologies. They will need to do this in order to maintain their position of authority and their relevance to both their nation's citizens and the international community at large. In part, this need to control new technologies, and especially AI technologies, will stem from traditional military planning needs. Traditional methods of war planning may become quickly obsolete as super-intelligent computers produce new strategies, weapon designs and sensing technologies at an exponential rate.[32] However, the need for a state to control access to AI technology goes beyond military needs. As our rate of technological progress increases exponentially, the types of developments, innovations and breakthroughs we might see in everything from environmental science to poverty reduction to health care will be enormous. Any state that has access to this information and technology will be well-positioned to dominate the international system politically, socially and economically.

Although the precise nature of the post-Singularity world is impossible to characterize, it is a safe assumption that nation states and insurgent groups alike would be keen to control access to the by-products of AI technology. Controlling AI and enhancement technologies will become an important tool of leverage in geopolitics, and if the nation state is to remain relevant in the post-Singularity world, it will be imperative that it maintains control over the technology.

## Human enhancement and transnational security

Human enhancement technologies could affect the transnational security dimension in a variety of ways. On the one hand, it may influence the ability of nations to cooperate or not to cooperate, as the case may be. Nations that are equally matched in terms of technology would probably have an easier time working together as their playing field would be more or less level, and one country would not feel disadvantaged by the other country's comparative advantage.

On the other hand, transnational security may be somewhat undermined by enhancement technologies, especially if a black market opens up in the trade of certain drugs and procedures. This would reinforce existing transnational threats of organized crime and the cross-border drugs trade.

Another consideration is the possible inequalities that may come about as certain people engage in enhancements to increase their intelligence, physical capabilities, and so on. This could give enhanced peoples an economic advantage over unenhanced peoples, causing economic instability and financial crises. Such instability would not necessarily be limited to one specific nation state but could easily spill across borders, causing dissent and violence. Human trafficking of non-enhanced individuals by enhanced individuals for the purposes of slavery or exploitation may also eventually become a problem.

From a health perspective, human enhancements could actually improve transnational security. Transhumanists argue that the central purpose of human enhancement technologies is to improve the human condition and quality of life, and increase our longevity. New advances in genetic sciences, pharmacology, nanotechnology and biotechnology mean that our quality of health care and medical treatments will improve dramatically in the coming decades. Infectious diseases are likely to be better contained or even eliminated, and societies will become more productive and economically stable as a result.

## Human enhancement and transcultural security

At the heart of transcultural security and human enhancement is the question of eugenics. As is noted above, eugenics deals with the selective breeding of people, and the promotion of specific desirable traits and the simultaneous repression of those traits seen as negative or undesirable. This is an issue related to transcultural security because the transcultural security dimension deals specifically with the integration of diverse cultures and geocultural domains. In some instances, this type of integration can be carried out fairly seamlessly or peacefully, but in more volatile, politically unstable regions such integration is a difficult criterion to meet.

In many instances, different cultural groups and geocultural domains can be distinguished from other groups by physical features. As human enhancement technologies begin to offer the possibility of eliminating certain characteristics and promoting others, societies will have to grapple with the question of whether only certain types of people receive enhancements, which enhancements are made publicly available and whether certain people are pushed to alter themselves according to what is deemed 'socially acceptable' – even if this means going against the established norms and values of a subculture or culture.

In this instance, a hypothetical example may be instructive. Imagine a society in which there is a minority culture whose members can be physically identified by their curly hair. It is not hard to imagine a scenario in which a government, seeking to consolidate power and reinforce the unity of the state, might deem the curly-haired minority as a provocative and unnecessary threat to the stability and longevity of the society as a whole. Through genetic modification and the like, it is conceivable that curly haired people could be made into straight haired people over the course of a generation or two. While the motivation to do this might be argued to be somewhat altruistic (e.g., to reduce conflict between a straight-haired majority and a curly-haired minority), the ramifications that such actions would have on cultural diversity, the self-esteem of the curly-haired people, cultural identity, and so on, would be enormous. Additionally, the very long term would raise the possibility of eliminating all physical differences among cultures, pushing towards a vast civilizational homogeneity at the cost of diversity. Under such a scenario, the *Brave New World* vision again comes to mind.

Similar scenarios could present themselves even in cases where there is no specific government intervention but where pressure to conform to society's ideals nonetheless exists. As the US President's Council on Bioethics argues, enhancement technologies that are 'freely permitted and widely used may, under certain circumstances, become practically mandatory'.[33] For example:

If most children are receiving memory enhancement or stimulant drugs, failure to provide them to your child might be seen as a form of child neglect. If all the defensive linemen [a term for those who are usually the largest players in American football] are on steroids, you risk mayhem if you go against them chemically pure. And, a point subtler still, some critics complain that, as with cosmetic surgery, Botox, and breast implants, many of the enhancement technologies of the future will very likely be used in slavish adherence to certain socially defined and merely fashionable notions of 'excellence' or improvement, very likely shallow and conformist.[34]

Human enhancement, whether under an authoritarian or libertarian government structure, could eliminate cultural differences. This could happen as the result of a malicious, intentional policy but, scarier still, it could just as likely happen as the result of unintended consequences.

Transcultural security is a challenge because in many instances cultural identity stems from characteristics that could actually be directly and fundamentally altered by emerging strategic and human enhancement technologies. We often take it for granted that, at least on some level, diversity is to be fostered and appreciated. While differences among different groups, races and nationalities can cause tensions among cultural groups, they also contribute to the richness of life experience. Human enhancement could, in the wrong hands, lead to the elimination of these valued forms of diversity. Although this might eliminate – or be misrepresented as a way to eliminate – transcultural security problems as we know them today, new types of humans and transhumans might represent even greater, unintended cultural divides. This is just one of many reasons why eugenics should be regarded with scepticism and concern.

Another challenge presents itself when religious inclinations are factored into the enhancement process. Many thinkers and scholars have already expressed discomfort at the similarities between the use of human enhancement technology and playing God. Another concern is that the power to enhance ourselves will ultimately blur this line even further, reducing the philosophical and ideological diversity in the world.

At the core of my '*multi-sum security principle*' is the concept of global justice, a fundamental element of international peace, stability and security. Human enhancement needs to be evaluated in the context of what is just and applied accordingly. Without justice, individuals, groups and cultures will inevitably feel marginalized by society and the international system, and this will cause instability and conflict. To the extent that human enhancement can offset or alleviate these feelings, it has enormous potential to improve international security across all five dimensions. Unfortunately, human enhancement has just as much potential to undermine and destroy the foundations of global justice. I have outlined above my belief that, for better or worse, human enhancement is an inevitable development. Below, I weigh the potential benefits of enhancement against the potential dangers. In so doing, I articulate my views on human enhancement and the destiny of the human race.

## Moving beyond the *'multi-sum security principle'*: human enhancement as an existential threat

So far, this chapter has looked at the potential implications of human enhancement for specific elements of global security. This section moves beyond the five substrates of global security to look at the cumulative effect which enhancement technologies may have on the destiny of the human race – specifically, the idea that human enhancement is a potential existential threat to the human race as a whole.

In the light of emerging and converging strategic technologies and all the new opportunities that are being created for humans to alter, enhance and fundamentally change the way their minds and bodies function, it is my conviction that human enhancement, combined with the fundamental characteristics of human nature, will in all likelihood lead the human race to lose control of its destiny.

This is a bold statement and one that deserves a little more explanation. Essentially, it is my fear that rapid technological advances and converging technologies will alter human nature to a point where we will no longer be human. Recall my description of human nature and its close entwinement with neurochemistry, genetics and emotions. Cognitive science, nanotechnology, biotechnology and information sciences all have the potential to change, disrupt or eliminate these fundamental components of human nature. Ironically, it is precisely because of the interplay of these core components of our human nature that we will seek these enhancements out to begin with.

Take, for example, antidepressants. These drugs, designed to help those who suffer from chronic depression, are classic examples of an already widely accepted mental enhancer. For the time being, we will leave aside the questionable distinction between therapy and enhancement and consider the two concepts as overlapping and somewhat interchangeable. If we continue altering our neurochemistry with mood-altering drugs, memory enhancers or happy pills, or any number of potential neuropharmacological innovations, I believe that, over the long term, the human race will cease to exist as we know it today. Right now, a depressed person who takes a mood-modulating drug has the potential to alter their mood, but only when they are physically taking the pill. In the future, we will be able to make more long term and permanent enhancements to our moods and brains. Emotions will become something to be controlled by drugs and treatment. They will no longer be tied to the human survival instinct but instead to the newest technological innovations and capabilities. Emotions have traditionally been a defining feature of the human urge to survive, but human enhancement will separate human emotions from our survival instinct. On a more elemental scale, human enhancements, if taken to the extreme, could ultimately result in the loss of core human characteristics such as compassion, loving and nurturing relationships, charity, dignity and equality. Although it is highly unlikely that these traits would be eliminated deliberately, the potential for human enhancement to lead to unintended consequences cannot be overstated. Such alterations would undermine concepts of justice and transform our notions of security and of what it means to be human. The precise implications that this will have for human nature and the future of our race are uncertain but potentially grave. Perhaps technology can improve our overall quality of life without sacrificing the key components that make up our human nature and dignity. Or, perhaps, we will completely reprogram our brains so that our motivations and nature are altered beyond the point of recognition. Even if the end results of this are unclear, the shift is certain to happen.

Beyond purely enhancement technologies, the creation of AI, the likelihood of humanity reaching the point of Singularity, and the merging of biology with technology also pose potentially existential threats to the human race. AI, or machines that are smarter than humans, will, presumably, come into existence in the next few decades. When this happens, these machines will be intelligent enough to design even more intelligent machines, and the rate of technological progress will expand exponentially.

In the context of human destiny, we are running the real risk of making our human intelligence and ourselves obsolete. Once smarter-than-human AI comes into existence, what is to stop this entity from completely replacing us? Some philosophers and scholars who study and speculate on the Singularity and the future of AI maintain that this question is simply a matter of ensuring that AI is created with pro-human tendencies.[35] If, however, we are creating an entity with greater than human intelligence that is capable of designing its own newer, better successors, why should we assume that human-friendly programming traits will not eventually fall by the wayside? What if these AI machines turn on our race? Again, the dangerous implications of emerging strategic technologies on our human destiny cannot be understated.

This possible doomsday scenario of human enhancement and strategic technologies is not meant to be confused with arguments against enhancement. Instead, it is a case for strong regulation. It is also a call for action to define international moral and ethical guidelines on what technologies and enhancements are acceptable to our human race and on what terms. The urgency surrounding these issues is high. Although it is easy to dismiss AI and permanent brain enhancement or alteration as purely futuristic concepts, we as individuals, the international community, and, indeed, the human race must begin thinking about how to approach these issues, dangers and temptations now so that we are prepared for new technology when – not if – it comes along. In some ways, we are already woefully behind on this front. Germ-line genetic engineering, cloning and neuropharmacology are already very advanced sciences, and yet we as a society still have ambiguous, often conflicting ideas on how these technologies should be adapted and regulated in our day-to-day lives.

Ironically, technology will lengthen our lives before it possibly eliminates them. Life expectancy has increased steadily across Europe since the 1800s, and the trend is for lives to be longer and longer.[36] As Tom Kirkwood argues, 'The ongoing increase in human life expectancy is without doubt one of the greatest changes to affect humanity in the last two hundred years [...] Although dangerous if we ignore it, [it] is the product of quite extraordinary success' in science and technology.[37] Technology can, in some scenarios, make us immortal, but the question is, in extending our lives and altering the aging process, are we losing a fundamental component of our humanness? Again, it is a slippery slope.

I outlined above my views on human dignity and the minimum criteria that must be met in order to protect this vital aspect of human life and social interaction. When it comes to the application and adoption of emerging and converging strategic technologies, we face a human decision on which parts of our existence are the most important in terms of guiding and regulating enhancement. In the short term, we may have choices on this front. We may be able to prioritize pleasure and minimize pain, and we may be able to learn from the trial and error of non-permanent enhancements. However, in the medium term, these choices will become blurred, and alterations and enhancements will take on a more permanent nature. Over the long term, we will have lost all control.

I strongly maintain that human enhancement is not just likely: it is inevitable. The inevitability of enhancement also means the inevitability of several tragedies for the human race. Chief among them is the fact that there will be conflict stemming from inequality between enhanced and unenhanced humans. In this competition, unenhanced beings will almost certainly lose, and our human race as we currently know it will be replaced with trans- and post-humans. Similarly, enhancement technologies are costly and likely to remain so for the foreseeable future. This means rich countries and rich individuals will have disproportionate abilities to enhance themselves, potentially reinforcing already problematic gaps between the rich and the poor and between developing and developed countries. We need to consider how to preserve equality and justice in the face of these challenges. Again, this will irrevocably change the destiny of our race.

These dangers to human nature, dignity and destiny are real, but that is absolutely not to say that technological advancement should be halted or stifled. Instead, there needs to be some reconciliation of the need to allow and even encourage scientific innovation to progress with the need to balance concerns over ethics, morals and the potential abuse of new and emerging technologies. To do this effectively, all the relevant players in science, government, society, civil liberties groups, businesses, cultural groups, religious groups, nation states and the international community must come together. Maintaining compassion for the human spirit and a sense of justice for all are important parts of achieving this goal. As I have argued before, global justice is the foundation of global peace and security. If we are to successfully and ethically regulate human enhancement and emerging strategic technologies, justice and equality must be top priorities. Good governance with a focus on global justice is imperative to protect human destiny. Chapter 15 focuses specifically on this topic.

# 15
## Criteria for a Regulatory Framework of Human Enhancement

Given all the potential dangers and opportunities related to human enhancement technologies, the issue of regulating these technologies is one that we must address urgently as an international community. As is the case with most transnational issues, the best way to comprehensively address and regulate human enhancement technologies will be through an international forum. However, with human enhancement technologies, the preference for regulating these emerging technologies at the international level is not merely a desirable situation – given the nature of the technologies and the elements of human destiny that are at stake, it is an imperative.

### The United Nations

As the most comprehensive and inclusive international regulatory body, the United Nations is a natural starting point for setting standards on human enhancement and defining the nature of human dignity and human destiny. Already, the United Nations has outlined basic guidelines on and definitions of human dignity in the Universal Declaration of Human Rights. This landmark document acknowledges as a given that the 'recognition of the inherent dignity and of the equal and inalienable rights of all members of the human family' are 'the foundation of freedom, justice and peace in the world'.[1] Its specific focus on human dignity and justice have helped set the tone for international relations and human rights standards in the post-World War II world.

The United Nations is an important starting point for human enhancement regulations because, as is noted above, standards and rules on human enhancement technologies must be universal. If this is not the case, people in countries where there are strict regulations against human enhancement will simply migrate or seek out enhancements in countries where restrictions are more lax. In such a situation, the key concerns of inequalities between enhanced and non-enhanced peoples, the challenge of who has access to the technology, and the question of preserving human nature and human dignity in the face of enhancement would be magnified. Unless ethical and legal regulations are standard across borders and cultures, there will be an opportunity to exploit and abuse human enhancement technologies. Because our human nature will naturally lead us down this path, the demise of humanity will be accelerated in the absence of unified international regulations and standards.

The United Nations Declaration on Human Rights was an important starting point in this area. As a next step, the United Nations should consider a Declaration on Human Dignity. Such a document would reflect my nine necessary minimum criteria of human dignity (reason, security, the promotion and protection of human rights, accountability, transparency, justice, opportunity, innovation and inclusiveness).[2] With these criteria formally acknowledged by the international community, it would be easier to open up a discussion on logical, humane and rational boundaries for human enhancement. So far, the closest thing the United Nations has to this declaration is the UNESCO Universal Declaration on the Human Genome and Human Rights. Although not legally binding, this document urged nations to ban practices that are contrary to or that violate human dignity. As an example, the declaration specifically cites reproductive cloning.[3] This is a positive step towards creating international standards and norms.

## The major technological players

While comprehensive international regulation is the top priority for protecting human destiny in the long term, the policies of the world's leading technological countries and regions will be just as important for setting the tone of international regulations in the short run. Countries such as the United States and regions such as the European Union (EU) are among the world's most advanced players in the development of emerging strategic technologies in general and in human enhancement technologies more specifically. As the international community debates the framework for a universal agreement on human enhancement technologies, their limits and their potential, individual countries will continue to regulate these issues at the national level. What these countries deem acceptable and how they approach enhancement is important, as it is likely to set the tone for international debates over the issue. Whether it is because these practices become customary international law or because they are merely the easiest starting point for an international discussion, leaders in technological development will play a large role in laying the foundation for comprehensive regulation.

So far, the United States under the Bush Administration has taken a rather conservative approach. Reflecting the general sentiment of many US citizens, one of United States' leading bioethicists, George Annas, has advocated a Convention on the Preservation of the Human Species that 'would make efforts to enhance human beings by making heritable changes in people's genomes a crime against humanity'.[4]

Given President Obama's stated aim to 'restore science to its rightful place',[5] the Obama Administration is likely to adopt a somewhat different approach. In 2009, Obama founded the Presidential Commission for the Study of Bioethical Issues with a mandate to 'advise the President on Bioethical issues and related areas of science and technology'.[6] As opposed to its predecessor, it is not composed of leading mainstream bioethicists. Therefore, it is expected to focus more on practical, specifically medical policy issues.[7] However, since the first meeting took place only in July 2010,[8] it remains to be seen what the concrete outputs of this newly established commission will be. There is currently no established overarching legal policy in place on this front in the United States.

The EU is even further behind than the United States. According to Martijntje Smits of the Netherlands Rathenau Institute: 'The EU has no platform for monitoring and discussing human enhancement. At the same time, these issues touch upon matters that have relevance at the EU level, such as health budgets, research policies, and economic issues. Differences between member states will likely pose problems in the future'.[9] The EU debate over enhancement regulation has also seen many references to the EU Charter of Fundamental Rights, which covers provisions on bioethics as well as the values of medicine and biomedicine.[10] As Roberto Mordacci has noted, 'This declaration is useful in this regard [...]Human dignity, autonomy and integrity are definitely shared values in the European tradition'.[11] Formalizing these values and gaining consensus on guidelines and regulations for human enhancement technologies are the next steps for the EU.

## Relevant agencies

In a similar vein to the technologically advanced nations, a number of think tanks and private sector organizations focused on bioethics and technological progress will also play a role in establishing the ground rules for using emerging strategic and enhancement technologies. One of the most high-profile groups is the US President's advisory Commission on bioethical issues. Called US Presidential Council on Bioethics under the Bush Administration, this Council has published a large body of research on the topics of human dignity, human enhancement and government regulation of related technologies. Bioconservatives such as Francis Fukuyama and Leon Kass have served on the board of the organization, and many of the Council's recommendations leaned towards the conservative side of things.

The Council's 2003 report, *Beyond Therapy: Biotechnology and the Pursuit of Happiness*, frames human enhancement as a double edged sword. On the one hand, it envisages 'a world in which many more human beings – biologically better-equipped, aided by performance-enhancers, liberated from the constraints of nature and fortune – can live lives of achievement, contentment and high self-esteem, come what may'.[12] On the other hand, the report cautions that such technologies and improvements do not come without cost. Essentially echoing my concerns, although for slightly different reasons, the President's Council calls for regulation of human enhancement technologies in order to preserve the future of the human race. As mentioned above, the Council has been succeeded by the Presidential Commission for the Study of Bioethical Issues created by US President Obama.

In Europe, there is the European Group on Ethics in Science and New Technologies, which plays a similar role to the Presidential Commission for the Study of Bioethical Issues, albeit with a wider mandate. The European Group has addressed issues related to nanotechnology and other potential enhancement technologies. With regard to nanotechnology, for example, the Group focuses on the importance of establishing the safety of certain nanomedicines and treatments before they are made widely available to society.[13]

Groups unaligned with any particular government are also proving to be influential in the debate over human enhancement. In the United States, centres such as the Hastings Center and Georgetown University's Kennedy Institute of Ethics also often weigh in on issues related to human nature and our growing potential to alter or tamper with it. Such governmental and non-governmental organizations will, like states, be instrumental in shaping the debate over human enhancement at the international level.

## What should be permissible?

Having looked at the existing frameworks, ideas and institutions that could govern international human enhancement, the next natural question is: how should these groups regulate this issue? This is a sticky question full of contradictions and ethical obstacles.

Positions taken in the debate advocate everything from a research moratorium to strong regulation to minimal regulation.[14] Some, such as Fukuyama and Habermas, call for heavy controls, or even a for the prohibition of human enhancement technologies. Habermas, for instance, holds that negative, or therapeutic genetic interventions may be allowed under certain circumstances, while genetic enhancements, or what he calls positive eugenics, are to be prohibited.[15] This is seen as indispensable to shelter society from unpredictable and undesirable effects, including the creation of inequalities among enhanced and non-enhanced humans. Others, like Fritz Allhoff, advocate no more than minimal regulation in order to safeguard the individual's ability to choose whether it wants to be enhanced, and how it wants to design its life.[16]

I believe the central issue in this debate is the need to foster innovation and scientific advancement while keeping the inadvertently self-destructive elements of human nature firmly in check. Any practices that infringe on our human dignity negatively impact our vital human needs, and thus the fundamental concept of human dignity is a good place to start when evaluating what should be permissible.

First, we must consider the key components of human dignity, as these are the core elements we must protect and prioritize as we decide how to regulate human enhancement. Any enhancement that violates any dimension of human dignity simply cannot be considered an enhancement; indeed, any violations of dignity are a setback for the human race. In *Sustainable History and the Dignity of Man*, I outline nine minimum criteria of human dignity, which I define as 'the status of human beings that warrants respect'.[17] These nine criteria are reason, security, the promotion and protection of human rights, accountability, transparency, justice, opportunity, innovation and inclusiveness.[18]

At the national level, the state has numerous obligations to preserve the dignity of its citizens. These obligations include the provision of basic needs, inclusiveness and participation, socio-economic justice, gender equality, human rights, and the protection of the environment and ecological balance.[19] From a global perspective, ensuring dignity means avoidance of conflict, provision of basic needs, participation, socio-economic justice, gender equality, the promotion of human rights, and the protection of the environment and ecological balance. Obviously, there is some overlap between national and global responsibilities. Regardless of whether we are speaking of national or global regulations, these criteria at least provide a benchmark for judging how well governments and global society are doing when it comes to protecting the dignity of each individual and global society as a whole.

Looking at human enhancement technology in this context can be instructive. For example, if the avoidance of conflict is a key measure for human dignity, then enhancements that would encourage conflict or aggression should be prohibited. This could apply, for example, to a neuropharmacological product that promotes aggression or fraudulent ambition. If such a drug were used on the battlefield, it would be likely to increase conflict or make certain individuals more prone to violence. Thus, based on the criteria of human dignity, such enhancing technologies should be made illegal. Similarly, enhancing one society, culture or population while denying the same opportunities to another would lead to tremendous inequalities and potential conflicts between enhanced and unenhanced individuals. Again, by using the human dignity criteria, we at least have some guidelines and overarching principles for the regulation of these emerging strategic technologies. In many senses, the human dignity guidelines reflect and reinforce what we already intuitively know about where to draw the line in the human enhancement debate.

Continuing with this line of thought, accountability and transparency are two key components of human dignity that presumably would prevent governments from manipulating their populations or from forcing enhancements – particularly in the context of eugenics – on their citizens. If governments are truly held accountable for their actions and if they behave clearly and openly, they would, for example, be prevented from promoting one set of genes at the cost of another because, if their actions were open and transparent, they would face a natural backlash, particularly from the group whose genes were being negatively manipulated. Similarly, states would be held accountable to their citizens and international law if they engaged in practices such as forced abortions of foetuses deemed abnormal. The lines in this situation are not black and white but by being transparent, states and societies are better prepared to identify the tipping point for when a fundamental moral line that might infringe on human dignity has been crossed.

Another of my criteria for human dignity is the provision of basic human needs, such as food, housing, clothing, health care, education and security. Certain types of enhancement technology could help to fulfil these basic needs, particularly health care but also education – because of technologies such as cognitive enhancement, memory improvement and increased concentration. If a person who has lost a leg is unable to secure work, for example, society must bear the costs of providing unemployment and benefits for that individual. With the help of an enhancement such as a prosthesis, however, that person would have an easier time finding a job, thus causing less of a burden to society. In this sense, human enhancement plays an important role in advancing human dignity.

However, a challenge to human dignity arises with regard to opportunity and innovation. Innovation is one of my nine core elements of human dignity, and I believe innovation should be pursued, as it generally tends to improve the quality of life and the calibre of society as a whole. Unfortunately, innovation has a dark side and, especially in science, new discoveries can have negative applications. Bioweapons, addressed in Part I, are one high-profile example of this, but there are equally dangerous innovations in the field of human enhancement. I have outlined above how enhancements may change our core human nature. We may choose drugs that help us maximize pleasure and minimize pain; we may seek brain treatments that help us forget bad memories; or we may alter our genes to avoid certain illnesses. The question then becomes at what point do innovation and progress in enhancement technologies cross the line and become dangerous. An enhancement innovation that makes a soldier violent and aggressive while eliminating the soldier's capacity for compassion would be quite deadly, but this is an extreme hypothetical example. Before we get to such an extreme, we must already have an understanding of the point at which innovation, a key element of human dignity, begins to undermine or challenge human dignity.

Certain enhancement technologies also pose a challenge to the question of human rights. This is especially true in reproductive technologies where the rights of embryos and foetuses are hotly debated. Embryonic stem cells can be used in a variety of medical research endeavours and could potentially lead to breakthroughs in the treatment of everything from cancer to Parkinson's disease. If embryos are classified as humans, however, as many people maintain, then the dismemberment of these embryos in the name of scientific research and human enhancement would be a gross violation of human rights. Again, such issues are fraught with moral questions and challenges, but it is up to us as a society to begin to address them directly.

Androgyny is a theme that occurs frequently in certain circles of transhumanism literature, and this feeds into the issue of gender equality, another component of human dignity. Those who adhere to this school of thought maintain that 'technology is eroding the biological, physiological and social role of gender'.[20] For many, this is a positive step. Post-genderists, as they are called, argue that 'dyadic gender roles and sexual dimorphisms are generally to the detriment of individuals and society'.[21] Enhancement, these post-genderists would argue, is beneficial to human dignity, as it may blur the lines between genders. However, whether this undermines other components of human dignity is an open question.

Perhaps some of the most central issues in this discussion of human dignity and human enhancement are equality and inclusiveness. Equality has been a common theme throughout this book, whether it is in terms of equality of access to new and emerging technologies; equal consideration in the development of new technologies; or growing inequalities brought about by differences between enhanced and unenhanced individuals, it is a crucial concept that must be aggressively tackled by policymakers. For any enhancement technology truly to be used and adopted in a moral fashion, it must be made equally available and accessible to all members of the global society. This is a tall order even in the best of circumstances, but we have a moral obligation to at least aim our efforts in this direction. While no one should be forced to enhance themselves, all people must at least have the option of adopting the technologies deemed acceptable on whatever level they deem appropriate for themselves.

Certain technologies may be deemed too dangerous for human use, and they will be universally barred. This is acceptable from a dignity perspective, as the ban will be complete, and thus everyone is equally prevented from utilizing the technology. In this situation, there can be no exceptions for the rich, the well-connected or the especially ill. If we deem a technology such as a brain implant to be morally unacceptable for one group of people, that same logic must extend to all. Thus, the central question in terms of dignity and equality is not whether we allow enhancement technology. As I have argued, the adoption of such technology is inevitable. The real question focuses on *which* technologies we allow and how we make these acceptable technologies equally available and accessible. Once those questions are answered, we will have gone a long way towards assuring the preservation of human dignity. The decision-making process on this front ought to be inclusive, with opinions and input from all sectors of society. This issue is addressed in greater depth below.

As a humanist scholar who is interested in the betterment of humanity, I do not place specific moral restrictions on which scientific innovations should be allowed to occur. Provided that scientific innovations do not undermine any of the nine core criteria of human dignity that I have outlined, I am inclined to encourage scientific innovation and its adaptation to the human body and mind as is individually appropriate. From my perspective, the priority for global policymakers is to *urgently* develop strict, ethical guidelines at the global level, as I am most concerned about the creation of an international black market for enhancement technologies. Even in the best possible regulatory situation, human enhancement technologies have a high chance of getting out of our control and negatively affecting human dignity, possibly leading to the extinction of the human race as we know it today. If enhancement technologies were allowed to develop in an unregulated environment, the disastrous effects would be both augmented and accelerated. As policymakers look to regulate enhancement technologies, they should refer to the nine core criteria of human dignity as guidelines. While there are few absolute truths in this field, human dignity is an indispensable moral compass.

Because of the powerful attraction of human enhancements and their drug-like nature – drugs, after all, are enhancers of pleasure and feelings – rogue elements of society will ultimately seek to access and control these technologies. Without strong, clear, enforceable and global regulations, these groups are likely to be successful. Thus, we need to begin by defining where the proverbial line in the sand is. What is the point in human enhancement that is inviolable, the point after which everything else becomes taboo? Is it reproductive cloning as the EU has suggested? Or perhaps it is germ-line genetic engineering. Even once we define this tipping point, it still does not mean that everything on the other side of the line is acceptable. Rather, as Jorgo Chatzimarkakis has said, 'We need a broader discussion, and we need a normative framework. Yes, we need a red line. That does not mean that no discussion is needed up until that line – it only means that what lies beyond the line is taboo. Up until the red line, there should be discussion on a case-by-case basis'.[22] The process is a difficult one with many potential pitfalls, but it is one we must tackle aggressively.

## Who should decide what is permissible?

Despite the ambiguous boundaries and rules that need to be created with regard to which types of human enhancement and related technologies are permissible, and socially and morally acceptable, the question of who should make these types of decisions is somewhat more straightforward. Overall, for any enhancement-related regulation to be effective or valid, all facets of society must have a stake or a voice in establishing guidelines. This includes governments, parents, responsible kin, religious groups, ethical organizations and legal experts. In a diverse world such as the one we live in, each of these group's is bound to have dramatically different opinions on the topic of human dignity, human enhancement and the role of technology in society. However, if we neglect to at least take account of the broad spectrum of ideas or the needs of specific groups (e.g., women and unborn children), we run the risk of creating regulations that are ineffective or that do not reflect the realities of the world we live in.

I believe that health and education must never be governed by economic issues and that these are two basic human needs that must be available to all regardless of the cost to the state. In terms of education, ability must be the only distinguishing factor, but in terms of human enhancement, which will have consequences for both the health and the educational aspects of our lives, the solution is simply that everyone must have access to approved enhancement technologies.

One area that I am particularly concerned about deals with the question of enhanced and dramatically lengthened physical longevity in the absence of comparable cognitive enhancements. We run the risk of having people who are physically capable of living well into their 100s but suffer from dementia and mental deterioration that has its onset in such people's 70s or 80s. This will not only burden social security systems around the world but also present society and carers with challenges that we are currently not equipped to deal with.

Another important area of regulation is pre-emptively preventing abuses or at least reducing the possibility of runaway technologies taking over our lives, decisions and human destiny as we know it. As if this does not set the bar high enough for setting regulatory standards, regulations must also take account of social, cultural, ethnic, religious and economic aspects. Regulations that fail to incorporate these many facets of global life will run the risk of being unenforceable and non-inclusive.

Overall, humanity is faced with a challenge to its values, and it is finally time to decide what we as a race prioritize. Is it selfish needs such as profit, permanency, power and pleasure, or is it a number of other things? Such decisions and prioritizations are crucial as we decide which enhancements are acceptable and in what form.

# 16
# Conclusions of Part II

Part II examined the intersection between emerging strategic technologies (ESTs), human dignity and human destiny. While Part I looked at technologies individually, Part II looked at how different ESTs are combining and converging to fundamentally alter people as human beings.

Such enhancement and alteration is not merely a futuristic concept relegated to science fiction. Already, human enhancement technologies are quite widespread – involving everything from hearing aids to artificial insemination to genetic testing for predisposition to diseases – and while we have some control over how we adopt these technologies into our lives in the short run, in the medium and long term we are going to lose control. Indeed, one of the greatest risks of emerging strategic technologies being applied to human enhancement and human improvement is the possibility that we will adopt these technologies to such an extent that the human race as we know it today will be eliminated. Again, while such a development seems far away, the reality is that we have already started down the slippery slope to this destination.

Although the dangers of human enhancement are very real and the effects on our human destiny are significant, there are strong arguments to be made both for and against enhancement. On the pro side, ESTs in the context of human enhancement can help to cure diseases and improve the quality of life for many people, especially those who suffer from disabilities or so-called abnormalities. Human enhancement can be portrayed as a moral issue because such technologies have the potential to alleviate human suffering and therefore, some argue, we have an *obligation* to take advantage of the opportunities that ESTs present us. Those against human enhancement argue that in using technologies to improve our physical and mental status, we run the risk of altering the fundamental essence of our human nature. Although the original intentions behind enhancing ourselves may be well-founded, the unintended consequences of such behaviour, especially when evaluated at the aggregate level, could be severe and destructive. In the worst case scenario, it could lead to conflict between enhanced and unenhanced peoples or to the end of humanity.

My position on human enhancement does not start from the perspective of advocating for or against such applications of science and technology as, ultimately, I believe that human nature will inevitably drive the human race to embrace such technological developments as they become available, and this will happen regardless of whether it is good or bad for the human race. This inevitability is rooted in the facets of human nature that I outline in my book *Emotional Amoral Egoism* and which have been described throughout this book. Human enhancements offer us the chance to satisfy the neurochemically based motivating factors that drive our survival. Whether it is relieving pain or stimulating pleasure, human enhancement technology can fulfil our most visceral human demands, and that will ensure that we as a race will seek out these technologies.

The technologies that humans will adopt to fulfil their neurochemically driven motivations are diverse. There are strictly physical enhancements such as prostheses, contact lenses, reconstructive surgery and anabolic steroids; and mental enhancements such as mood modulating drugs, brain prostheses and cybernetic implants. Also falling under the umbrella of enhancement technologies is the field of reproductive science, specifically as it relates to fertility drugs, in-vitro fertilization and the screening of embryos for diseases.

As these technologies are adopted on a bigger and bigger scale, human beings will gradually undergo a sort of evolutionary process from human beings to transhumans to post-humans. 'Transhuman' refers to an intermediary form of human, which has several of the key characteristics of humans but has also embraced traits such as the intensive use of cosmetic surgery and performance-enhancing drugs on a large scale. 'Post-human' refers to the point where humans have adapted technology and merged their biology with technology to the point where they are no longer identifiable as humans. At this point, our core nature will have changed and we will have evolved into an entirely different species.

Such an evolution will have dramatic effects on our human destiny. Using my '*multi-sum security principle*', I argue that human enhancement has implications for each of the five dimensions of global security. In the context of human security, human enhancement presents the dual challenges of equality and justice, particularly between enhanced and unenhanced individuals. One of the greatest security risks of enhancement is that enhanced individuals and societies discriminate against unenhanced peoples or use the benefits of enhancement to gain leverage and power over unenhanced people. The counter-argument to that point is that human enhancement technology can dramatically improve the quality of life. Some even argue that widespread human enhancement will happen so quickly that any inequalities will be short-lived.

The second dimension of my *'multi-sum security principle'* is environmental security and, although it may seem counterintuitive, human enhancement technologies affect this dimension of global security. Whether it is because of waste disposal issues related to enhancement technology or the greater strain we as humans put on the environment by increasing our population and living longer, the impacts that a collective change in the nature of the human race could have on the environment, natural resources and how we use them is not to be dismissed.

With regard to national security, I argue that human enhancement will be a way for more scientifically advanced states to gain an advantage over those without the technology or the resources to invest in such technology. States with broad access to human enhancement technology will also benefit from less tangible but nonetheless important benefits such as increased national pride and potentially stronger economies. The latter is particularly true if a country is able to develop a monopoly or strong competitive advantage in one form of technology or another.

The next dimension of global security I addressed was transnational security, where equality or lack thereof in enhancement technologies may affect the way nations cooperate. Again, inequality continues to be a theme of human enhancement, and in this context it becomes especially relevant if one entire society, nation or culture adopts enhancement techniques that undermine the safety, well-being or dignity of another. However, health-related human enhancement technologies could improve the health of societies, increasing stability and reducing the challenges brought about by infectious diseases or physical disabilities.

Finally, I looked at human enhancement in the context of transcultural security, arguing that eugenics is a particular concern. If governments threaten to use enhancement technologies to promote a 'desirable' gene or to eliminate an 'undesirable' one, the dignity of entire cultural domains could be at risk. Such threats do not necessarily have to be related to government action. Indeed, the pursuit of things such as breast augmentation or rhinoplasty in pursuit of cultural ideals of beauty could, if taken to the extreme, force us to completely inhuman notions of excellence. Again, such developments would attack the basis of human dignity and pose a significant threat to us as a human race.

Overall, human enhancement technologies are an inevitable part of our future, but they may also mean the future demise of human nature as we know it. As they are inevitably adopted on a bigger and bigger scale, I maintain that human enhancement technologies may cause us to lose control of our own human destiny, especially if we do not put the appropriate regulatory structures in place early enough. Whether it is undermining the core, evolutionary aspects of our human nature that have allowed us to survive for so many thousands of years, or the complete merging of our human biology with technology, enhancement will quite possibly render our human intelligence obsolete. Figuring out how to encourage scientific innovation while preserving human dignity and equality is a major challenge for policymakers. Although it is often an unpalatable issue for lawmakers, it is one that must be tackled aggressively in the near future if we are to preserve the long-term existence of the human race.

# General Conclusions

This book has addressed emerging strategic technologies and their potential impacts on geopolitics and human destiny with an eye to scientific sophistication and comprehensiveness. Using the Science and Technology Studies framework evaluated through the lens of *neuro-rational physicalism*, it has shown how society and technology are growing and developing in tandem, with each factor driving the development of the other. Knowledge, which has a physical, neuro-biological basis and is based on reason and sense data subject to interpretation, is at the core of this technological innovation.

Beginning with information and communications technologies (ICT), we saw how rapidly new strategic technologies in general and information technologies in particular are growing and spreading. Whether it is the Internet, Web 2.0, blogs, and mobile phones; or more futuristic aspects such as smart search capabilities in search engines or radio frequency identification tags, ICT pervades our modern lives.

From a geopolitical perspective, ICT is affecting social dynamics and cultural groups, and it is also influencing more traditional political issues. Moldova has witnessed a so-called Twitter Revolution, and Russia is widely suspected of having led cyber attacks against Estonia and Georgia. From these examples, we can see that ICT is playing an important role in both peaceful resistance and more traditional military warfare. ICT is a challenge and an opportunity for policymakers. On the one hand, the rapidity with which information spreads in this day and age results in time compression issues, or situations in which policymakers are expected to respond to developments in real time – without the opportunity for fact-checking or deeper reflection on the options. On the other hand, ICT is giving policymakers, politicians and civil society new ways to engage with each other and to share information on related causes and issues.

Unfortunately, the greater engagement offered by ICT is not yet truly global. As is the case with other strategic technologies, ICT suffers from a gap between what rich and poor countries have. And for those who do have access to ICT, there are constant concerns over privacy and cyber security. The protection of private information, the Internet and individual computer networks from cyber crime and other attacks are major geostrategic challenges. Governments and civil society have been slow in addressing these concerns, and our global vulnerabilities on this front should not be underestimated.

Chapter 3 moved on to look at the broad category of emerging strategic technologies that relate to global climate change, energy and the environment. Truly, these are two of the most significant global geostrategic challenges we face, and the sense of urgency around developing less-polluting and pollution-offsetting technologies is high. The two challenges of the looming global energy crunch and the ongoing damage we are doing to our environment are closely related, and similar technological advances can be applied to both problems. Scientists and politicians need to look to new, ideally renewable, energy sources to meet the growing needs and demands of their populations. The two general genres of emerging strategic technologies that can be adopted in response to these problems are renewable energy technologies and technologies that can improve the efficiency of non-renewable resources or existing, polluting technologies. As is noted above, even with advancements in renewable energy resources, fossil fuels will continue to feed our energy needs, and so the impetus to improve the efficiency of and reduce the emissions from these fuels is high.

Governance of energy, the environment and related technologies is an international challenge. The 2009 UN Climate Change Conference in Copenhagen failed to negotiate a legally binding agreement to address climate change, and a successor treaty to the Kyoto Protocol is still to be negotiated. From the food for fuel trade-off to the inherent unfairness of forcing the same emissions cuts on both developing and developed countries, energy, climate change and the environment present some of the most significant and truly among the most global challenges of our generation.

Chapter 4 examined another central geopolitical challenge: global health care and the emerging strategic technologies that may help us to improve the quality and length of people's lives. The worldwide health care industry is worth trillions of dollars and is expected to grow in the coming decades. From a technological perspective, the industry is being driven by developments in the fields of genomics, ICT, antimicrobial drugs and nanotechnology. These technologies can influence multiple facets of global health, including minimizing the spread of infectious diseases, facilitating the development and growth of impoverished countries and areas which are otherwise debilitated by diseases such as AIDS and malaria, and increasing longevity and the quality of life. Collectively, the impacts of these emerging strategic technologies can positively affect global civil society and stability in highly positive ways. Naturally, there are also some challenges at the intersection of strategic technologies and health care, chief among them being the familiar problem of providing access to these technologies to developing and impoverished countries. This inequality is one of the major geopolitical issues of health care and technology.

After looking broadly at emerging strategic technologies in the health care context, we proceeded to look at some more specific technologies that have obvious applications in health care but that are also affecting other facets of geopolitics and geostrategy. Biotechnology is a leading example of this. In medicine, biotechnology and related sciences are helping to treat ailments such as non-Hodgkin's lymphoma and aiding in processes such as early cancer detection. For the biotechnology industry, one of the major challenges is the issue of dual-use technologies and the question of balancing regulation and safety issues against the need to foster innovation and progress. Although it has tremendous potential to help with geopolitical challenges such as global food supply, clean energy, and improved medicines and vaccines, many of the technologies and products of biotechnology can also be manipulated into dangerous products that can cause harm. Bioweapons and chemical products, especially if they fall into the hands of terrorists, can undermine global security and alter our traditional thinking of military conflicts. Unfortunately, global regulation on this front falls far short of technological capabilities. Technology is simply advancing too fast for bureaucratic regulatory agencies to keep up – a reality we see across the realm of emerging strategic technologies and their integration into society. Other geostrategic implications of biotechnology that must be taken into account are ethical issues over highly promising but morally contentious stem cell research and the growing need for better international research governance. A common theme throughout this book – the need to balance regulation and safety with fostering scientific innovation – applies equally to biotechnology and is, in general, one of society's biggest collective challenges.

The next technology addressed was genomics. The successful mapping of the human genome and related developments, research and discoveries are perhaps the most high-profile scientific achievements in this field, and advances and developments continue. Increasingly, we have the ability to manipulate and alter DNA and genetics. As the science progresses, such manipulation is highly likely to be done for our benefit – to reduce the probability of inheriting a genetic disease or to improve qualities such as intelligence or athleticism. Genetics can also be applied to security issues, particularly in terms of identifying criminals or limiting the opportunities for impersonation and identity theft. However, there is a large chance that abuse or misuse of these technologies could irrevocably alter our essence as human beings. Applications of genomics are not just limited to human health. Genetically modified foods are potentially an important component of solving global food shortages and improving the quality, freshness and nutritional value of agricultural crops. Of course, the flip side of this issue is the question of the safety of genetically modified foods, something that needs to be studied in much greater depth at a global level before any conclusions are drawn about the long-term viability of these products. Equality of access to genomics technologies presents another geostrategic challenge. Two further geostrategic challenges in genomics that I touch on are privacy issues in genome mapping and how these might affect the provision of health insurance, and evolutionary metabolism.

The next technology was nanotechnology, which is one of the emerging strategic technologies with the most potential to change our world and the ways in which we interact with it. It is a rapidly growing industry with enormous potential. Working with objects less than one micrometer in size, nanotechnology gives us the chance to alter matter and its qualities on a nano-scale. Ultimately, it is hoped that nanotechnology will enable us to manipulate molecular and atomic structures and to create materials from scratch. Nanotechnology will affect everything from medicine to the military, but most of the biggest dangers surrounding this emerging technology are unknown. When we manipulate materials on a nano-level, we do not know how the materials will respond, as often materials behave differently on the nanoscale than they do in their normal scales. Especially if nano-particles become self-producing, there could be numerous global challenges related to the unintended consequences of technology gone awry. Indeed, nanotechnology and nano-particles may be able to reduce pollution in the air, but another type of nano-particle may have just the opposite effect. From a military perspective, nanotechnology has an array of geostrategic implications, including the ability to facilitate surveillance and possibly the development of smaller scale nuclear weapons. One key theme in nanotechnology is nanoconvergence. Nanotechnology is converging with technologies such as biotechnology, cognitive science and ICT, and it is widely believed that the future of technological innovation will be defined not by one single scientific field or area of study, but by all these distinct areas converging and collaborating together. Indeed, it can be said that convergence is the true future of technological revolution.

Materials science, a field closely related to nanotechnology, has similar implications and applications, including better protection for military forces and improving the efficiency and safety of international space exploration. Materials science technology includes innovations in smart materials, organic electronics, pharmaceutical materials, composites and coatings. A truly interdisciplinary field, materials science is difficult to regulate but, on a positive note, many countries have shown a willingness to cooperate across borders on materials-related research. In addition to military and space applications, materials science is also influencing health care, telemedicine and pharmaceutical and drug delivery.

Finally, I addressed an issue that is increasingly relevant to both geopolitics and our future as human beings: artificial intelligence (AI). AI is especially important with regard to the Singularity – or the point where technological change will become so rapid that it will irrevocably transform human life and the world as we know it. In essence, we are approaching the point where our human biology merges with technology. AI will play a key role in this technological revolution because computers have the potential to think, process and communicate far more quickly and efficiently than we as humans ever could. Once a smarter-than-human machine is created, it will presumably have the capacity to design a machine smarter than itself, and that machine will be able to design an even smarter machine, and so on and so forth. Such intelligence will give us as humans resources and information to solve problems that currently seem unsolvable. Whether it is global warming or global hunger, the potential for strong AI to improve our daily lives is enormous, and AI could presumably solve some of humanity's greatest challenges in a matter of seconds. Smarter-than-human AI could also turn against humans and seek to destroy our race. We would potentially be unable to offset this threat to our human destiny.

Such themes of human destiny are the focus of Part II of this book for as great as the impact of emerging strategic technologies is on geopolitics and global civil society, they also have the potential to touch our lives in extremely personal ways. Collectively, these individual or personal experiences with and decisions vis-à-vis emerging strategic technologies are likely to alter our entire species, quickly transforming what can be construed as questions of individual choice into challenges and opportunities that transcend borders and extend to all corners of the globe. Part II began by defining some of the key concepts in the discussion of emerging strategic technologies, human enhancement, human evolution, human nature, human dignity and our *'sustainable history'*.

I defined what could be dubbed different phases of humanity or the potential stages of human evolution. Transhumans are intermediary forms of human beings that have undergone some enhancements but are not yet transformed entirely beyond the human form. Post-humans are creatures that belong to a race of beings fundamentally different from our own. They are so different from what we consider human beings that they can no longer be considered human, even if their evolutionary roots are in humanity. When described in this context, it is much easier to see the issue of human enhancement as one that potentially affects our entire human destiny.

The next concept to define was human nature, as much of this book's framework for addressing enhancement technologies and human destiny builds on my detailed analyses of human nature, human dignity and *'sustainable history'*. Speaking of human nature, the core concept to remember is that we, as humans, are defined by *'emotional amoral egoism'*. Our nature is rooted in biology and our neurochemically driven need to protect and promote our own self-interest. The central idea of this concept of human nature in the context of emerging strategic technologies and human enhancement is that humans will *inevitably* pursue technologies to improve or enhance their bodies and minds. We will do this even if such technologies are potentially harmful to us collectively or in the long run.

Perhaps the greatest potential harm that human enhancement could cause would be the undermining of our human dignity. In the discussion and debate over human enhancement, we cannot lose sight of the importance of human dignity. In the short term, we may have choices and options regarding whether to enhance, but in the medium and long terms these choices will be blurred and enhancements will take on a more personal nature. Although human dignity can be a difficult concept to define, my definition centres on the minimum criteria that must be met to ensure human dignity. These nine criteria: reason, security, promotion and protection of human rights, accountability, transparency, justice, opportunity, innovation and inclusiveness are important benchmarks for deciding how to regulate emerging strategic technologies and their applications to human enhancement.

The concept of *'sustainable history'* was also central to laying the foundation for a discussion of emerging strategic technologies and human destiny. To me, *'sustainable history'* means an enduring and progressive trajectory for human life coupled with the guarantee of human dignity for all people and at all times. I emphasized that good governance is at the core of *'sustainable history'*, and I argued that technologies have the power to alter, enhance or derail our pursuit of *'sustainable history'* and civilizational triumph.

After defining human nature, human dignity and *'sustainable history'*, I outlined the cases for and against the application of technologies to human enhancement. In this discussion, my fundamental conclusion was again that regardless of the pros and cons, our human nature will take us down the path of enhancement. I analysed the perspectives of all the major groups in the human enhancement debate, ranging from bioconservatives to libertarian transhumanists. I then presented my own assessment of the future of human enhancement, something I have dubbed *'inevitable transhumanism'*. This idea reflects the fact that human nature will inevitably lead us to adopt technologies for enhancement purposes. Thus our focus, as a society, needs to be on carefully considered and strong regulation. Banning these technologies completely will be impossible, so we must direct our energies to developing global legal and ethical frameworks to deal with the most dramatic, significant or morally challenging technologies.

After outlining and defining the key terms and nature of the debate in this increasingly important and urgent arena, Part II examined the emerging strategic technologies first discussed in Part I, which technologies may have specific enhancement applications, and their current and future potential. These technologies encompass everything from physical enhancements, such as prosthetic limbs and cosmetic surgery, to reproductive enhancements, such as in-vitro fertilization. In the future, genetic engineering of embryos is likely to become common practice, and we as a society will have to grapple with the ethical and strategic implications of issues such as germ-line genetic engineering and the question of designer children.

Over the course of the descriptions of human enhancements and potential outcomes, I focused on the fundamental questions posed by these technologies in combination with human nature. I made the bold statement that human enhancement, combined with the core characteristics and traits of human nature, may ultimately lead us, as a human race, to lose control of our destiny. Emerging strategic technologies such as cognitive science, nanotechnology, biotechnology and information science all have the potential to individually or collectively alter the core elements of human nature. From mood altering drugs that undermine our emotions that have evolved in response to specific human challenges over time, to the suppression or encouragement of certain human responses, technologies can take away or alter key components of our 'humanness'. Ultimately, we run the risk of having our emotions become controlled by drugs and treatment. In this process, we will lose the emotionality tied to our survival instinct. The full implications of such a shift cannot be guessed at, but it is likely that even incremental changes on this front will ultimately have very dramatic and significant consequences for the human race.

Lest the reader think that the questions of emerging strategic technologies and human enhancement are limited to questions of our human identity and dignity, I finish the book by tying these issues directly to my *'multi-sum security principle'*. This approach to these issues is new and completely unique, but I believe that it will be highly instrumental to global policymakers. In the globalized world we live in, global security can no longer be seen as simply a collection of different nations acting on behalf of their own national security interests. Rather, I maintain that global security has five dimensions: human, environmental, national, transnational and transcultural security. Global security and the security of any state or culture cannot be achieved without good governance at all levels that guarantees security through *justice* for *all* individuals. I apply the challenges and issues of emerging strategic technologies, human enhancement and human destiny to this framework to provide policymakers with fresh perspectives on revolutionary challenges.

After defining each dimension of the *'multi-sum security principle'* in general terms, I apply them to human enhancement and human security. Possible security risks on this front include the risk that an enhanced human will perceive unenhanced humans as inferior and as a target of potential exploitation. This could be true from a mental or a physical standpoint, and it reminds us that human enhancement could augment existing inequalities among peoples and races – or even create new ones. In addition, potential violations of human dignity are and will continue to be a core concern.

The next component of the *'multi-sum security principle'* discussed was environmental security. Human enhancement affects this substrate in several ways. For example, human enhancement and enhancement technologies mean that people are likely to lead longer and healthier lives. As the world's population grows organically and people start to live longer and longer, the potential strain on our environment's natural resources will be enormous. Alternatively, we as humans could become so integrated with technology that we would be able to upload our thoughts, mental processes, memories and 'essential selves' on to a computer, thereby eliminating the need for food and other resources. In this scenario, human enhancement taken to radical levels would reduce our dependency on natural resources and, in this sense, would be good for environmental security. Another challenge of human enhancement for environmental security relates to the fact that the technologies used in human enhancement may not be biodegradable or may have negative consequences for air and water quality. These are serious geopolitical challenges that policymakers must consider.

The third substrate of my *'multi-sum security principle'* is national security, one of the most traditional dimensions of international security. Countries with access to a wide spectrum of enhancement technologies will have the opportunity to improve their country's overall prestige and possibly to sell such technologies for economic profit and advantage. Similarly, enhanced military and army personnel are likely to have an advantage on the battlefield, whether it is increased endurance or greater aggression, thereby altering the traditional balance of power and methods of conventional warfare. The issue of eugenics also arises in the discussion of human enhancement and national security, especially since eugenics is often associated with state policies. In some of the scarier situations I outlined, it is possible to imagine a world in which a so-called ideal race of peoples is pursued and engineered thanks to strategic technologies. This is something policymakers and citizens need to keep close watch on.

The fourth substrate of my multi-sum security principle examined human enhancement and its relation to transnational security. On the one hand, human enhancement could help nations cooperate, especially if the countries are matched in their technological capabilities and not directly competing on that front. On the other hand, human enhancement technologies could undermine transnational security, especially if certain countries restrict certain technologies while other countries do not. In this case, a black market for enhancement technologies could emerge and, transnationally, there could be a large spike in organized crime, smuggling and even human trafficking.

Finally, Part II addressed human enhancement technologies in the context of *transcultural security*. Once again, eugenics is a key transcultural concern as it threatens to undermine diverse cultures and civilizational forms, which are the foundation of transcultural security. In many instances, different cultures have unique physical features that add to their culture's richness and history. The idea of eugenics may give some policymakers the illusion that they are able to control or eliminate such features to gain stronger control over their populations, and this may lead to abuse or attempts at ethnic cleansing. Similarly, pressure to conform to society's ideals could push people to trend enhancements, eliminating diversity. Meanwhile, parents who choose not to pursue enhancements for their children run the risk of being accused of negligence. The idea that religious factors could eventually play a role in enhancement decisions, for better or worse, is also a challenge from the transcultural perspective.

To wrap up the discussion on human enhancement, we moved beyond the five substrates of global security to look at the destiny of the human race as a whole. Because the path to human enhancement is one that we will inevitably take, it is imperative that we carefully consider the enormous risks that such improvements may pose to our species. Such risks are not intended to be interpreted as arguments against enhancement. Rather, they are cautionary points that should remind us of the importance of clear, well thought out regulation.

So far, we as a global society are falling short on issues of human enhancement regulation. The United Nations could potentially be well-positioned to address these issues, but beyond the Universal Declaration of Human Rights, the UN has done little to tackle issues directly or tangentially related to human enhancement technologies. I also note that the countries that lead in the development of enhancement-related technologies should also play a role in standards-setting and in defining best practices for the use of emerging strategic technologies that alter or improve our bodies and minds. Certain non-governmental agencies and institutions can and should also play a role in this process, and already some universities and panels of experts are making recommendations and generating global discussion over our options and how to manage them.

Human enhancement is a complex, morally contentious issue, and it is no surprise that governments and policymakers – often more concerned with the challenges of the next election than the long term fate of the human race – have been slow to formulate specific guidelines on this front. Beyond the obvious questions of what is permissible and who should decide what is permissible, it is hard to make concrete policies in the face of so much uncertainty and speculation. However, I maintain that the above-mentioned nine core minimum criteria of human dignity are important benchmarks for policymakers looking for guidance on this front. At the national and global levels, these goals can be advanced through the provision of basic needs, inclusiveness and participation, socio-economic justice, gender equality, human rights and the protection of the environment and ecological balance. All elements of society, from governments to civil society groups, and from lawyers to individuals, must collaborate and work together to decide which enhancements promote human dignity and should be made globally acceptable.

This book has presented and analysed some of the most significant emerging strategic technologies of our day. Looking specifically at geopolitics and human enhancement, I have shown the numerous dimensions of our global lives that are being influenced, and often driven, by emerging strategic technologies. These issues and technologies are important because they have such enormous potential to change the course of individual, global and human lives in the near and distant future.

To end this book with a launching point for future discussion and debate over technology and geopolitics, I consider the challenges and opportunities of emerging strategic technologies and provide some general guidelines that national and global policymakers should take into account as they address the regulatory, moral and scientific issues that society faces in response to new and ongoing technological developments.

At the top of global policymakers' agendas should be the issue of global justice, which in this context translates to equality of access for all peoples to emerging strategic technologies that qualify as both legal and ethical. We will not be able to achieve global justice, and technology will fall way short of its potential to ameliorate some of the world's biggest challenges if emerging strategic technologies are only available to wealthy countries or to those individuals with thick wallets. Sharing information, technology, scientific data and resources is the single best way to ensure that the global challenges we face – from global warming to malaria to safe communications – are adequately and comprehensively addressed and, more importantly, solved.

A second consideration for policymakers is the importance of balancing scientific innovation with safety and privacy. As a scientist myself, I strongly favour regulatory frameworks that encourage and promote scientific innovation across all disciplines. That said many emerging strategic technologies have potentially deadly or dangerous dual uses.

In an ideal world, we would have freedom of innovation and the financial and technological resources to fully pursue the multitude of opportunities that science offers us both as individuals and as global citizens. However, we do not live in an ideal world, and this must be taken into consideration when developing regulations. I outlined in Part II of this book my theory of human nature, explaining that we as humans are defined by our *'emotional amoral egoism'*. I stressed that this nature will *inevitably* drive us to use technologies to improve and enhance our minds and bodies. By virtue of being human, we will need on a visceral level to take advantage of opportunities to better ourselves. This is an important fundamental to remember when discussing government regulation of emerging strategic technologies. Because our human nature will drive us to adopt strategic technologies to improve ourselves – potentially to the point where we cease to be really human and where we evolve into trans and post-humans, we need strict government regulation of the development and application of such technologies, which will act as a successful counterbalance to our human nature. As we have seen countless times in the news, people, whether financiers, bankers, doctors or lawyers, are always on the lookout for loopholes in rules that will allow them to benefit or profit. Again, such an inclination is a basic part of human nature and an inherent part of ourselves. Regulation of technology should err on the side of being strict in order to limit any potential loopholes for exploitation. Technology presents a slippery slope that could ultimately culminate in the destruction of the human race as we know it. Strong regulations are an important counterbalance to our natural human tendencies; preservation and the promotion of human dignity should be a central focus in this process.

Another important point to make when speaking about regulations is that the regulation of emerging strategic technologies and human enhancement technologies must be universally applied to all corners of the world. The reason for this is that if one country or region fails to adopt strict regulations on potentially dangerous technologies, any technologies created, developed or applied there could, for all intents and purposes, be applied anywhere. Take, for example, the practical, contemporary example of nuclear weapons and uranium enrichment. Nuclear non-proliferation is a top priority for many countries around the world, including major powers such as the United States and Russia. When countries such as Iran or North Korea engage in uranium enrichment, however, they are creating technology that could easily be sold or transferred across borders. Just because large portions of the global community consider such uranium enrichment and weapons programmes illegal does not mean that one rogue state or terrorist group cannot take advantage of the technologies. Through the black market and the forces of globalization, once a technology exists somewhere it can, in theory, exist anywhere, presenting a possible threat to global peace and security. Only with uniform, consistent regulations and monitoring across borders and regions will we be able to ensure that one region does not gain a competitive advantage over another region in potentially dangerous technologies.

Whatever the regulatory frameworks for emerging strategic technologies in general, and human enhancement technologies in particular, ultimately look like, it is important to underscore in no uncertain terms that such regulations must keep pace with rapidly evolving technologies. This sets the bar extremely high, especially since we are already woefully behind on this front, but it is an important goal that must be achieved.

Technology promises to be one of the major forces in our lives throughout the twenty-first century. This will be true for individuals, societies, governments and the global community as a whole. Because of technology's dual potential to help and to harm, and because of its vast, far-reaching impact, how we approach these issues will shape our geopolitical and human futures. This book provides a framework for tackling these issues and the key concepts for debating the role of technology in our lives.

# Notes

## General Introduction

1. 'Emerging Technologies' (2010), Bitpipe Website, http://www.bitpipe.com/tlist/Emerging-Technologies.html, date accessed 17 August 2010.
2. 'Emerging Technologies' (2010), BusinessDictionary Website, http://www.businessdictionary.com/definition/emerging-technologies.html, date accessed 17 August 2010.
3. J. Erlendsson (2004), 'Emerging Technology', University of Iceland: Scientific and Technical Information Services Website, 20 April, http://www3.hi.is/~joner/eaps/emergtd.htm, date accessed 17 August 2010.
4. J.H. Barton (2006), 'Scientific and Technical Information for Developing Nations' in Secretariat of the International Task Force on Global Public Goods (eds) *Expert Paper Series, Expert Paper Six: Knowledge* (Washington, D.C.: The Secretariat of the International Task Force on Global Public Goods), p. 1.
5. Ibid.
6. International Task Force on Global Public Goods (2006), *Meeting Global Challenges: International Cooperation in the National Interest* (Stockholm: Erlanders Infologistics Väst AB).
7. Barton (2006), 'Scientific and Technical Information for Developing Nations', p. 1.
8. Ibid., p. 2.
9. N.R.F. Al-Rodhan (2009), *Sustainable History and the Dignity of Man: A Philosophy of History and Civilisational Triumph* (Berlin: LIT), p. 125.
10. Ibid., p. 127.
11. Ibid.
12. Ibid., p. 108.
13. Ibid., p. 101.
14. Ibid., p. 131.
15. Ibid., p. 127.
16. Ibid., pp. 126–127.
17. Ibid., p. 131.
18. Ibid.
19. Ibid., p. 437.
20. Ibid., p. 125.
21. Ibid.
22. Ibid., p. 101.
23. S. Fuller (2007), *New Frontiers in Science and Technology Studies* (Cambridge: Polity Press), p. 1.
24. W.K. Bauchspies, J. Croissant and S. Restivo (2006), *Science, Technology, and Society* (Oxford: Blackwell Publishing), p. vii.
25. Ibid., p. viii.

26. Ibid., p. 4.
27. Ibid.
28. Al-Rodhan (2009), *Sustainable History and the Dignity of Man: A Philosophy of History and Civilisational Triumph*, p. 133.
29. Ibid., p. 130.
30. Cf. N. R.F. Al-Rodhan (2007), *The Five Dimensions of Global Security: Proposal for a Multi-sum Security Principle* (Berlin: LIT).

# 1  Introduction

1. 'Fascinating Facts about the Invention of the Wheel' (2005), IdeaFinder Website, March, http://www.ideafinder.com/history/inventions/wheel.htm, date accessed 4 August 2010.
2. 'Starfish Prime' (2010), Absolute Astronomy Website, http://www.absoluteastronomy.com/topics/Starfish_Prime, date accessed 17 August 2010.
3. R.G. Lipsey, K.I. Carlaw and C.T. Bekar (2005), *Economic Transformations: General Purpose Technologies and Long-Term Economic Growth* (Oxford: Oxford University Press), p. 99.
4. Ibid., p. 131.
5. Ibid.
6. Ibid., p. 133.

# 2  Information and Communications Technology (ICT)

1. Cf. B. Gates (2007), 'The Tech Revolution Has Just Begun: The Big Picture', *PC Magazine*, Vol. 27, No. 1/2, http://www.pcmag.com/article2/0,2817,2238181,00.asp, date accessed 2 February 2011.
2. N.L. Rudenstein (2001), *Pointing Our Thoughts: Reflections on Harvard and Higher Education, 1991–2001* (Cambridge, MA: Harvard University), p. 140.
3. E.J. Wilson III (1998), *Globalization, Information Technology, and Conflict in the Second and Third Worlds: A Critical Review of the Literature* (New York: Rockefeller Brothers Fund), p. 6.
4. J.P. Singh (2002), 'Introduction: Information Technologies and the Changing Scope of Global Power and Governance' in J.N. Rosenau and J.P. Singh (eds) *Information Technologies and Global Politics: The Changing Scope of Power and Governance* (Albany, NY: State University of New York Press), p. 2.
5. R. Silberglitt, P.S. Anton, D.R. Howell, A. Wong, N. Gassman, B.A. Jackson, E. Landree, S. L. Pfleeger, E.M. Newton and F. Wu (2006), *The Global Technology Revolution 2020 – In-Depth Analyses: Bio/Nano/Materials/Information Trends, Drivers, Barriers, and Social Implications* (Santa Monica, CA: RAND), p. 14.
6. 'Internet Usage Statistics' (2009), Internet World Stats Website, 31 March, http://www.internetworldstats.com/stats.htm, date accessed 10 August 2010.
7. 'Global ICT Spending Tops $3.5 Trillion' (2008), *JCN Newswires*, 20 May, http://www.japancorp.net/Article.Asp?Art_ID=18281, date accessed 10 August 2010.
8. Ibid.

9. Organisation for Economic Cooperation and Development (OECD) (2008), 'OECD Information Technology Outlook 2008 Highlights', www.oecd.org/sti/ito, date accessed 4 August 2010.

10. R. MacManus (2007), '10 Future Web Trends', Read Write Web Website, 5 September, http://www.readwriteweb.com/archives/10_future_web_trends.php, date accessed 4 August 2010.

11. B.M. Leiner, V.G. Cerf, D.D. Clark, R.E. Kahn, L. Kleinrock, D.C. Lynch, J. Postel, L.G. Roberts, and S. Wolff (2003), 'A Brief History of the Internet', Internet Society Website, 10 December, http://www.isoc.org/internet/history/brief.shtml, date accessed 4 August 2010.

12. Ibid.

13. T. Berners-Lee (2010), 'Frequently Asked Questions', World Wide Web Consortium Website, http://www.w3.org/People/Berners-Lee/FAQ.html#InternetWeb, date accessed 4 August 2010.

14. 'The Difference Between the Internet and the World Wide Web' (2008), Webopedia Website, 29 February, http://www.webopedia.com/DidYouKnow/Internet/2002/Web_vs_Internet.asp, date accessed 4 August 2010.

15. Berners-Lee, 'Frequently Asked Questions'.

16. J.R. Okin (2005), *The Internet: The Not-for-Dummies Guide to the History, Technology, and Use of the Internet* (Winter Harbor, ME: Ironbound Press), p. 24.

17. Ibid., p. 26.

18. A. Paris (2008), 'Cyberethics: The Emerging Codes of Online Conduct', *Policy Innovations, A Publication of the Carnegie Council*, 9 April, http://www.policyinnovations.org/ideas/briefings/data/000046, date accessed 4 August 2010.

19. A.G.K. Solomon (2005), *Technology Futures and Global Power, Wealth, and Conflict* (Washington, D.C.: Center for Strategic and International Studies), p. vi.

20. OECD (2007), 'Social and Economic Factors Shaping the Future of the Internet, Workshop Proceedings, DSTI/ICCP(2007),12/FINAL', 25 July, http://www.oecd.org/dataoecd/43/13/38818332.pdf, date accessed 10 August, p. 5.

21. Okin (2005), *The Internet: The Not-for-Dummies Guide to the History, Technology, and Use of the Internet*, p. 4.

22. Federal Communications Commission (FCC) (2009), 'What Is Broadband?' http://www.fcc.gov/broadband/, date accessed 18 August 2010.

23. Ibid.

24. P. Anderson (2007), 'What is Web 2.0? Ideas, Technologies and Implications for Education', JISC Technology & Standards Watch Website, February, http://www.jisc.ac.uk/whatwedo/services/techwatch/reports/horizonscanning/hs0701.aspx, date accessed 17 August 2010, 5.

25. T. O'Reilly 'Web 2.0: Compact Definition?', O'Reilly Radar Website, http://radar.oreilly.com/archives/2005/10/web-20-compact-definition.html, date accessed 4 August 2010

26. D. Tapscott and A.D. Williams (2008), *Wikinomics: How Mass Collaboration Changes Everything*, expanded edn (London: Atlantic Books), p. 19.

27. B. Evangelista (2010), 'Facebook Directs More Users Online than Google', SF Gate Website, 15 February, http://articles.sfgate.com/2010-02-15/

business/17876925_1_palo-alto-s-facebook-search-engine-gigya, date accessed 10 August 2010.

28. N.R.F. Al-Rodhan (2007), *The Emergence of Blogs as a Fifth Estate and Their Security Implications* (Geneva: Slatkine), p. 29.
29. Ibid., p. 16.
30. Ibid., p. 29.
31. Ibid.
32. N. Aranda (2007), 'A Brief History of Mobile Computing', Ezine Articles Website, 27 March, http://ezinearticles.com/?A-Brief-History-of-Mobile-Computing&id=505215, date accessed 4 August 2010.
33. 'What is Wi-Fi?' (2010), Webopedia Website, http://www.webopedia.com/TERM/W/Wi_Fi.html, date accessed 4 August 2010.
34. 'Internet's Future in 2020 Debated' (2006), *BBC News*, 24 September, http://news.bbc.co.uk/2/hi/technology/5370688.stm, date accessed 4 August 2010.
35. International Telecommunication Union (ITU) (2009), *Confronting the Crisis: ICT Stimulus Plans for Economic Growth*, 2nd edn (Geneva: ITU), p. 62.
36. R. MacManus (2007), '10 Future Web Trends', Read Write Web Website, 5 September, http://www.readwriteweb.com/archives/10_future_web_trends.php, date accessed 4 August 2010.
37. 'Reality, Improved' (2009), *The Economist*, 3 September, http://www.economist.com/node/14299602, date accessed 10 August 2010.
38. Ibid.
39. Ibid.
40. Ibid.
41. Ibid.
42. 'Definitions', Cloud Computing Website, http://searchcloudcomputing.techtarget.com/sDefinition/0,,sid201_gci1287881,00.html, date accessed 10 August 2010.
43. 'Press Release: Gartner Identifies the Top 10 Strategic Technologies for 2009' (2009), Gartner Website, 12–16 October, http://www.gartner.com/it/page.jsp?id=777212, date accessed 10 August 2010.
44. E. Knorr and G. Gruman (2009), 'What Cloud Computing Really Means', Info World Website, 2 May, http://www.infoworld.com/d/cloud-computing/what-cloud-computing-really-means-031, accessed 4 August 2010.
45. D. Talbot (2009), 'TR10: HashCache', *MIT Technology Review*, March/April, http://www.technologyreview.com/printer_friendly_article.aspx?id=22119&channel=specialsections&section=tr10, date accessed 17 August 2010.
46. Ibid.
47. Ibid.
48. Global Strategy Institute, Centre for Strategic & International Studies (CSIS) 'Revolution 3 – Technology', http://gsi.csis.org/index.php?option=com_content&task=view&id=24&Itemid=53, date accessed 4 August 2010.
49. Ibid.
50. 'Moore's Law' (2008), Webopedia Website, 29 February, http://www.webopedia.com/TERM/M/Moores_Law.html, date accessed 4 August 2010.
51. E. Schmidt (2008), 'Inspiring Innovation and Exploration', NASA's 50th Anniversary Lecture Series, 17 January, www.nasa.gov/50th/NASA_lecture_series/schmidt.html, date accessed 4 August 2010.

52. J. Stokes (2008), 'Understanding Moore's Law', *Ars Technica*, 27 September, http://arstechnica.com/hardware/news/2008/09/moore.ars/5, date accessed 4 August 2010.
53. 'How Important Is a Fast CPU?' (2009), Help Desk Geek Website, 1 August, http://helpdeskgeek.com/windows-xp-tips/importance-of-cpu-in-compute/, date accessed 10 August 2010.
54. 'The Top 20 Applications for an Infinitely Fast Computer' (2009), Skytopia Website, 13 July, http://www.skytopia.com/project/cpu/cpu.html, date accessed 10 August 2010.
55. Ibid.
56. 'Memories Are Made of This' (2009), *The Economist*, 3 September, Vol. 392, http://www.economist.com/node/14299550, date accessed 10 August 2010, 14.
57. Ibid.
58. Ibid.
59. W. Drake (2004), 'Reframing Internet Governance Discourse: Fifteen Baseline Propositions. Memo #2 for the Social Science Research Council's Research Network on IT and Governance, Paper Based on Presentation at the Workshop on Internet Governance, ITU, Geneva, 26–27 February and the UN ICT Task Force Global Forum on Internet Governance', New York City, 25–26 March, http://www.un-ngls.org/orf/drake.pdf, date accessed 17 August 2010, p. 5.
60. Europe's Information Society (2009), 'Internet Governance', 18 June, http://ec.europa.eu/information_society/policy/internet_gov/index_en.htm, date accessed 4 August 2010.
61. Drake (2004), 'Reframing Internet Governance Discourse: Fifteen Baseline Propositions. Memo #2 for the Social Science Research Council's Research Network on IT and Governance', p. 5.
62. Ibid.
63. D. McGuire (2003), 'U.N. Summit to Focus on Internet', *Washington Post*, 5 December, http://pqasb.pqarchiver.com/washingtonpost/access/477322551.html?FMT=ABS&FMTS=ABS:FT&date=Dec+5%2C+2003&author=David+McGuire&pub=The+Washington+Post&edition=&startpage=E.05&desc=U.N.+Summit+to+Focus+on+Internet%3B+Officials+to+Discuss+Shifting+of+Control+to+International+Body, date accessed 10 August 2010.
64. 'Can ICANN Cope?' (2001), *Foreign Policy*, No. 126, September/October, 99.
65. M.L. Mueller, J. Mathiason and H. Klein (2007), 'The Internet and Global Governance: Principles and Norms for a New Regime', *Global Governance*, Vol. 13, No. 2, April–June, 238.
66. 'ICANN Be Independent' (2009), *The Economist*, 26 September, Vol. 392, No. 8650, http://www.economist.com/node/14517430, date accessed 10 August 2010.
67. 'ICANN Approves Internet Addresses in Arabic' (2010), Agence France Presse Website, 22 January, http://www.google.com/hostednews/afp/article/ALeqM5hdOGa_b355dSfA2Lhasd0mAYRzuw, date accessed 4 August 2010.
68. W. J. Drake (2001), 'Communications' in P.J. Simmons and C. de Jonge Oudraat (eds) *Managing Global Issues: Lessons Learned* (Washington, D.C.: Carnegie Endowment for International Peace), p. 41.
69. Ibid.

70. Ibid.
71. E.A. Fischer (2005), 'Creating a National Framework for Cybersecurity: An Analysis of Issues and Options' in L.V. Choi (ed.) *Cybersecurity and Homeland Security* (New York: Nova Science Publishers, Inc.), p. 1.
72. J.J. Hamre, P. Ambegaonkar and K.C. Zuback (2006), 'Bringing International Governance to Cyber Space' in J.A. Lewis (ed.) *Cyber Security: Turning National Solutions into International Cooperation* (Washington, D.C.: Center for Strategic and International Studies), p. 116.
73. S.E. Goodman (2006), 'Toward a Treaty-Based International Regime on Cyber Crime and Terrorism' in Lewis (2006), *Cyber Security: Turning National Solutions into International Cooperation*, p. 74.
74. N. Miwa (2006), 'Informal, Non-Treaty-Based Multilateral Coordination' in Lewis (2006), *Cyber Security: Turning National Solutions into International Cooperation*, p. 103.
75. N.R.F. Al-Rodhan (ed.) (2006), 'Editorial of Policy Brief on Information Technology, Terrorism, and Global Security', *Policy Briefs on the Transcultural Aspects of Security and Stability* (Berlin: LIT), p. 181.
76. Ibid.
77. Ibid.
78. S. Ham and R.D. Atkinson (2001), *A Third Way Framework for Global E-Commerce* (Washington, D.C.: Progressive Policy Institute).
79. Ibid.
80. ITU 'About ITU', http://www.itu.int/net/about/index.aspx, date accessed 4 August 2010.
81. Ibid.
82. Ibid.
83. ITU 'The ITU Mission', http://www.itu.int/net/about/mission.aspx, date accessed 4 August 2010.
84. UN Global Alliance for ICT and Development (GAID) 'Mission and Objectives', http://www.un-gaid.org/About/OurMission/tabid/893/language/en-US/Default.aspx, date accessed 4 August 2010.
85. GAID, 'GAID Areas of Focus', http://www.un-gaid.org/Activities/PriorityAreas/tabid/862/language/en-US/Default.aspx, date accessed 4 August 2010.
86. B.S. Buckland, F. Schreider and T.H. Winkler (2010), *Democratic Governance Challenges of Cyber Security*, DCAF Horizon 2015 Working Paper Series 1 (Geneva: DCAF), p. 12.
87. Ibid.
88. F. Assandri and D. Martings (eds) (2009), *From Early Tang Court Debates to China's Peaceful Rise* (Amsterdam: ICAS/Amsterdam University Press), pp. 143–144.
89. S. Lafraniere (2010), 'China Moves to Tighten Data Controls', *The New York Times*, 27 April, http://www.nytimes.com/2010/04/28/world/asia/28china.html?_r=1&scp=1&sq=China%20moves%20to%20tighten%20Data%20Controls%E2%80%99&st=cse, date accessed 10 August 2010.
90. 'Google darf in China weitermachen' (2010), *Zeit Online*, 9 July, http://www.zeit.de/digital/internet/2010-07/google-china-zensur, date accessed 10 August 2010.

91. Center for Strategic and International Studies Commission on Cybersecurity for the 44th Presidency (2008), *Securing Cyberspace for the 44th Presidency* (Washington, D.C.: Center for Strategic and International Studies), pp. 1–3.

92. J. Markoff (2010), 'Step Taken to End Impasse Over Cybersecurity Talks', *The New York Times*, 16 July, http://www.nytimes.com/2010/07/17/world/17cyber.html?ref=john_markoff, date accessed 19 August 2010.

93. Buckland et al. (2010), *Democratic Governance Challenges of Cyber Security*, p. 29.

94. Drake (2001), 'Communications', p. 41.

95. Drake (2004), 'Reframing Internet Governance Discourse: Fifteen Baseline Propositions. Memo #2 for the Social Science Research Council's Research Network on IT and Governance', p. 5.

96. Malkoff (2010), 'Step Taken to End Impasse Over Cybersecurity Talks'.

97. S. Buckley (2000), 'Radio's New Horizons: Democracy and Popular Communication in the Digital Age', *International Journal of Cultural Studies*, Vol. 3, No. 2, 181.

98. Drake (2001), 'Communications', pp. 25–26.

99. J.S. Nye (2010), *Cyber Power* (Cambridge, MA: Harvard Kennedy School, Belfer Center), p. 11.

100. J.A. Lewis (2006), 'Introduction' in Lewis (2006), *Cyber Security: Turning National Solutions into International Cooperation*, p. xiii.

101. Ibid.

102. Goodman (2006), 'Toward a Treaty-Based International Regime on Cyber Crime and Terrorism', p. 74.

103. Drake (2001), 'Communications', p. 47.

104. Schmidt (2008), 'Inspiring Innovation and Exploration'.

105. E.B. Skolnikoff (2001), 'International Governance in a Technological Age' in J. De La Mothe (ed.) *Science, Technology and Governance* (Trombridge, Wilts: Crombrell Press), p. 4.

106. Ibid.

107. Ibid., p. 3.

108. Ibid.

109. European Commission, Information Society and Media (2006), 'EU–US Summit on Cyber Trust: System Dependability and Security, Workshop Report', 15–16 November, p. 4.

110. McAfee and Security and Defence Agenda (SDA) (2010), 'Cyber Security: A Transatlantic Perspective', SDA Evening Debate Report, 22 March, http://www.securitydefenceagenda.org/Portals/7/2010/Publications/Report_Cyber_security_Final.pdf, date accessed 10 August 2010, p. 5.

111. P. Biggs, 'Trends in Social Media & the Social Web', ITU.

112. C. Arthur (2010), 'Facebook Privacy Lets You See Where Strangers Plan to Go', *The Guardian*, 26 April, http://www.guardian.co.uk/technology/2010/apr/26/facebook-privacy-hole, date accessed 10 August 2010.

113. R. Waters (2010), 'Google's Buzz criticized by Privacy Regulators', *The Financial Times*, 20 April, http://www.ft.com/cms/s/0/63d763c4-4ca6-11df-9977-00144feab49a.html, date accessed 10 August 2010.

114. K. Allison (2007), 'Key to It All', *The Financial Times*, 24 September, http://www.ft.com/cms/s/0/82693014-6a32-11dc-a571-0000779fd2ac.html, date accessed 10 August 2010.

115. M.S. Smith (2003), 'Internet Privacy: Overview and Pending Legislation', CRS Report for Congress, 10 July, http://www.firstamendmentcenter.org/pdf/CRS.internet1.pdf, date accessed 4 August 2010, p. 2.

116. S. Hansell (2009), 'Agency Skeptical of Internet Privacy Policies', *The New York Times*, 12 February, http://www.nytimes.com/2009/02/13/technology/internet/13privacy.html?scp=1&sq=Agency%20Skeptical%20of%20Internet%20Privacy%20Policies&st=cse, date accessed 10 August 2010.

117. R.J. Deibert (2002), 'Circuits of Power: Security in the Internet Environment' in J.N. Rosenau and J.P. Singh (eds) *Information Technologies and Global Politics: The Changing Scope of Power and Governance* (Albany, NY: State University of New York Press), p. 126.

.    Ibid., p. 128.

119. Smith (2003), 'Internet Privacy: Overview and Pending Legislation', p. 4.

120. Ibid., p. 2.

121. Ibid.

122. S. Rodotà (2006), 'Europe and Cyber Security' in Lewis (2006), *Cyber Security: Turning National Solutions into International Cooperation*, p. 83.

123. Ibid., p. 83.

124. Allison (2007), 'Key to It All', p. 11.

125. Ibid.

126. Rodotà (2006), 'Europe and Cyber Security', p. 83.

127. J. Brodkin (2008), 'Gartner's Top 10 Strategic Technologies for 2008', Networked World Website, http://www.networkworld.com/news/2007/100907-10-strategic-technologies-gartner.html, date accessed 4 August 2010.

128. R. Stancich (2008), 'Green ICT: Banking on a Software Solution to Climate Change', Climate Change Corp Website, 22 October, http://www.climate-changecorp.com/content.asp?contentid=5727, date accessed 4 August 2010.

129. 'Smart 2020: Enabling the Low Carbon Economy in the Information Age: United States Addendum' (2008), Global E-Sustainability Initiative and the Boston Consulting Group Website, http://www.smart2020.org/_assets/files/Smart2020UnitedStatesReportAddendum.pdf, p. 6.

130. ITU (2009), 'Confronting the Crisis: Its Impact on the ICT Industry', pp. 65–67.

131. D.W. Drezner (2005), 'Weighing the Scales: the Internet's Effect on State-Society Relations', March, http://www.danieldrezner.com/research/scales.pdf, accessed 6 August 2010.

132. 'Moldova's "Twitter Revolutionary" Speaks Out' (2009), *BBC News*, 25 April, http://news.bbc.co.uk/2/hi/europe/8018017.stm, date accessed 4 August 2010.

133. N. Hodge (2009), 'Inside Moldova's Twitter Revolution', *Wired*, 8 April, http://www.wired.com/dangerroom/2009/04/inside-moldovas, date accessed 4 August 2010.

134. United Nations Educational, Scientific, and Cultural Organization (UNESCO) (2005), 'Women, Poverty and ICT: Mediating Social Change', 25 March, http://portal.unesco.org/ci/en/ev.php-URL_ID=18443&URL_DO=DO_TOPIC&URL_SECTION=201.html, date accessed 4 August 2010.

135. Ibid.

136. N.R.F. Al-Rodhan (2006), 'Editorial of Policy Brief on Xenophobia, Media Stereotyping, and Their Role in Global Insecurity' in Al-Rodhan (2006), *Policy Briefs on the Transcultural Aspects of Security and Stability*, p. 38.

137. Ibid.

138. G. Herd and N.R.F. Al-Rodhan (2006), 'Danish Cartoons: A Symptom of Global Insecurity' in Al-Rodhan (2006), *Policy Briefs on the Transcultural Aspects of Security and Stability*, p. 49.

139. Ibid.

140. Ibid., p. 50.

141. L.F. Baron Porras (2003), 'IC(K)Ts, Civil Society and New Social Debates', Center of Research & Popular Education (CINEP), March, p. 2.

142. M. Vatis (2006), 'International Cyber-Security Cooperation: Informal Bilateral Models' in Lewis (2006), *Cyber Security: Turning National Solutions into International Cooperation*, p. 10.

143. D. Ventre (2009), *Information Warfare* (Hoboken, NJ, London: Wiley-ISTE), p. 213.

144. McAfee and SDA (2010), *Cyber Security: A Transatlantic Perspective*, p. 11.

145. Fischer (2005), 'Creating a National Framework for Cybersecurity: An Analysis of Issues and Options', p. 13.

146. World Economic Forum (WEF) (2008), *Global Risks 2008: A Global Risk Network Report, A World Economic Forum Report in Collaboration with Citigroup, Marsh & McLennan Companies (MMC), Swiss Re, Wharton School Risk Center, Zurich Financial Services* (Cologny/Geneva: WEF), p. 22.

147. M. Vatis (2006), 'The Next Battlefield', *Harvard International Review*, Vol. 28, No. 3, 60.

148. Ibid.

149. Nye (2010), *Cyber Power*, p. 1.

150. J.A. Lewis (2007), 'Cyber Attacks Explained', CSIS: Commentary, 15 June, http://csis.org/files/media/csis/pubs/070615_cyber_attacks.pdf , date accessed 10 August, p. 1.

151. B. Griggs (2008), 'US at Risk of Cyberattacks, Experts Say', *CNN*, 18 August, http://edition.cnn.com/2008/TECH/08/18/cyber.warfare/index.html, date accessed 4 August.

152. S. Gorman (2008), 'Cyberattacks on Georgian Web Sites Are Reigning a Washington Debate', *The Wall Street Journal*, 14 August, http://online.wsj.com/article/SB121867946115739465.html?mod=googlenews_wsj, accessed 4 August 2010.

153. J.A. Lewis (2009), 'Crisis in Cyberspace' in N.R.F. Al-Rodhan (ed.) (2009), *Potential Global Strategic Catastrophes: Balancing Transnational Responsibilities and Burden-sharing with Sovereignty and Human Dignity* (Berlin: LIT), p. 189.

154. Gorman (2008), 'Cyberattacks on Georgian Web Sites Are Reigning a Washington Debate'.

155. S. Hoffman (2008), 'Russian Cyber Attacks Shut Down Georgian Websites', ChannelWeb Website, 12 August, http://www.crn.com/security/210003057, date accessed 4 August 2010.

156. Ventre (2009), *Information Warfare*, pp. 209–210.

157. Ibid., p. 212.

158. Cf. McAfee and SDA (2010), 'Cyber Security: A Transatlantic Perspective'.

159. Buckland et al. (2010), *Democratic Governance Challenges of Cyber Security*, p. 10.
160. Gorman (2008), 'Cyberattacks on Georgian Web Sites Are Reigning a Washington Debate'.
161. Ibid.
162. Ibid.
163. 'War in the Fifth Domain: Are the Mouse and the Keyboard the New Weapons of Conflict?' (2010), *The Economist*, 1 July, http://www.economist.com/node/16478792, date accessed 6 August 2010.
164. Fischer (2005), 'Creating a National Framework for Cybersecurity: An Analysis of Issues and Options', p. 2.
165. 'Cyber Crime', Privacy International Website, http://www.privacyinternational.org/issues/cybercrime/index2.html, date accessed 6 August 2010.
166. T. Burghardt (2009), 'The Launching of the US Cyber Command (CYBERCOM) Offensive Operations in Cyberspace', Global Researcher: Centre for Research on Globalisation Website, 1 July, http://www.globalresearch.ca/index.php?context=va&aid=14186, date accessed 6 August 2010.
167. S. Baker (2008), 'Cyber-Security: A Hard Sell', *Business Week Online*, 9 December, http://www.businessweek.com/technology/content/dec2008/tc2008128_182619.htm, date accessed 6 August 2010.
168. R.C. Hodgin (2009), 'FBI Ranks Cyber Attacks Third Most Dangerous behind Nuclear War and WMDs', *TG Daily*, 7 January, http://www.tgdaily.com/security-features/40861-fbi-ranks-cyber-attacks-third-most-dangerous-behind-nuclear-war-and-wmds, date accessed 6 August 2010, p. 1.
169. T. Nakatomi (2001), 'Threats to the Information Society', *The OECD Observer*, January, No. 224, http://www.oecdobserver.org/news/fullstory.php/aid/410/Threats_to_the_information_society.html, date accessed 10 August 2010.
170. Vatis (2006), 'The Next Battlefield', 60.
171. Fischer (2005), 'Creating a National Framework for Cybersecurity: An Analysis of Issues and Options', p. 7.
172. Lewis (2009), 'Crisis in Cyberspace', p. 181.
173. Vatis (2006), 'The Next Battlefield', 60.
174. P. Mukundan (2006), 'Laying the Foundations for a Cyber-Secure World' in Lewis (2006), *Cyber Security: Turning National Solutions into International Cooperation*, p. 31.
175. Ibid. p. 32.
176. Lewis (2006), 'Introduction', p. xiv.
177. Ventre (2009), *Information Warfare*, p. xviii.
178. Ibid., pp. 252–266
179. G. Bruno (2008), 'The Evolution of Cyber Warfare', Council on Foreign Relations Website, 27 February, http://www.cfr.org/publication/15577/, date accessed 6 August.
180. Ibid.
181. Fischer (2005), 'Creating a National Framework for Cybersecurity: An Analysis of Issues and Options', p. 8.
182. Ibid., p. 9.
183. Ibid.

184. J. Markoff (2009), 'Do We Need a New Internet?' *The New York Times*, 14 February, http://www.nytimes.com/2009/02/15/weekinreview/15markoff. html?_r=1&scp=1&sq=internet percent20security&st=cse, date accessed 6 August 2010.

185. Fischer (2005), 'Creating a National Framework for Cybersecurity: An Analysis of Issues and Options', p. 2.

186. Markoff (2009), 'Do We Need a New Internet?'

187. Ibid.

188. Ibid.

189. Lewis (2009), 'Crisis in Cyberspace', p. 180.

190. Fischer (2005), 'Creating a National Framework for Cybersecurity: An Analysis of Issues and Options', p. 2.

191. Lewis (2009), 'Crisis in Cyberspace', p. 181.

192. Ibid. p. 182.

193. ITU (2008), 'The ICT Opportunity Index: The Evolution of the Digital Divide', http://www.itu.int/ITU-D/ict/statistics/ict_oi.html, date accessed 6 August 2010.

194. United Nations Conference on Trade and Development (UNCTAD) (2006), *The Digital Divide Report: ICT Diffusion Index 200* (New York and Geneva: United Nations), p. iii.

195. Ibid.

196. ITU (2009), *Confronting the Crisis: ICT Stimulus Plans for Economic Growth*, p. 55.

197. A. Cane (2006), 'Digital Divide Stops Growing', *The Financial Times*, 4 December, http://www.ft.com/cms/s/0/9341842e-833b-11db-a38a-0000779e2340.html, date accessed 6 August 2010.

198. 'Digital Divide: What It Is and Why It Matters', DigitalDivide Website, http://www.digitaldivide.org/digitaldivide.html, date accessed 6 August 2010.

199. Ibid.

200. Cane (2006), 'Digital Divide Stops Growing', 6.

201. R.D. Atkinson and A.S. McKay (2007), 'Digital Prosperity: Understanding the Economic Benefits of the Information Technology Revolution', The Information Technology & Innovation Foundation Website, March, http://archive.itif.org/index.php?id=34, date accessed 18 August 2010.

202. 'Global Digital Divide "Narrowing"' (2005), *BBC News*, 25 February, http://news.bbc.co.uk/2/hi/technology/4296919.stm, date accessed 6 August 2010.

203. K. Banks (2008), 'Mobile Phones and the Digital Divide', *PC World*, 29 July, http://www.pcworld.com/article/149075/mobile_phones_and_the_digital_divide.html, date accessed 6 August 2010.

204. E. Hansberry (2009), 'Are Mobile Devices Closing the Digital Divide?' *Information Week*, 30 July, http://www.informationweek.com/blog/main/archives/2009/07/are_mobile_devi_1.html;jsessionid=BGBO1EYZ1HTDT QE1GHPCKHWATMY32JVN, date accessed 6 August 2010.

205. Al-Rodhan (2007), *The Emergence of Blogs as a Fifth Estate and Their Security Implications*, p. 57.

206. Ibid.

207. E. MacAskill (2005), 'Tighter Restrictions on Military Blogs Angers U.S. Soldiers', *The Guardian*, 5 May, http://www.guardian.co.uk/technology/2007/may/05/news.usnews, date accessed 6 August 2010.
208. Al-Rodhan (2007), *The Emergence of Blogs as a Fifth Estate and Their Security Implications*, p. 149.
209. Ibid., pp. 150–151.
210. Ibid., p. 154.
211. Ibid., p. 155.
212. Ibid., p. 156.

## 3  Energy and Climate Change

1. Intergovernmental Panel on Climate Change (IPCC) (2007), 'Summary for Policymakers' in M.L. Parry, O.F. Canziani, J.P. Palutikof, P.J. van der Linden and C.E. Hanson (eds) *Climate Change 2007: Impacts, Adaptation and Vulnerability. Contribution of Working Group II to the Fourth Assessment Report of the Intergovernmental Panel on Climate Change* (Cambridge: Cambridge University Press), p. 21.
2. T. Ries (2009), 'Global Warming' in N.R.F. Al-Rodhan (ed.) *Potential Global Strategic Catastrophes: Balancing Transnational Responsibilities and Burden-sharing with Sovereignty and Human Dignity* (Berlin: LIT), pp. 127–128.
3. K.A. Baumert (2005), 'The Challenge Of Climate Protection: Balancing Energy and Environment' in J.H. Kalicki and D.L. Goldwyn (eds) *Energy & Security: Toward a New Foreign Policy Strategy* (Washington, D.C.: Woodrow Wilson Center Press), p. 486.
4. US Department of Commerce, 'Trends in Atmospheric Carbon Dioxide', NOAA Research Website, http://www.esrl.noaa.gov/gmd/ccgg/trends/, date accessed 30 July 2010.
5. N. Stern (2006), *The Stern Review: The Economics of Climate Change* (London: HM Treasury), http://www.hm-treasury.gov.uk/media/3/2/Summary_of_Conclusions.pdf, date accessed 6 August 2010, p. ix.
6. J. Podesta and P. Ogden (2007/08) 'The Security Implications of Climate Change', *The Washington Quarterly*, Vol. 31, No. 1, 115.
7. M. Wahlström (2007), 'Before the Next Disaster Strikes: The Humanitarian Impact of Climate Change', *UN Chronicle Online Edition*, Issue 2, http://www.un.org/wcm/content/site/chronicle/cache/bypass/home/archive/issues2007/pid/4829?ctnscroll_articleContainerList=1_0&ctnlistpaginati on_articleContainerList=true, date accessed 10 August 2010.
8. The CAN Corporation (2007), *National Security and the Threat of Climate Change* (Alexandria, VA: The CAN Corporation), p. 15.
9. N.R.F. Al-Rodhan (ed.) (2007), 'Editorial of Policy Brief on Natural Disasters, Globalization, and the Implications for Global Security', *Policy Briefs on the Transnational Aspects of Security and Stability* (Berlin: LIT), p. 163.
10. Ibid.
11. The CAN Corporation (2007), *National Security and the Threat of Climate Change*, p. 13.
12. World Health Organization (WHO) (2005), 'Climate and Health: Fact Sheet', July, http://www.who.int/globalchange/news/fsclimandhealth/en/index.html, date accessed 6 August 2010.

13. Ibid.
14. Podesta and Ogden (2007/08) 'The Security Implications of Climate Change', pp. 115–116.
15. Ibid., p. 117.
16. S. Dalby (2009), 'The Relevance of Environmental Security', Presentation at the 8th International Security Forum, Geneva, Switzerland, 18 May.
17. A. Gore (2007), 'Nobel Lecture', The Nobel Peace Prize 2007, 10 December, http://nobelprize.org/nobel_prizes/peace/laureates/2007/gore-lecture_en.html, date accessed 6 August 2010.
18. German Advisory Council on Global Change (WBGU) (2007), *Climate Change as a Security Risk* (London and Sterling: Earthscan), p. 1.
19. International Energy Agency (IEA) (2009), *World Energy Outlook 2009 Fact Sheet* (Paris: OECD/IEA), p. 1.
20. Ibid.
21. IEA (2007), *World Energy Outlook 2007, Executive Summary: China and India Insights* (Paris: OECD/IEA), p. 4.
22. IEA (2009), *World Energy Outlook 2009 Fact Sheet*, p. 1.
23. E. Crooks (2008), 'Energy: Center of Power Is on the Move', *The Financial Times*, 23 January, http://www.ft.com/cms/s/0/49d8d9ce-c7c1-11dc-a0b4-0000779fd2ac,dwp_uuid=73b9f204-baa7-11dc-9fbc-0000779fd2ac.html, date accessed 6 August 2010.
24. Ibid.
25. Cf. IEA (2009), *World Energy Outlook 2009 Fact Sheet*.
26. Crooks (2008), 'Energy: Center of Power Is on the Move'.
27. Ibid.
28. World Economic Forum (WEF) in Partnership with Cambridge Energy Research Associates (2006), *The New Energy Security Paradigm* (Geneva: WEF), p. 20.
29. 'The Power and the Glory' (2008), *The Economist*, 19 June, http://www.economist.com/specialreports/displaystory.cfm?story_id=11565685, accessed 6 August 2010.
30. Ibid.
31. P. Taylor (2008), 'Europe Makes Pitch for Green Leadership', *International Herald Tribune*, 28 January, http://www.iht.com/articles/2008/01/28/business/rtrinside29.php, date accessed 6 August 2010.
32. Podesta and Ogden (2007/08) 'The Security Implications of Climate Change', p. 131.
33. Europa Press Releases (2008), 'Memo on the Renewable Energy and Climate Change Package, MEMO/08/33 D', Brussels, 23 January, http://europa.eu/rapid/pressReleasesAction.do?reference=MEMO/08/33, date accessed 6 August 2010.
34. K. Mallon (2006), *Renewable Energy Policy and Politics: A Handbook for Decision-Making* (London: EarthScan Publications Ltd), p. 149.
35. S. Lozanova (2008), 'Renewable Energy: How Storage Can Make It Cheaper and More Reliable', CleanTechnica Website, 21 August, http://cleantechnica.com/2008/08/21/renewable-energy-how-storage-can-make-it-cheaper-more-reliable/, date accessed 6 August 2010.
36. US Department of Energy: Energy Efficiency and Renewable Energy (2010), 'Biomass Program', http://www1.eere.energy.gov/biomass, date accessed 6 August 2010.

37. IEA (2007), *Energy Technologies at the Cutting Age: International Energy Technology Collaboration, IEA Implementing Agreements*, p. 75.
38. 'Quick Guide: Biofuels' (2007), *BBC News*, 24 January, http://news.bbc.co.uk/2/hi/science/nature/6294133.stm, date accessed 6 August 2010.
39. Committee on Science, United States House of Representatives (2005), 'Testimony of Dr. Gal Luft, Executive Director, Institute for the Analysis of Global Security (IAGS)', 9 February, http://www.internationalrelations.house.gov/archives/109/luf072705.pdf, date accessed 20 August 2010, p. 4.
40. National Renewable Energy Laboratory (NREL) 'Biomass Research', http://www.nrel.gov/biomass/, date accessed 18 August 2010.
41. Biotechnology Industry Organization (BIO) (2008), *Guide to Biotechnology 2008* (Washington, D.C.: BIO), http://bio.org/speeches/pubs/er/Biotech-Guide2008.pdf, date accessed 17 August 2010, p. 65.
42. M. Scott (2008), 'Business and the Environment: Innovative Plans May Be Key for Green Future', *The Financial Times*, 17 April, http://www.ft.com/cms/s/0/afa073fa-0b6b-11dd-8ccf-0000779fd2ac,dwp_uuid=cfa40b38-0b6d-11dd-8ccf-0000779fd2ac.html, date accessed 6 August 2010.
43. Ibid.
44. US Department of Energy: Energy Efficiency and Renewable Energy 'Solar Energy Technologies Program', http://www1.eere.energy.gov/solar/, date accessed 9 August 2010.
45. IEA (2007), *Energy Technologies at the Cutting Age: International Energy Technology Collaboration, IEA Implementing Agreements* (Paris: OECD/IEA), p. 82.
46. Ibid.
47. US Department of Energy: Energy Efficiency and Renewable Energy (2008), 'Geothermal Technologies Program', http://www1.eere.energy.gov/geothermal/geothermal_basics.html, date accessed 9 August 2010.
48. IEA (2007), *Energy Technologies at the Cutting Age: International Energy Technology Collaboration, IEA Implementing Agreements*, p. 77.
49. IEA (2008), *World Energy Outlook 2008* (Paris: OECD / IEA), p. 170.
50. 'Geothermal Power', Clean Energy Ideas Website, http://www.clean-energy-ideas.com/geothermal_power.html, date accessed 9 August 2010.
51. Ibid.
52. US Department of Energy: Energy Efficiency and Renewable Energy (2010), 'Wind and Hydropower Technologies Program', http://www1.eere.energy.gov/windandhydro/wind_basics.html, date accessed 9 August 2010.
53. Ibid.
54. M. Scott (2008), 'Business and the Environment: Innovative Plans May Be Key for Green Future'.
55. Ibid.
56. 'Tilting in the Breeze' (2009), *The Economist*, 3 September, http://www.economist.com/sciencetechnology/tq/displaystory.cfm?story_id=E1_TQNJJGQQ, date accessed 10 August 2010.
57. Ibid.
58. Canadian Hydropower Association, 'About Hydro', http://www.canhydropower.org/hydro_e/p_hyd_b.htm, date accessed 9 August 2010.
59. IEA (2008), *World Energy Outlook 2008*, p. 165.

60. International Hydropower Association, International Commission on Large Dams, Implementing Agreement on Hydropower Technologies and Programmes/ International Energy Agency, Canadian Hydropower Association (2000), 'Hydropower and the World's Energy Future: The Role of Hydropower in Bringing Clean, Renewable, Energy to the World', November, http://www.ieahydro.org/reports/Hydrofut.pdf, date accessed 9 August 2010, p. 2.
61. Ibid.
62. R. Lafitte, 'World Hydro Power Potential', http://www.uniseo.org/hydropower.html, date accessed 9 August 2010.
63. Canadian Hydropower Association, 'About Hydro'.
64. M. Anastasio, M. Kluse, S. Aronson, T. Mason, S. Chu, G.H. Miller, J. Grossenbacher, R. Rosner, T. Hunter and S. Bhattacharyya (2008), 'A Sustainable Energy Future: The Essential Role of Nuclear Energy', US Department of Energy: Nuclear Energy, http://www.ne.doe.gov/pdffiles/rpt_sustainableenergyfuture_aug2008.pdf, date accessed 19 August 2010, p. 3.
65. US Environmental Protection Agency, 'Nuclear Energy', http://www.epa.gov/cleanenergy/energy-and-you/affect/nuclear.html, date accessed 9 August 2010.
66. Ibid.
67. C.D. Ferguson (2007), 'Nuclear Energy: Balancing Benefits and Risks', Council on Foreign Relations, Council on Foreign Relations (CSR) No. 28, http://www.cfr.org/content/publications/attachments/NuclearEnergyCSR28.pdf, date accessed 9 August 2010, p. 3.
68. Ibid., p. 5.
69. Ibid., p. 6.
70. Ibid.
71. M. Wald (2009), 'TR10: Traveling-Wave Reactor', *Technology Review*, March/April, http://www.technologyreview.com/energy/22114/, date accessed 9 August 2010.
72. Ibid.
73. A.E. Sieminski (2005), 'World Energy Futures' in J.H. Kalicki and D.L. Goldwyn (eds) *Energy & Security: Toward a New Foreign Policy Strategy* (Washington, D.C.: Woodrow Wilson Center Press), p. 47.
74. IEA (2007), *Energy Technologies at the Cutting Age: International Energy Technology Collaboration, IEA Implementing Agreements*, p. 78.
75. Ibid.
76. Sieminski (2005), 'World Energy Futures', p. 27.
77. IEA (2007), *Energy Technologies at the Cutting Age: International Energy Technology Collaboration, IEA Implementing Agreements*, p. 79.
78. Ibid., p. 46.
79. IEA (2006), *Mobilising Energy Technology: Activities of the IEA Working Parties and Expert Groups* (Paris: OECD/IEA), p. 44.
80. IEA (2007), 'IEA Energy Technology Essentials', p. 1.
81. IEA (2008), *World Energy Outlook*, p. 171.
82. H. Whiteman and I. Hart (2008), 'Carbon Capture and Storage: How Does It Work?' *CNN*, 7 August, http://edition.cnn.com/2008/TECH/science/07/29/carbon.capture/index.html, date accessed 9 August 2010.

83. Cf. H.N. Soud (2000), *Developments in FGD* (London: IEA Clean Coal Centre).
84. The World Bank (2000), 'Wet Flue Gas Desulfurization' (FGD).
85. Cf. H.N. Soud (2000), *Developments in FGD*.
86. The World Bank (2000), 'Wet Flue Gas Desulfurization (FDG)'.
87. Ibid.
88. Ibid.
89. United States Department of Energy (2007),'How Coal Gasification Power Plants Work', 11 September, http://fossil.energy.gov/programs/powersystems/gasification/howgasificationworks.html, date accessed 10 August 2010.
90. Ibid.
91. P. Fairley (2007), 'China's Coal Future I', *Technology Review*, 1 January, http://www.technologyreview.com/energy/18069/, date accessed 5 September 2010.
92. Ibid.
93. Ibid.
94. Cf. Development, Security, and Cooperation Policy and Global Affairs, National Research Council of the National Academies, National Academy of Engineering of the National Academies, Chinese Academy of Engineering, Chinese Academy of Sciences (2004), *Urbanization, Energy, and Air Pollution in China: The Challenges Ahead – Proceedings of a Symposium* (Arlington, VA: The National Academies Press).
95. Fairley (2007), 'China's Coal Future I'.
96. IEA Clean Coal Centre, 'Clean Coal Technologies: Integrated Gasification Combined Cycle', http://www.iea-coal.org.uk/site/ieacoal_old/clean-coal-technologies-pages/clean-coal-technologies-integrated-gasification-combined-cycle-igcc?, date accessed 19 August 2010.
97. Ibid.
98. Whiteman and Hart (2008), 'Carbon Capture and Storage: How Does It Work?'.
99. Ibid.
100. University of California, Berkeley, 'Hybrid Vehicles', http://www.ocf.berkeley.edu/~coreyp/hybridhome3.html, date accessed 10 August 2010.
101. 'Buying a Small Hybrid Car' (2008), *BBC News*, 30 April, http://www.bbc.co.uk/bloom/actions/hybridcar.shtml#quickjump, date accessed 10 August 2010.
102. Ibid.
103. University of California, Berkeley, 'Hybrid Vehicles'.
104. B. Vlasic and N. Bunkley (2010), 'A Future that Doesn't Guzzle', *The New York Times*, 11 January, http://www.nytimes.com/2010/01/12/automobiles/autoshow/12electric.html, date accessed 10 August 2010.
105. K. Nice and C.W. Bryant, 'How Catalytic Converters Work', HowStuffWorks Website, http://auto.howstuffworks.com/catalytic-converter.htm, date accessed 10 August 2010.
106. M. Gerson (2007), 'Hope on Climate Change? Here's Why', *The Washington Post*, 15 August, http://www.washingtonpost.com/wp-dyn/content/article/2007/08/14/AR2007081401327.html, date accessed 20 August 2010.

107. 'Keeping a Grip' (2009), *The Economist*, 3 September, http://www.econo-mist.com/sciencetechnology/tq/displaystory.cfm?story_id=E1_TQNJJGSQ, date accessed 10 August 2010.
108. T. Ries (2007), 'Global Warming' in Al-Rodhan (2007), *Potential Global Strategic Catastrophes: Balancing Transnational Responsibilities and Burden-sharing with Sovereignty and Human Dignity*, p. 133.
109. Ibid., p. 132.
110. B. Ki-Moon (2007), 'Now is the Time: We Must Find a Global Response to the Most Global Problems', *UN Chronicle Online Edition*, Issue 2, http://www.un.org/wcm/content/site/chronicle/cache/bypass/home/archive/issues2007/pid/4818?ctnscroll_articleContainerList=1_0&ctnlistpaginati on_articleContainerList=true, date accessed 10 August 2010.
111. United Nations Environment Programme (UNEP) (1972), 'Report of the United Nations Conference on the Human Environment', http://www.unep.org/Documents.Multilingual/Default.asp?DocumentID=97, date accessed 10 August 2010.
112. UNEP, 'About UNEP', http://www.unep.org/Documents.Multilingual/Default.asp?DocumentID=43&ArticleID=3301&l=en, date accessed 10 August 2010.
113. Intergovernmental Panel on Climate Change (IPCC), 'History', http://www.ipcc.ch/organization/organization_history.htm, date accessed 17 August 2010.
114. Earth Summit (1992), 'United Nations Conference on Environment and Development (UNCED)' http://www.un.org/geninfo/bp/enviro.html, date accessed 10 August 2010.
115. A. Steiner (2007), 'The UN Role in Climate Change Action: Taking the Lead Towards a Global Response', *UN Chronicle Online Edition*, Issue 2, http://www.un.org/wcm/content/site/chronicle/cache/bypass/home/archive/issues2007/pid/4827?ctnscroll_articleContainerList=1_0&ctnlistpaginati on_articleContainerList=true, date accessed 10 August 2010.
116. United Nations Industrial Development Organization (UNIDO), 'Montreal Protocol', http://www.unido.org/index.php?id=7854, date accessed 10 August 2010.
117. United Nations Framework Convention on Climate Change (UNFCC), 'Kyoto Protocol', http://unfccc.int/kyoto_protocol/items/2830.php, date accessed 10 August 2010.
118. M. Hulme (2010), 'Moving Beyond Climate Change', *Environment*, Vol. 52, No. 3, 15–19.
119. Ibid.; F. Schreier (2010), *Trends and Challenges in International Security: An Inventory*, Geneva Centre for Democratic Control of Armed Forces (DCAF) Occasional Paper Number 19 (Geneva: DCAF), p. 43.
120. G. Dvorsky (2010), 'Five Reasons the Copenhagen Climate Conference Failed', Institute for Ethics and Emerging Technologies (IEET) Website, 8 January, http://ieet.org/index.php/IEET/more/3639, date accessed 1 August 2010.
121. Podesta and Ogden (2007/08) 'The Security Implications of Climate Change', p. 115.
122. Stern (2006), *The Stern Review: The Economics of Climate Change*, p. vii.
123. Ibid.

124. 'Lightly Carbonated' (2007), *The Economist*, 4 August, http://www.econo-mist.com/node/9587705, date accessed 10 August 2010.
125. Ibid.
126. N. Stern (2006), *The Stern Review: The Economics of Climate Change* (London: HM Treasury), http://www.hm-treasury.gov.uk/media/3/2/Summary_of_Conclusions.pdf, p. vii.
127. Ibid.
128. M.E. Porter and F.L. Reinhardt (2007), 'Grist: A Strategic Approach to Climate', *Harvard Business Review*, October, http://www.erb.umich.edu/News-and-Events/news-events-pics/HBR-Oct07.pdf, date accessed 5 September 2010.
129. Ibid.
130. Ibid.
131. Ibid.
132. Global Reporting Initiative (GRI), 'About GRI', http://www.globalreport-ing.org/AboutGRI/, date accessed 10 August 2010.
133. C. Bortz (2007), 'Conversation: Alyson Slater, Global Reporting Initiative's Director of Strategy, On How Disclosing Emissions Benefits Companies', *Harvard Business Review*, October, http://www.erb.umich.edu/News-and-Events/news-events-pics/HBR-Oct07.pdf, date accessed 5 September 2010.
134. F. Harvey (2008), 'Environment: Airlines Under Pressure to Turn Skies Green', *The Financial Times*, 14 July, http://www.ft.com/cms/s/0/0677f170-4ede-11dd-ba7c-000077b07658,dwp_uuid=f0cf5eee-4708-11dd-876a-0000779fd2ac.html, date accessed 10 August 2010.
135. K. Done and J. Boxell (2007), 'Sector Soars Above Green Fears', *The Financial Times*, 18 June, http://www.ft.com/cms/s/1/c7f50f64-1b44-11dc-bc55-000b5df10621,dwp_uuid=dacfc068-1769-11dc-86d1-000b5df10621.html, date accessed 10 August 2010.
136. D. Cameron (2007), 'Environmental Issues: Industry to Redress Imbalance', *The Financial Times*, 18 June, http://www.ft.com/cms/s/1/c691bd3e-1b44-11dc-bc55-000b5df10621,dwp_uuid=dacfc068-1769-11dc-86d1-000b5df10621.html, date accessed 10 August 2010.
137. Ibid.
138. Harvey (2008), 'Environment: Airlines Under Pressure to Turn Skies Green'.
139. M. Palmer (2008), 'Information Technology: Cool Is Hot for Data Storage', *The Financial Times*, 17 April, http://www.ft.com/cms/s/0/a9f1b0c2-0b6b-11dd-8ccf-0000779fd2ac,dwp_uuid=cfa40b38-0b6d-11dd-8ccf-0000779fd2ac.html, date accessed 10 August 2010.
140. Ibid.
141. Ibid.
142. Ibid.
143. Ibid.
144. Ibid.
145. IPCC (2007), 'Summary for Policymakers' in Parry et al. (2007), *Climate Change 2007: Impacts, Adaptation and Vulnerability. Contribution of Working Group II to the Fourth Assessment Report of the Intergovernmental Panel on Climate Change*, p. 20.

146. International Task Force on Global Public Goods (2006), *Meeting Global Challenges: International Cooperation in the National Interest* (Stockholm: Erlanders Infologistics Väst AB), p. 42.

147. M.A. Levy (2009), 'Climate Change and the Future of Geopolitics', Presentation at the 8th International Security Forum, Geneva, Switzerland, 18 May.

148. 'London Summit: Leader's Statement' (2009), G20 Communiqué, Climate Café Website, 20 April, http://www.climatecafe.org/g20_communique_020409.pdf, date accessed 18 August 2010.

149. 'Food vs. Fuel' (2007), *Business Week Online*, 5 February, http://www.businessweek.com/magazine/content/07_06/b4020093.htm?chan=search, date accessed 10 August 2010.

150. A. Faiola (2008), 'The New Economics of Hunger', *The Washington Post*, 27 April, http://www.washingtonpost.com/newssearch/search.html?st=%E2%80%98The+New+Economics+of+Hunger&fn=&sfn=&sa=&cp=1&hl=true&sb=-1&sd=undefined&ed=20100810&blt=&bln=&sdt=1987+-+Current&dpp=10&scoa=1987+-+Current&addedNav=, date accessed 10 August 2010.

151. S. Mufson (2008), 'Siphoning Off Corn to Fuel Our Cars', *The Washington Post*, 30 April, http://www.washingtonpost.com/newssearch/search.html?st=Siphoning+Off+Corn+to+Fuel+Our+Cars%2C&fn=&sfn=&sa=&cp=1&hl=true&sb=-1&sd=undefined&ed=20100810&blt=&bln=&sdt=1987+-+Current&dpp=10&scoa=1987+-+Current&addedNav=, date accessed 10 August 2010.

152. A. Elobeid and C. Hart (2007), 'Ethanol Expansion in the Food versus Fuel Debate: How Will Developing Countries Fare?', *Journal of Agricultural & Food Industrial Organization*, Vol. 5, No. 2, 3.

153. A. Chakrabortty (2008), 'Secret Report: Biofuel Caused Food Crisis', *The Guardian*, 3 July, http://www.guardian.co.uk/environment/2008/jul/03/biofuels.renewableenergy, date accessed 10 August 2010.

154. C.W. Calomiris (2007), 'Food for Fuel? Debating the Tradeoffs of Corn-Based Ethanol', *Foreign Affairs*, Vol. 86, No. 5, September/October, 157.

155. 'Food vs. Fuel' (2007), *Business Week Online*.

156. 'Developing Countries Lack Means To Acquire More Efficient Technologies' (2008), *Science Daily*, 24 December, http://www.sciencedaily.com/releases/2008/12/081209125931.htm, date accessed 10 August 2010.

157. 'Developing Country Commitments? Ask George Bush (Senior)', The Center for International Environmental Law (CIEL) Website, http://www.ciel.org/Climate/devcountrycommit.html, date accessed 10 August 2010.

158. United States Energy Information Administration (EIA) (2009), *International Energy Outlook 2009* (Washington, D.C.: EIA), http://www.eia.doe.gov/oiaf/archive/ieo09/index.html, date accessed 10 August 2010.

159. M. ElBaradei (2008), 'Addressing the Global Energy Crisis', International Atomic Energy Agency (IAEA), 6–8 October, http://www.iaea.org/NewsCenter/Transcripts/2008/cfm061008.pdf, date accessed 18 August 2010, p. 1.

160. M.B. Zuckerman (2007), 'The Energy Emergency', *US News & World Report*, Vol. 143, No. 8, 10 September, 72.

161. N. Snow (2008), 'Shell Officials Outline Routes to World's Energy Future', *Oil & Gas Journal*, Vol. 106, No. 13, 7 April, 30.
162. J. Raloff (2008), 'US Must Invest in Technologies to Avoid Energy Crisis: Interview with Steven Chu', *Science News*, Vol. 174, No. 9, 25 October, 32.
163. Ibid.
164. ElBaradei (2008), 'Addressing the Global Energy Crisis', p. 1.
165. Ferguson (2007), 'Nuclear Energy: Balancing Benefits and Risks', p. 3.
166. M.A. Kenderdine and E.J. Moniz (2005), 'Technology Development and Energy Security' in J.H. Kalicki and D.L. Goodwyn (eds) *Energy & Security: Toward a New Foreign Policy Strategy* (Washington, D.C.: Woodrow Wilson Center Press), p. 427.
167. Ferguson (2007), 'Nuclear Energy: Balancing Benefits and Risks', p. 14.
168. Ibid., p. 36.
169. 'World Energy Needs and Nuclear Power' (2010), World Nuclear Association Website, June, http://www.world-nuclear.org/info/inf16.html, date accessed 10 August 2010.
170. Kenderdine and Moniz (2005), 'Technology Development and Energy Security', p. 449.
171. Ibid.
172. Baumert (2005), 'The Challenge Of Climate Protection: Balancing Energy and Environment', p. 485.

## 4  Health Care

1. International Task Force on Global Public Goods (2006), *Meeting Global Challenges: International Cooperation in the National Interest* (Stockholm: Erlanders Infologistics Väst AB), p. 9.
2. N.R.F. Al-Rodhan (ed.) (2007), 'Editorial of Policy Brief on Changing Health Paradigms, Globalization, and Global Security' , *Policy Briefs on the Transnational Aspects of Security and Stability* (Berlin: LIT), p. 179.
3. P.B. Mansourian (2005), 'The Promise of Technology' in S.W.A. Gunn, P.B. Mansourian, A.M. Davies, A. Piel, and B.McA. Sayers (eds) *Understanding the Global Dimensions of Health* (New York: Springer), p. 126.
4. World Health Organization (WHO) 'Health Technologies – the Backbone of Health Services', http://www.who.int/eht/en/Backbone.pdf, date accessed 10 August 2010, p. 1.
5. Mansourian (2005), 'The Promise of Technology', p. 130.
6. Ibid.
7. Ibid.
8. Al-Rodhan (2007), 'Editorial of Policy Brief on Changing Health Paradigms, Globalization, and Global Security', p. 181.
9. Ibid.
10. Organisation for Economic Co-operation and Development (OECD) (2010), 'Growing Health Spending Puts Pressure on Government Budgets, According to OECD Health Data 2010', 29 June, http://www.oecd.org/do cument/11/0,3343,en_2649_34631_45549771_1_1_1_37407,00.html, date accessed 10 August 2010.

11. 'Medical Industry Overview', The Medica Website, http://www.themedica. com/industry-overview.html, date accessed 10 August 2010.
12. Ibid.
13. 'Executive Summary: Health Care'08: Global Trends and Best Practices' (2008), American Association of Retired Persons (AARP) International Website, 1 May, www.aarpinternational.org/resourcelibrary/resourcelibrary_ show.htm?doc_id=714368, date accessed 10 August 2010.
14. Ibid.
15. A.D. Barker (2005), 'The Need for Harmonization in Biobanking to Realize the Potential of 21st Century Medicine', IBM World Wide Biobank Summit IV, National Cancer Institute, National Institutes of Health, 8 November.
16. Ibid.
17. Ibid.
18. WHO (2007), *World Health Statistics 2007* (Geneva: WHO), http://www.who. int/whosis/whostat2007.pdf, accessed 10 August 2010, p. 12.
19. Mansourian (2005), 'The Promise of Technology', p. 128.
20. Biotechnology Industry Organization (BIO) (2008), *Guide to Biotechnology 2008* (Washington, D.C.: BIO), http://bio.org/speeches/pubs/er/Biotech-Guide2008.pdf, date accessed 17 August 2010, p. 31.
21. Mansourian (2005), 'The Promise of Technology', p. 128.
22. Ibid.
23. T. Pang (2002), 'The Impact of Genomics on Global Health', *American Journal of Public Health*, Vol. 92, No. 7, July, http://www.pubmedcentral.nih. gov/articlerender.fcgi?artid=1447192, date accessed 10 August 2010.
24. Ibid.
25. The Genomics Working Group of the Science and Technology Task Force of the United Nations Millennium Project (2004), 'Genomics and Global Health', http://belfercenter.ksg.harvard.edu/files/genomics.pdf, date accessed 10 August 2010, p. 1.
26. Centre for Ecology & Hydrology (CEH) (2009), 'Bio-Linux Goes Global', 12 January, http://www.ceh.ac.uk/news/news_archive/2009_news_item_01. html, date accessed 10 August 2010.
27. Medicare Payment Advisory Commission (MedPAC) (2004), 'Report to the Congress: New Approaches in Medicare', June, http://www.medpac.gov/ publications, date accessed 10 August 2010, p. 6.
28. Sir C. Davis (2008), 'Data and Technology Could Save Lives', *The Financial Times*, 16 May, http://www.ft.com/cms/s/0/e7526326-16b0-11dd-bbfc-0000779fd2ac.html, date accessed 10 August 2010.
29. WHO, 'Information Technology in Support of Health Care', http://www. who.int/eht/en/InformationTech.pdf, date accessed 10 August, p. 1.
30. L.M. Etheredge (2007), 'A Rapid-Learning Health System', *Health Affairs*, Vol. 26, No 2, 108.
31. National Cancer Institute (2010), 'About caBIG', https://cabig.nci.nih.gov/ overview/, date accessed 10 August 2010.
32. Ibid.
33. D.L. Heymann (2005), 'Dealing with Global Infectious Disease Emergencies' in Gunn et al. (2005), *Understanding the Global Dimensions of Health*, p. 173.
34. Ibid.

35. 'Global Health Care Information Technology (2009–2014)', Markets and Markets Website, September, http://www.marketsandmarkets.com/Market-Reports/healthcare-information-technology-market%20-136.html, date accessed 10 August 2010.
36. M. Pearson (2010), 'E-ffective healthcare', *OECD Observer*, http://www.oecdobserver.org/news/fullstory.php/aid/3231/E-ffective_healthcare.html, date accessed 10 August 2010.
37. American Association for the Advancement of Science (AAAS) (2005), *Vision2033: Linking Science and Policy for Tomorrow's World* (Washington, D.C.: AAAS), p. 121.
38. Ibid.
39. Ibid.
40. 'Nanotechnology May Hold Promise for Health Care Design Materials' (2005), *Health Facilities Management*, Vol. 18, No. 9, 6.
41. F. Harvey (2004), 'Can We Overcome Nano-fear?' *The Financial Times*, 15 January, 12.
42. D.C. Rickerby (2006), 'Societal and Policy Aspects of the Introduction of Nanotechnology in Healthcare', *International Journal of Healthcare Technology & Management*, Vol. 7, No. 6, 463.
43. WHO (2009), 'About WHO', http://www.who.int/about/en, date accessed 10 August 2010.
44. WHO (2009), 'The WHO Agenda', http://www.who.int/about/agenda/en/index.html, date accessed 10 August 2010.
45. WHO (2009), 'Online Q&A', http://www.who.int/features/qa/39/en/index.html, date accessed 10 August 2010.
46. Ibid.
47. WHO (2008), *The World Health Report 2008, Primary Health Care* (Geneva: WHO).
48. P. Clevestig (2009), 'Pandemics and Bio-catastrophes' in N.R.F. Al-Rodhan (ed.) *Potential Global Strategic Catastrophes: Balancing Transnational Responsibilities and Burden-sharing with Sovereignty and Human Dignity* (Berlin: LIT), p. 85.
49. Ibid.
50. J.B. Tucker (2001), 'Contagious Fears: Infectious Disease and National Security', *Harvard International Review*, Vol. 23, No. 2, 82.
51. Ibid.
52. Ibid.
53. 'Sub Saharan Africa: HIV & AIDS Statistics' (2009), Avert Website, 7 July, http://www.avert.org/subaadults.htm, date accessed 10 August 2010.
54. M. Manciaux and T.M. Flender (2005), 'World Health, A Mobilizing Utopia?' in Gunn et al. (2005), *Understanding the Global Dimensions of Health*, p. 77.
55. AAAS (2005), *Vision2033: Linking Science and Policy for Tomorrow's World*, p. 116.
56. Ibid, p. 117.
57. WHO 'Prevention of Health Care-Associated HIV Infection', http://www.who.int/eht/en/HealthCareHIV.pdf, date accessed 11 August 2010, p. 2.
58. Ibid.
59. D.P. Fidler (2003), 'Public Health and National Security in the Global Age: Infectious Diseases, Bioterrorism, and Realpolitik', *George Washington International Law Review*, Vol. 35, No. 4, 787.

60. S. Gay Stolberg (2009), 'Swine Flu Diary: Caught in a Beijing Dragnet', *The New York Times*, 27 July, http://www.nytimes.com/2009/07/28/health/28flu. html?_r=1, date accessed 11 August 2010.
61. D. Brown (2009), 'System Set Up After SARS Epidemic Was Slow to Alert Global Authorities', *The Washington Post*, 30 April, http://www.washingtonpost.com/wp-dyn/content/article/2009/04/29/AR2009042904911.html, date accessed 11 August 2010.
62. A. Buelva (2006), 'Technology Plays Key Role in Fight vs H1N1', *The Philippine Star*, 6 July, http://www.philstar.com/Article.aspx?articleid=484115, date accessed 10 July 2010.
63. S.S. Morse (2007), 'Global Infectious Disease Surveillance and Health Intelligence', *Health Affairs*, Vol. 26, No. 4, July/August, 1069.
64. 'Genomics and Biology Join Forces to Fight HIV' (2008), Swissinfo Website, 8 May, http://www.swissinfo.ch/eng/front/Genomics_and_biology_join_ forces_to_fight_HIV.html?siteSect=105&sid=9062679&rss=true&ty=st, date accessed 19 August 2010.
65. Ibid.
66. 'The Global Fund to Fight AIDS, Tuberculosis, and Malaria' (2006), Avert Website, 16 July, http://www.avert.org/global-fund.htm, date accessed 11 August 2010.
67. The Global Fund to fight AIDS, Tuberculosis and Malaria 'Who We Are What Do We Do?', http://www.theglobalfund.org/documents/publications/ brochures/whoweare/gf_brochure_07_full_high_en.pdf, date accessed 11 August 2010, p. 6.
68. 'The Global Fund to Fight AIDS, Tuberculosis, and Malaria' (2009), Avert Website.
69. The Global Fund to fight AIDS, Tuberculosis and Malaria 'Who We Are What We Do?', p. 6.
70. Ibid.
71. 'The Global Fund to Fight AIDS, Tuberculosis, and Malaria', Avert Website.
72. Ibid.
73. Global Alliance for Vaccines and Immunisation (GAVI Alliance) 'Governance', http://www.gavialliance.org/about/governance/index.php, date accessed 11 August 2010.
74. J. Martens (2007), 'Multistakeholder Partnerships – Future Models of Multilateralism?' Dialogue on Globalization, Occasional Papers, Friedrich Ebert Stiftung Berlin, No. 29, http://library.fes.de/pdf-files/iez/04244.pdf, date accessed 10 August 2010, p. 26.
75. WHO (2009), 'Health Systems Development: GAVI Alliance – Health Systems Strengthening', http://www.searo.who.int/en/Section1243/Section2448. htm, date accessed 11 August 2010.
76. WHO (2010), 'EHT Advocacy folder', http://www.who.int/eht/advocacy_ folder/en/index.html, date accessed 10 August 2010.
77. WHO (2010), 'Essential Health Technologies', http://www.who.int/eht/eht_ intro/en/index.html, date accessed 10 August 2010.
78. 'Genomics and Biology Join Forces to Fight HIV' (2008), Swissinfo Website, 8 May, http://www.swissinfo.ch/eng/front/Genomics_and_biology_join_ forces_to_fight_HIV.html?siteSect=105&sid=9062679&rss=true&ty=st, date accessed 11 August 2010.

79. Cf. World Economic Forum (WEF) (2002), *Global Health Initiative Resource Paper* (Geneva: WEF).
80. WEF 'The Global Health Initiative: A Catalyst for Partnership in Global Public Health', http://www.weforum.org/pdf/Initiatives/GHI_Brochure.pdf, date accessed 11 August 2010.
81. Helsinki Process on Globalization and Democracy (2005), *Empowering People at Risk: Human Security Priorities for the 21st Century* (Helsinki: Finish Ministry for Foreign Affairs), p. 15.
82. United Nations Children's Fund (UNICEF) (2009), *The State of the World's Children 2009: Maternal and Newborn Health*, http://www.unicef.org/sowc09/docs/SOWC09-FullReport-EN.pdf, date accessed 11 August 2010, p. 2.
83. WHO, *'Investing in Health: A Summary of Findings of the Commission on Macroeconomics and Health* (Geneva: WHO) http://www.who.int/macrohealth/infocentre/advocacy/en/investinginhealth02052003.pdf, date accessed 10 August 2010, p. 8.
84. Ibid.
85. Ibid., p. 10.
86. Ibid.
87. International Task Force on Global Public Goods (2006), *Meeting Global Challenges: International Cooperation in the National Interest*, p. 67.
88. J.H. Barton (2006), 'Scientific and Technical Information for Developing Nations' in Secretariat of the International Task Force on Global Public Goods (eds) *Expert Paper Series, Expert Paper Six: Knowledge* (Washington, D.C.: The Secretariat of the International Task Force on Global Public Goods), p. 26.
89. International Task Force on Global Public Goods (2006), *Meeting Global Challenges: International Cooperation in the National Interest*, p. 67.
90. Ibid.
91. D. Gardner (2009), ' "There's No Reason Only Poor People Should Get Malaria": The Moment Bill Gates Released Jar of Mosquitoes at Packed Conference', *Daily Mail*, 6 February, http://www.dailymail.co.uk/news/worldnews/article-1136463/Theres-reason-poor-people-malaria-The-moment-Bill-Gates-released-jar-mosquitoes-packed-conference.html, date accessed 11 August 2010.
92. D. Sridhar (2006), 'Inequality in the United States Healthcare System', Human Development Report Office Occasional Paper (New York: UNDP), http://hdr.undp.org/en/reports/global/hdr2005/papers/hdr2005_sridhar_devi_36.pdf , date accessed 18 August 2010, pp. 1–2.
93. The White House, President Obama, 'Health Care', http://www.whitehouse.gov/Issues/health-Care, date accessed 29 July 2010.
94. J.H. Barton (2006), 'Knowledge' in Secretariat of the International Task Force on Global Public Goods (eds) *Expert Paper Series, Expert Paper Six: Knowledge* (Washington, D.C.: The Secretariat of the International Task Force on Global Public Goods), p. 11.
95. Barton (2006), 'Scientific and Technical Information for Developing Nations', p. 26.
96. Barton (2006), 'Knowledge', p. 11.
97. International Task Force on Global Public Goods (2006), *Meeting Global Challenges: International Cooperation in the National Interest*, p. 68.

98. Ibid.
99. Ibid., p. 69.
100. 'Goals', The Grand Challenges in Global Health Website, http://www. grandchallenges.org/Pages/BrowseByGoal.aspx, date accessed 29 August 2010.
101. Ibid.

## 5 Biotechnology

1. European Commission, 'Competitiveness in Biotechnology: Introduction', http://ec.europa.eu/enterprise/phabiocom/comp_biotech_intro.htm, date accessed 17 August 2010.
2. D. Steele (2004), 'Danger Lurks in a Biotech World', *The Aquarian*, Spring, http://www.aquarianonline.com/Sci-Tech/Steele_Biotech.html, date accessed 11 August 2010.
3. G.A. Koehler (1996), 'Bioindustry: A Description of California's Bioindustry and Summary of the Public Issues Affecting Its Development, An Overview of Biotechnology and Bioindustry', http://www.library.ca.gov/crb/96/07/ BIOT_CH1.html, date accessed 11 August 2010.
4. Biotechnology Industry Organization (BIO) (2008), *Guide to Biotechnology 2008* (Washington, D.C.: BIO), http://bio.org/speeches/pubs/er/ BiotechGuide2008.pdf, date accessed 17 August 2010, p. 1.
5. European Commission (2002), *Life Sciences and Biotechnology – A Strategy for Europe* (Luxembourg: Office for Official Publications of the European Communities), http://ec.europa.eu/biotechnology/pdf/com2002-27_ en.pdf, date accessed 11 August 2010, p. 10.
6. 'Briefing: Genetically Modified Crops and Food' (2003), Friends of the Earth (FOE) Website, http://www.foe.co.uk/resource/briefings/gm_crops_ food.pdf, date accessed 11 August 2010, p. 3.
7. Ibid.
8. Confederation of Indian Industry and the Department of Technology of the Ministry of Science and Technology of the Government of India, 'Facts about Biotechnology', http://www.docstoc.com/docs/2308176/Facts- about-Biotechnology, date accessed 11 August 2010.
9. 'Food vs. Fuel' (2007), *Business Week Online*, 5 February, http://www.busi- nessweek.com/magazine/content/07_06/b4020093.htm?chan=search, date accessed 16 August 2010.
10. EuropaBio (2008), 'How Industrial Biotechnology Can Tackle Climate Change', p. 6.
11. BIO (2008), *Guide to Biotechnology 2008*, p. 69.
12. Katholieke Universiteit Leuven 'BioSCENTer – Context and Motivation', http://homes.esat.kuleuven.be/~bioiuser/bioscenter/contextandmotiva- tion.php#trends, date accessed 11 August 2010.
13. T. Taylor (2010), 'Biotechnology Industry', Presentation at the Strategic Technologies and Our Global Future Course, Geneva Centre for Security Policy, Geneva, Switzerland, 15 April.
14. 'History of Biotechnology' (2010), Dupont Website, http://www2.dupont. com/Biotechnology/en_US/intro/history.html, date accessed 11 August 2010.

15. BIO (2008), *Guide to Biotechnology 2008*, p. 2.
16. Massachusetts Institute of Technology (MIT) (1997), 'Inventor of the Week', http://web.mit.edu/invent/iow/boyercohen.html, date accessed 11 August 2010.
17. BIO (2008), *Guide to Biotechnology 2008*, p. 2.
18. Ibid.
19. National Research Council Committee on Advances in Technology and the Prevention of Their Application to Next Generation Biowarfare Threats (2006), *Globalization, Biosecurity, and the Future of the Life Sciences* (Washington, D.C.: The National Academies Press), p. 86.
20. S. Singh (2008), 'Biotech Growth Slows for the First Time in 5 Years', LiveMint.com, *The Wall Street Journal*, 15 July, http://www.livemint.com/2008/07/15215237/Biotech-growth-slows-for-first.html, date accessed 11 August 2010.
21. Ibid.
22. 'European Biotech Industry Shows Double Digit Revenue Growth' (2007), Ernst & Young Website, 16 April. http://www.prnewswire.com/news-releases/global-biotechnology-makes-historic-advances-58331937.html, date accessed 11 August 2011.
23. Ibid.
24. A. Pollack and D. Wilson (2010), 'Safety Rules Can't Keep Up with Biotech Industry', *The New York Times*, 27 May, http://www.nytimes.com/2010/05/28/business/28hazard.html?ref=biotechnology, date accessed 25 July 2010.
25. BIO (2007), *Milestones 2006–2007* (Washington, D.C.: BIO), http://www.bio.org/speeches/pubs/milestone07/Milestones2007.pdf, date accessed 26 August 2010, p. 23.
26. BIO (2010), 'Battelle / Bio State Bioscience Initiatives 2010', http://www.bio.org/local/battelle2010/m, date accessed 11 August 2010.
27. North Carolina State University 'What Is Bioprocess Engineering?' http://www.bae.ncsu.edu/undergrad/biopro_eng_info.htm, date accessed 11 August 2010.
28. BIO (2008), *Guide to Biotechnology 2008*, p. 18.
29. Ibid.
30. 'Recombinant DNA Technology', Encyclopedia Britannica Website, http://www.britannica.com/EBchecked/topic/493667/recombinant-DNA-technology, date accessed 11 August 2010.
31. BIO (2008), *Guide to Biotechnology 2008*, p. 18.
32. Ibid., p. 19.
33. Ibid.
34. Ibid., p. 20.
35. Ibid.
36. Ibid., p. 21.
37. Ibid.
38. Committee on Developing Biomarker-Based Tools for Cancer Screening, Diagnosis, and Treatment (2007), *Cancer Biomarkers: The Promises and Challenges of Improving Detection and Treatment* (Washington, D.C.: National Academies Press), p. 1.
39. 'Definition of Biomarker', Medicine.net Website, http://www.medterms.com/script/main/art.asp?articlekey=6685, date accessed 11 August 2010.

40. Committee on Developing Biomarker-Based Tools for Cancer Screening, Diagnosis, and Treatment (2007), *Cancer Biomarkers: The Promises and Challenges of Improving Detection and Treatment*, p. 1.
41. 'Biomarkers', Environmental Health Research Network Website, http://www.rrse.ca/en/research/biomarkers-intro.htm, date accessed 11 August 2010.
42. BIO (2008), *Guide to Biotechnology 2008*, p. 22.
43. D.F. Betsch (1994), 'DNA Fingerprinting in Human Health and Society', Access Excellence Resource Center Website, http://www.accessexcellence.org/RC/AB/BA/DNA_Fingerprinting_Basics.php, date accessed 11 August 2010.
44. Ibid.
45. K. Leutwyler (2001), 'New Micromachined Cantilever Quickly Detects Disease Markers', *Scientific American*, 4 September, http://www.scientificamerican.com/article.cfm?id=new-micromachined-cantile, date accessed 11 August 2010.
46. Ibid.
47. International Society for Stem Cell Research (ICCSR) (2008), *Stem-cell Facts: The Next Frontier?* (Deerfield, IL: ICCSR), http://www.isscr.org/public/ISSCR08_PubEdBroch.pdf, date accessed 11 August 2010, p. 2.
48. Ibid.
49. Ibid.
50. Ibid., p. 3.
51. Stanford University, 'Overview: Biomemisis', http://www-cdr.stanford.edu/biomimetics/, date accessed 11 August 2010.
52. Ibid.
53. 'Electrical Potential' (2009), *The Economist*, 10 December, http://www.economist.com/node/15048719, date accessed 11 August 2010.
54. Ibid.
55. The University of Melbourne, 'About Biomedicine', http://www.bbiomed.unimelb.edu.au/bachelor_of_biomedicine/about_biomedicine, date accessed 17 August 2010.
56. Ibid.
57. Ibid.
58. European Commission (2005), *Synthetic Biology: Applying Engineering to Biology, Report of a NEST High-Level Expert Group* (Brussels: European Commission), p. 5.
59. N. Wade (2010), 'Researchers Say They Created a 'Synthetic Cell', *The New York Times*, 20 May, http://www.nytimes.com/2010/05/21/science/21cell.html, date accessed 11 August 2010.
60. S. Benner (2010), 'Q&A: Life, Synthetic Biology and Risk', *BMC Biology*, Vol. 8, 78.
61. The White House (2010), 'Letter from President Obama', Bioethic.gov Website, 20 May, http://www.bioethics.gov/documents/Letter-from-President-Obama-05.20.10.pdf, date accessed 11 August 2010.
62. A.H. Cordesman (2005), *The Challenge of Biological Terrorism* (Washington, D.C.: CSIS Press), p. 91.
63. 'Follow-up of Deaths Among US Postal Service Workers Potentially Exposed to Bacillus Anthracis'(2003), Center for Disease Control (CDC) Website, 3

October, http://www.cdc.gov/mmwr/preview/mmwrhtml/mm5239a2.htm, date accessed 11 August 2010.

64. Cordesman (2005), *The Challenge of Biological Terrorism*, pp. 12–17.
65. The United Nations Office at Geneva (UNOG) 'Disarmament: The Biological Weapons Convention', http://www.unog.ch/bwc, date accessed 11 August 2010.
66. 'Syria Profile: Biological Overview', The Nuclear Threat Initiative (NTI) Website, http://www.nti.org/e_research/profiles/index_4470.html, date accessed 18 August 2010.
67. Cf. J. Parachini (2002), 'Control Biological Weapons, but Defend Biotechnology', *RAND Review*, Vol. 26, No. 2, 34–35.
68. M.D. Kellerhals (2009), 'United States Introduces New Biological Weapons Security System', America.gov Website, 9 December, http://www.america.gov/st/peacesec-english/2009/December/20091209130215dmslahrellek0.1 077997.html, date accessed 11 August 2010.
69. Cordesman (2005), *The Challenge of Biological Terrorism*, p. 93.
70. Ibid., p. 185.
71. United Nations Security Council (2004), 'Resolution 1540, S/RES/1540', April 28, http://www.treas.gov/offices/enforcement/ofac/legal/unscrs/1540. pdf, date accessed 11 August 2010.
72. Committee on Research Standards and Practices to Prevent the Destructive Application of Biotechnology (2004), *Biotechnology Research in an Age of Terrorism* (Washington, DC: National Academies Press), p. 43.
73. Ibid., p. 44.
74. European Union (EU) (2008), 'European Union Legislation and Recommendations Related to Biosafety and Biosecurity, Working Paper submitted by Germany on behalf of the EU 2008 Meeting of Experts', http://www.unog.ch/80256EDD006B8954/(httpAssets)/C4BD641FF55084 A1C12574A2004242B0/$file/Germany+EU+legislation+and+recommend ations+to+implement+and+improve+biosafety+and+biosecurity+WP.pdf, date accessed 11 August 2010.
75. Cordesman (2005), *The Challenge of Biological Terrorism*, p. 95.
76. M. Stanley (2002), 'Scientific Publishing: Knowledge is Power', 30 September, http://www.econ.ucsb.edu/~tedb/Journals/morganstanley.pdf, date accessed 11 August 2010, p. 2.
77. 'Scientific Publishing: Access All Areas' (2004), *The Economist*, 5 August, http://www.economist.com/science/displaystory.cfm?story_id=3061258, date accessed 11 August 2010.
78. Cordesman (2005), *The Challenge of Biological Terrorism*, p. 96.
79. Pollack and Wilson (2010), 'Safety Rules Can't Keep Up with Biotech Industry'.
80. T. Taylor (2010), 'Healthcare Industry', Presentation at the Strategic Technologies and Our Global Future Course, Geneva Centre for Security Policy, Switzerland, 15 April.
81. Committee on Research Standards and Practices to Prevent the Destructive Application of Biotechnology (2004), *Biotechnology Research in an Age of Terrorism*, p. 16.
82. Ibid.
83. Cf. J.D. Steinbruner, E.D. Harris, N. Galladher and S.M. Okutani (2005), *Controlling Dangerous Pathogens* (College Park, MD: Center for International

and Security Studies at Maryland) http://www.cissm.umd.edu/papers/files/ pathogens_project_monograph.pdf, date accessed 11 August 2010.

84. Committee on Advances in Technology and the Prevention of Their Application to Next Generation Biowarfare Threats, Development, Security, and Cooperation Policy and Global Affairs Division, Board on Global Health, Institute of Medicine and National Research Council of the National Academies (2006), *Globalization, Biosecurity, and the Future of the Life Sciences* (Washington, D.C.: The National Academies Press), pp. 246–250.

85. N.R.F. Al-Rodhan, L. Nazaruk, M. Finaud and J. Mackby (2008), *Global Biosecurity: Towards a New Governance Paradigm* (Geneva: Éditions Slatkine), p. 225.

86. A.H. Zakiri, S. Johnston and B. Tobin (2005), 'The Biodiplomacy Initiative: Informing Equitable and Ethical Decision-Making for Present and Future Generations', *Work in Progress*, Vol. 17, No. 2, Summer, http://www.unu.edu/ hq/ginfo/wip/wip17-2-summer2005.pdf, date accessed 11 August 2010, 3.

87. 'Stem Cell Research at the Crossroads of Religion and Politics' (2008), The Pew Forum on Religion & Public Life Website, 17 July, http://pewforum.org/ docs/?DocID=316, date accessed 11 August 2010.

88. T.F. Budinger and M.D. Budinger (2006), *Ethics of Emerging Strategic Technologies: Scientific Facts and Moral Challenges* (Hoboken, NJ: John Wiley & Sons), p. 342.

89. B. Agnew (2001), 'The Politics of Stem Cells: Legislative Uncertainties Hinder Research in the US', Genome News Network Website, 21 February, http:// www.genomenewsnetwork.org/articles/02_03/stem.shtml, date accessed 11 August 2010.

90. R.S. Schwartz (2006), 'The Politics and Promise of Stem-Cell Research', *The New England Journal of Medicine*, Vol. 355, No. 12, 21 September, http:// content.nejm.org/cgi/content/full/355/12/1189, date accessed 20 August 2010.

91. 'Obama Ends Stem Cell Funding Ban' (2009), *BBC News*, 9 March, http:// news.bbc.co.uk/2/hi/americas/7929690.stm, date accessed 10 August 2010.

92. G. Harris (2010), 'U.S. Judge Rules against Obama's Stem Cel Policy', *The New York Times*, 23 August, http://www.nytimes.com/2010/08/24/health/ policy/24stem.html?ref=stemcells, date accessed 27 August 2010.

93. S.G. Stolberg and G. Harris (2010), 'Stem Cell Ruling Will Be Appealed', *The New York Times*, 24 August, http://www.nytimes.com/2010/08/25/health/ policy/25stem.html?ref=stemcells, date accessed 27 August 2010.

94. 'Stem Cell Research Around the World' (2008), The Pew Forum on Religion and Public Life Website, 17 July, http://pewforum.org/docs/?DocID=318, date accessed 18 August 2010.

95. Ibid.

96. A. Bonnicksen (2007), 'Therapeutic Cloning: Politics and Policy' in B. Steinbock (ed.) *The Oxford Handbook of Bioethics* (Oxford: Oxford University Press), p. 444.

97. B. Agnew (2003),'The Politics of Stem Cells: Legislative Uncertainties Hinder Research in the US', Genome News Network Website, 21 February, http:// www.genomenewsnetwork.org/articles/02_03/stem.shtml, date accessed 10 August 2010.

98. Bonnicksen (2007), 'Therapeutic Cloning: Politics and Policy', p. 444.

99.  J. Kuzma (2004), 'Global Challenges and Biotechnology', Economic Perspectives, October, 4; K. Pawar, S. P. Pawar, V. A. Patel, H. V. Patel (2010), 'Promotion of Global Health through Oral Immunotherapy Using Edible Vaccines', *Pharmaceutical Reviews*, Vol. 8, No. 1, http://www.pharmainfo.net/reviews/promotion-global-health-through-oral-immunotherapy-using-edible-vaccines, date accessed 10 August 2010.
100. BIO (2008), *Guide to Biotechnology 2008*, p. 32.
101. Ibid.
102. Ibid.
103. Ibid., p. 35.
104. Ibid., p. 36.
105. Ibid., p. 37.
106. Committee on Advances in Technology and the Prevention of Their Application to Next Generation Biowarfare Threats et al. (2006), *Globalization, Biosecurity, and the Future of the Life Sciences*, p. 32.
107. Ibid., p. 49.
108. Ibid.
109. Ibid., p. 52.
110. The Central Intelligence Agency (CIA) (2003), 'The Darker Bioweapons Future', 3 November, http://www.fas.org/irp/cia/product/bw1103.pdf, date accessed 11 August 2010.
111. Ibid.
112. Ibid.
113. M.G. Kortepeter and G.W. Parker (1999), 'Potential Biological Weapons Threats', *Emerging Infectious Diseases*, 1 July, http://www.cdc.gov/ncidod/eid/vol5no4/kortepeter.htm, date accessed 11 August 2010.
114. 'Biological Warfare' (2004), *CBC News Online*, 18 February, http://www.cbc.ca/news/background/bioweapons/, date accessed 1 August 2010.
115. Ibid.
116. Kortepeter and Parker (1999), 'Potential Biological Weapons Threats'.
117. 'Biological Warfare' (2004), *CBC News Online*.

# 6   Genomics

1.  'T. Acharya, A.S. Daar, E. Dowdeswell, P.A. Singer and H. Thorsteinsdóttir (2004), *Genomics and Global Health: A Report of the Genomics Working Group of the Science and Technology Task Force of the United Nations Millennium Project* (Toronto: Joint Center for Bioethics) http://belfercenter.ksg.harvard.edu/files/genomics.pdf, date accessed 12 August 2010, p. 6.
2.  US Department of Energy Human Genome Project (2001), 'Other Anticipated Benefits of Genetic Research', http://www.ornl.gov/sci/techresources/Human_Genome/publicat/primer2001/7.shtml, date accessed 12 August 2010.
3.  Acharya et al. (2004), *Genomics and Global Health: A Report of the Genomics Working Group of the Science and Technology Task Force of the United Nations Millennium Project*, p. 9.
4.  P. Baldi and G.W. Hatfield (2002), *DNA Microarrays and Gene Expression: From Experiments to Data Analysis and Modeling* (Cambridge: Cambridge University Press), p. 2.

5. Committee on Advances in Technology and the Prevention of Their Application to Next Generation Biowarfare Threats, Development, Security, and Cooperation Policy and Global Affairs Division, Board on Global Health, Institute of Medicine, Institute of Medicine and National Research Council of the National Academies (2006), *Globalization, Biosecurity, and the Future of the Life Sciences* (Washington, D.C.: The National Academies Press), p. 15.

6. '100 Greatest Discoveries', The Science Channel Website, http://science. discovery.com/convergence/100discoveries/big100/genetics.html, date accessed 12 August 2010.

7. G.J. Hannon (2002), 'RNA Interference', *Nature*, Vol. 418, July, http://www. ncbi.nlm.nih.gov/pubmed/12110901, date accessed 12 August 2010.

8. Cf. K. Nixdorff and W. Bender (2002), 'Biotechnology, Ethics of Research and Potential Military Spin-off', International Network of Engineers and Scientists Against Proliferation, *Information Bulletin*, No. 19, March.

9. Human Genome Project Information 'Frequently Asked Questions', http:// www.ornl.gov/sci/techresources/Human_Genome/faq/faqs1.shtml, date accessed 12 August 2010.

10. Human Genome Project Information 'Spinoff Projects Related to the Human Genome Project', http://www.ornl.gov/sci/techresources/Human_Genome/ research/spinoffs.shtml, date accessed 12 August 2010.

11. A. Harmon (2008), 'Gene Map Becomes a Luxury Item', *International Herald Tribune*, 4 March, http://www.iht.com/articles/2008/03/04/healthscience/ 04geno.php, date accessed 12 August 2010.

12. Ibid.

13. 'The Genomics Revolution' (2000), *BCC Research*, http://www.bccresearch. com/report/BIO026A.html, date accessed 12 August 2010.

14. P. Ball (2010), 'Bursting the Genomics Bubble', *Nature News*, 31 March, http://www.nature.com/news/2010/100324/full/news.2010.145.html, date accessed 12 August 2010; N. Wade (2010), 'A Decade Later, Genetics Maps Yields Few New Cures', *The New York Times*, 13 June, http://www. nytimes.com/2010/06/13/health/research/13genome.html, date accessed 12 August 2010; A. Pollack (2010), 'The Genome at 10: Awaiting the Genome Payoff, *The New York Times*, 15 June, http://www.nytimes. com/2010/06/15/business/15genome.html, date accessed 12 August 2010.

15. B. Salter and N. Perez-Solorzano (2003), 'Civil Society and the Governance of Human Genetics: Report to the King Baudouin Foundation Regarding Its Project "Citizen Participation in the European Public Debate on the Issue of Human Genetics"', April, p. 6.

16. C.J. Henry, R. Phillips, F. Carpanini, J.C. Corton, K. Craig, K. Igarashi, R. Leboeuf, G. Marchant, K. Osborn, W.D. Pennie, L.L. Smith, M.J. Teta and V. Vu (2002), 'Use of Genomics in Toxicology and Epidemiology: Findings and Recommendations of a Workshop', *Environmental Health Perspectives*, Vol. 110, No. 10, October, 1047.

17. Biotechnology Industry Organization (BIO) (2008), *Guide to Biotechnology 2008* (Washington, D.C.: BIO), http://bio.org/speeches/pubs/er/Biotech-Guide2008.pdf, date accessed 17 August 2010, p. 27.

18. Ibid., p. 30.

19. 'The Mechanism of RNA Interference (RNAi)', Applied Biosystems Website, http://www.ambion.com/techlib/append/RNAi_mechanism.html, date accessed 12 August 2010.

20. B. Ashbridge (2008), 'RNA Interferene Explained', Science Articles: The Naked Scientists Website, http://www.thenakedscientists.com/HTML/articles/article/rna-interference-explained, date accessed 12 August 2010.

21. Sixth Review Conference of the States Parties to the Convention on the Prohibition of the Development, Production and Stockpiling of Bacteriological (Biological) and Toxin Weapons and on Their Destruction, Secretariat (2006), 'Background Information Document on New Scientific and Technological Developments Relevant to the Convention' BWC/CONF. VI/INF.4, Geneva, 28 September.

22. Ibid.

23. C. Cookson (2008), 'Research: Three Technologies That Could Transform Patient Care', *The Financial Times*, 3 July, http://www.ft.com/cms/s/0/668c7654-4898-11dd-a851000077b07658.html?nclick_check=1, date accessed 12 August 2010.

24. Ibid.

25. US Department of Health and Human Services, 'Genome-Wide Association Studies (GWAS)', http://grants.nih.gov/grants/gwas/, date accessed 19 August 2010.

26. 'Gene Mapping', Web Books 2.0 Website, http://web-books.com/MoBio/Free/Ch10A.htm, date accessed 12 August 2010.

27. The National Human Genome Research Institute (2009), 'Genetic Mapping', 28 January, http://www.genome.gov/10000715, date accessed 12 August 2010.

28. 'Gene Profiling Reveals the Essence of "Stemness"' (2002), Howard Hughes Medical Institute (HHMI) Website, 12 September, http://www.hhmi.org/news/melton3.html, date accessed 12 August 2010.

29. National Research Council of the National Academies (2003), *Biotechnology Research in an Age of Terrorism* (Washington, D.C.: The National Academies Press), p. 16.

30. S.G. Uzogara (2000), 'The Impact of Genetic Modification of Human Foods in the 21st Century: A Review', *Biotechnology Advances*, No. 18, 180.

31. T. Taylor (2010), 'Healthcare Industry', Presentation at the Strategic Technologies and Our Global Future Course, Geneva Centre for Security Policy, Switzerland, 15 April.

32. Uzogara (2000), 'The Impact of Genetic Modification of Human Foods in the 21st Century: A Review', p. 182.

33. 'Genetic Testing', MedlinePlus Website, http://www.nlm.nih.gov/medlineplus/genetictesting.html, date accessed 12 August 2010.

34. Ibid.

35. Genetic Science Learning Center at the University of Utah, http://learn.genetics.utah.edu/content/tech/genetherapy/whatisgt/, date accessed 12 August 2010.

36. Ibid.

37. Genetics Home Reference as a Service of the US National Library of Medicine (2009), 'What Is Gene Therapy?', 10 July, http://ghr.nlm.nih.gov/handbook/therapy/genetherapy, date accessed 12 August 2010.

38. Ibid.
39. The National Human Genome Research Institute (2010), 'Cloning', http://www.genome.gov/25020028, date accessed 12 August 2010.
40. Ibid.
41. Ibid.
42. Human Genome Project Information, 'Pharmacogenomics' http://www.ornl.gov/sci/techresources/Human_Genome/medicine/pharma.shtml, date accessed 12 August 2010.
43. Ibid.
44. National Center for Biotechnology Information Website (NCBI), 'Bioinformatics', http://www.ncbi.nlm.nih.gov/About/primer/bioinformatics.html, date accessed 12 August 2010.
45. Ibid.
46. University of Texas, 'Bioinformatics', http://biotech.icmb.utexas.edu/pages/bioinfo.html, date accessed 12 August 2010.
47. L. Gravitz (2009), 'TR10: $100 Genome', *Technology Review*, March/April, http://www.technologyreview.com/read_article.aspx?ch=specialsections&s c=tr10&id=22112, date accessed 12 August 2010.
48. Ibid.
49. Ibid.
50. S. Hensley (2001), 'Proteins – Not Genes – Could Be Clue to Human Complexity, Disease', *The Wall Street Journal*, 13 February, B1.
51. Ibid.
52. BIO (2008), *Guide to Biotechnology 2008*, p. 27.
53. M. Wheelis (2002), 'Biotechnology and Biochemical Weapons', *The Nonproliferation Review*, Vol. IX, No. 1, 50.
54. P. Jenkins (2001), 'Map of Proteins Holds the Key to Disease', *The Financial Times*, 27 November, 6.
55. A. Kalia and R.P. Gupta (2005), 'Proteomics: A Paradigm Shift', *Critical Reviews in Biotechnology*, Vol. 25, 187.
56. US Department of Energy Human Genome Project (2001), 'Gene Testing, Pharamacogenomics, and Gene Therapy', http://www.ornl.gov/sci/techresources/Human_Genome/publicat/primer2001/6.shtml, date accessed 12 August 2010.
57. Kalia and Gupta (2005), 'Proteomics: A Paradigm Shift', p. 188.
58. Ibid., p. 194.
59. 'Proteomics' (2010), BIO Website, 16 July, http://www.bio.org/speeches/pubs/er/biotechtools.asp, date accessed 12 August 2010.
60. 'Metabolomics', The Human Genome Website, http://genome.wellcome.ac.uk/doc_WTD020768.html, date accessed 12 August 2010.
61. University of Wisconsin, Biological Magnetic Resonance Data Bank, 'Metabolomics/Metabonomics', http://www.bmrb.wisc.edu/metabolomics/, date accessed 12 August 2010.
62. 'Metabolomics', The Human Genome Website.
63. Ibid.
64. E. Harrell (2009), 'The Human Epigenome, Decoded', *Time Magazine*, 8 December, http://www.time.com/time/specials/packages/article/0,28804,1945379_1944416_1944420,00.html, date accessed 12 August 2010.

65. J. Cloud (2010), 'Why Your DNA Isn't Your Destiny', *Time Magazine*, 6 January, http://www.time.com/time/printout/0,8816,1951968,00.html, date accessed 12 August 2010.
66. Harrell (2009), 'The Human Epigenome, Decoded'.
67. Cloud (2010), 'Why Your DNA Isn't Your Destiny'.
68. Harrell (2009), 'The Human Epigenome, Decoded'.
69. Cloud (2010), 'Why Your DNA Isn't Your Destiny'.
70. Ibid.
71. Ibid.
72. C. Kuppuswamy (2007), *The International Legal Governance of the Human Genome* (New York: Routledge).
73. United Nations General Assembly (2005), 'United Nations Declaration on Human Cloning', A/RES/59/280, 23 March, http://daccess-dds-ny.un.org/doc/UNDOC/GEN/N04/493/06/PDF/N0449306.pdf?OpenElement, date accessed 12 August 2010.
74. United Nations General Assembly (1998), 'United Nations Universal Declaration on the Genome and Human Rights', A/RES 53/152, 9 December, http://www2.ohchr.org/english/law/genome.htm, date accessed 12 August 2010.
75. The National Human Genome Research Institute 'Welcome to the Genome Statute and Legislation Database', http://www.genome.gov/PolicyEthics/LegDatabase/pubsearch.cfm, accessed 2 August 2010.
76. Food and Agriculture Organization of the United Nations (FAO) (2009), *The State of Food Insecurity in the World 2009* (Rome: FAO), ftp://ftp.fao.org/docrep/fao/012/i0876e/i0876e.pdf, p. 11.
77. 'Biotechnology and the World Food Supply' (2009), Union of Concerned Scientists Website, 29 October, http://www.ucsusa.org/food_and_agriculture/science_and_impacts/impacts_genetic_engineering/biotechnology-and-the-world.html, date accessed 12 August 2010.
78. World Health Organisation (WHO) '20 Questions on Genetically Modified Foods', http://www.who.int/foodsafety/publications/biotech/20questions/en/, date accessed 12 August 2010.
79. Human Genome Project Information, 'Genetically Modified Foods and Organisms', http://www.ornl.gov/sci/techresources/Human_Genome/elsi/gmfood.shtml, date accessed 12 August 2010.
80. D.B. Whitman (2000), 'Genetically Modified Foods: Harmful or Helpful?', April, ProQuest Website, http://www.csa.com/discoveryguides/gmfood/overview.php, date accessed 12 August 2010.
81. Human Genome Project Information, 'Genetically Modified Foods and Organisms'.
82. K. Harmon (2009), 'Cracked Corn: Scientists Solve Maize's Genetic Maze: Boasting More Genes than Humans, the Corn Genome Proved Difficult to Decode', *Scientific American*, 19 November, http://www.scientificamerican.com/article.cfm?id=corn-genome-cracked, date accessed 12 August 2010.
83. Uzogara (2000), 'The Impact of Genetic Modification of Human Foods in the 21st Century: A Review', pp. 185–189.
84. 'Biotechnology and the World Food Supply' (2009), Union of Concerned Scientists Website.
85. Ibid.

86. P.S. Anton, R. Silberglitt and J. Schneider (2001), *The Global Technology Revolution: Bio/Nano/Materials Trends and their Synergies with Information Technology by 2015* (Santa Monica, CA: RAND), p. 9.
87. WHO '20 Questions on Genetically Modified Foods'.
88. P. Collier (2003), 'The Politics of Hunger', *Foreign Affairs*, Vol. 87, No. 6, November/December, 67.
89. WHO, '20 Questions on Genetically Modified Foods'.
90. Ibid.
91. Anton et al. (2001), 'The Global Technology Revolution: Bio/Nano/Materials Trends and their Synergies with Information Technology by 2015', p. 10.
92. WHO '20 Questions on Genetically Modified Foods'.
93. J. Pohlhaus and R. Cook-Degan (2008), 'Genomics Research: World Survey of Public Funding', *BMC Genomics*, Vol. 9, 472.
94. 'Bridging the Genomics Divide' (2003), Ludwig Institute for Cancer Research Website, 29 January, http://www.licr.org/C_news/archive.php/2003/01/29/bridging-the-genomics-divide/, date accessed 12 August 2010.
95. A. Langlois (2006), 'The Governance of Genomic Information: Will It Come of Age?', *Genomics, Society and Policy*, Vol. 2, No. 3, December, http://www.pubmedcentral.nih.gov/articlerender.fcgi?artid=2291236, date accessed 4 September 2010.
96. Ibid.
97. WHO, 'GPG Aspects of Genomics', http://www.who.int/trade/distance_learning/gpgh/gpgh5/en/index5.html, date accessed 12 August 2010.
98. WHO, 'Genomics and the Global Health Divide', http://www.who.int/genomics/healthdivide/en/index.html, date accessed 18 August 2010.
99. E. Dowdeswell, A. Daar, and P. Singer (2003), 'Bridging the Genomics Divide', *Global Governance*, Vol. 9, No. 1, 1.
100. Ibid., p. 3.
101. D. Weisbrot (2009), 'Rethinking Privacy in the Genetic Age' in P. Atkinson, P. Glasner and M. Lock (2009), *Handbook of Genetics and Society: Mapping the New Genomic Era* (New York: Routledge), p. 324.
102. Anton et al. (2001), 'The Global Technology Revolution: Bio/Nano/Materials Trends and their Synergies with Information Technology by 2015', p. 9.
103. S. McLean (2009), 'Genetics: Law and Regulation' in Atkinson et al. (2009), *Handbook of Genetics and Society: Mapping the New Genomic Era*, p. 267.
104. Anton et al. (2001), *The Global Technology Revolution: Bio/Nano/Materials Trends and their Synergies with Information Technology by 2015*, p. 9.
105. WHO, 'Genomics and the Global Health Divide'.
106. Ibid.
107. Weisbrot (2009), 'Rethinking Privacy in the Genetic Age', p. 328.
108. B. Bogin (2003), 'The Evolution of Human Nutrition', *The Anthropology of Medicine*, December, http://web.archive.org/web/20031203003838/http://citd.scar.utoronto.ca/ANTA01/Projects/Bogin.html, date accessed 12 August 2010.
109. Ibid.
110. B. Balzer, 'Introduction to the Paleolithic Diet', Earth360 Website, http://www.earth360.com/diet_paleodiet_balzer.html, date accessed 12 August 2010.
111. Ibid.

112. N.R.F. Al-Rodhan (2009), 'Evolutionary Metabolism and Health Security', unpublished paper.
113. Bogin (2003), 'The Evolution of Human Nutrition'.
114. Balzer, 'Introduction to the Paleolithic Diet'.
115. Ibid.
116. Ibid.
117. Bogin (2003), 'The Evolution of Human Nutrition'.
118. National Diabetes Information Clearinghouse 'American Indians and Alaska Natives and Diabetes', http://diabetes.niddk.nih.gov/dm/pubs/americanindian, date accessed 12 August 2010.
119. R. Martorell (2005), 'Diabetes and Mexicans: Why the Two Are Linked', *Preventing Chronic Disease: Public Health Research, Practice, and Policy* 2, No. 1, http://www.pubmedcentral.nih.gov/articlerender.fcgi?artid=1323307, date accessed 12 August 2010.
120. Ibid.

# 7 Nanotechnology

1. M. Horton, A. Khan and S. Maddison (2006), 'Delivering Nanotechnology to the Healthcare, IT and Environmental Sectors', *BT Technology Journal*, Vol. 24, No. 3, 175.
2. W.S. Bainbridge (2007), *Nanoconvergence* (Upper Saddle River, NJ: Prentice Hall), p. 6.
3. T.F. Budinger and M.D. Budinger (2006), *Ethics of Emerging Technologies: Scientific Facts and Moral Challenges* (Hoboken, NJ: John Wiley & Sons), p. 444.
4. National Nanotechnology Initiative 'What is Nanotechnology?', http://www.nano.gov/html/facts/whatIsNano.html, date accessed 18 August 2010.
5. T. Shelley (2006), *Nanotechnology: New Promises, New Dangers* (London: Zed Books Ltd.), p. 14.
6. G. De Micheli (2009), 'Designing Nano Systems for a Safer Tomorrow' in N.R.F. Al-Rodhan (ed.) *Potential Global Strategic Catastrophes: Balancing Transnational Responsibilities and Burden-sharing with Sovereignty and Human Dignity* (Berlin: LIT), p. 149.
7. Budinger and Budinger (2006), *Ethics of Emerging Technologies: Scientific Facts and Moral Challenges*, p. 444.
8. M.C. Roco and W.S. Bainbridge (eds) (2001), *Societal Implications of Nanoscience and Nanotechnology* (Arlington, VA: National Science Foundation), p. 4.
9. P. Harris (2007), 'Carbon Nanotubes', Centre for Advanced Microscopy, University of Reading Website, 1 March, http://www.personal.rdg.ac.uk/~scsharip/tubes.htm, date accessed 12 August 2010.
10. K. Bullis (2008), 'The Year in Nanotech', *Technology Review*, 3 January, http://www.technologyreview.com/Nanotech/19983/?a=f, date accessed 12 August 2010.
11. 'The Project on Emerging Nanotechnologies' (2009), Nanotech Project Website, http://www.nanotechproject.org, date accessed 12 August 2010.

12. Roco and Bainbridge (2001), *Societal Implications of Nanoscience and Nanotechnology*, p. 2.
13. Bainbridge (2007), *Nanoconvergence*, p. 8.
14. S. Lenhert, 'A Brief History of Nanotechnology', Nanoword Website, http://www.nanoword.net/pages/history.htm, date accessed 12 August 2010.
15. 'There's Plenty of Room at the Bottom', EconomicExpert Website, http://www.economicexpert.com/a/There:s:Plenty:of:Room:at:the:Bottom.htm, date accessed 12 August 2010.
16. Office of Science and Technology, Executive Office of the President (2010), 'Independent Review finds Federal Nanotechnology Initiative Highly Effective; Recommends Changes to Assure =ngoing US dominance', 25 March, http://www.whitehouse.gov/sites/default/files/microsites/ostp/nano-release.pdf, date accessed 12 August 2010.
17. G. De Micheli (2009), 'Material and Nano Technology', Presentation at the 8th International Security Forum, Geneva, Switzerland, 19 May; G. De Micheli (2010), 'Nano-Technology', Presentation at the 'Strategic Technologies and Our Global Future Course' at the Geneva Centre for Security Policy, Geneva, Switzerland, 15 April.
18. M. Holman (2010), cited in 'Amid Nanotechnology's Dazzling Promise, Health Risks Grow', AOL News Website, 24 March, http://www.aolnews.com/nanotech/article/amid-nanotechs-dazzling-promise-health-risks-grow/19401235, date accessed 19 August 2010.
19. The White House (2003), 'President Bush Signs Nanotechnology Research and Development Act', 3 December, http://georgewbush-whitehouse.archives.gov/news/releases/2003/12/20031203-7.html, date accessed 18 August 2010.
20. 'NNI Amendmnt Act 2009 Introduced by Senator John Kerry' (2009), Nanotech Development Blog, July, http://www.nanotechnologydevelopment.com/government/nni-amendment-act-2009-introduced-by-senator-john-kerry.html, date accessed 12 August 2010.
21. De Micheli (2009), 'Material and Nano Technology'.
22. R.G. Lipsey, K.I. Carlaw and C.T. Bekar (2005), *Economic Transformations: General Purpose Technologies and Long-Term Economic Growth* (Oxford: Oxford University Press), p. 216.
23. Ibid.
24. 'The Wonders and Dangers of Nanotechnology' (2006), *Safety Compliance Letter*, Issue 2465, May, 1.
25. Ibid.
26. I. Amato (2004), 'Nano's Safety Checkup', *Technology Review*, Vol. 107, No. 1, February, 22.
27. 'The Risk in Nanotechnology' (2007), *The Economist*, 24 November, http://www.economist.com/node/10171212?story_id=10171212, date accessed 18 August 2010.
28. C. Phoenix and M. Treder (2008), 'Nanotechnology as Global Catastrophic Risks' in N. Bostrom and M.M. Ćircović (eds) *Global Catastrophic Risks* (Oxford: Oxford University Press), p. 483.
29. Ibid., p. 484.
30. R. Kurzweil (2006), *The Singularity Is Near* (London: Gerald Duckworth & Co., Ltd.), p. 247.

31. 'The Wonders and Dangers of Nanotechnology'(2006), *Safety Compliance Letter*, p. 1.
32. B.J. Feder (2002), 'New Economy; Nanotechnology Has Arrived; a Serious Opposition Is Forming', *The New York Times*, 19 August, http://www.nytimes.com/2004/11/01/technology/01nano.html, date accessed 12 August 2010.
33. 'The Wonders and Dangers of Nanotechnology' (2006), *Safety Compliance Letter*, p. 1.
34. G. Gonzalez (2007), 'Nanotechnology Poses Potential Risks to Environment, Health', *Business Insurance*, Vol. 41, No. 29, 3 December, 21.
35. 'Grey Goo Is a Small Issue' (2003), Center for Responsible Nanotechnology Website, 14 December, http://crnano.org/BD-Goo.htm, date accessed 12 August 2010.
36. Feder (2002), 'New Economy; Nanotechnology Has Arrived; a Serious Opposition Is Forming'.
37. C. Cookson (2008), 'Tight Regulation Urged on Nanotechnology', *The Financial Times*, 12 November, http://www.ft.com/cms/s/1bbac05a-b05c-11dd-a795-0000779fd18c,Authorised=false.html?_i_location=http%3A%2F%2Fwww.ft.com%2Fcms%2Fs%2F0%2F1bbac05a-b05c-11dd-a795-0000779fd18c.html&_i_referer=http%3A%2F%2Fsearch.ft.com%2Fsearch%3FqueryText%3Dnanotechnology, date accessed 19 August 2010, 4.
38. Shelley (2006), *Nanotechnology: New Promises, New Dangers*, p. 15.
39. Ibid.
40. G.A. Hodge, D. Bowman and K. Ludlow (eds) (2007), *New Global Frontiers in Regulation* (Cheltenham: Edward Elgar Publishing), p. 355.
41. International Organization for Standardization (ISO), 'Nanotechnologies', http://www.iso.org/iso/iso_technical_committee?commid=381983, date accessed 12 August 2010.
42. Organisation for Economic Co-operation and Development (OECD), 'OECD Working Party on Nanotechnology', http://www.oecd.org/document/36/0,3343,en_2649_34269_38829732_1_1_1_1,00.html, date accessed 12 August 2010.
43. Ibid.
44. De Micheli (2009), 'Designing Nanosystems for a Safer Tomorrow', p. 154.
45. Ibid., p. 155.
46. Ibid.
47. De Micheli (2010), 'Nano-Technology'.
48. J. Altmann and M.A. Gubrud (2004), 'Military, Arms Control, and Security Aspects of Nanotechnology' in D. Baird, A. Nordmann and J. Schummer (eds) *Discovering the Nanoscale* (Amsterdam: IOS Press), p. 275.
49. Ibid.
50. J. Altmann (2008), 'Military Uses of Nanotechnology – Too Much Complexity for International Security', *Complexity and Security*, Vol. 14, No. 1, 69.
51. Hodge et al. (2007), *New Global Frontiers in Regulation*, p. 361.
52. Ibid., p. 365.
53. Ibid., p. 364.
54. Ibid.
55. M.C. Roco and W.S. Bainbridge (2002), *Converging Technologies for Improving Human Performance: Nanotechnology, Biotechnology, Information Technology and Cognitive Science* (Arlington: National Science Foundation), p. ix.

56. Ibid., p. 2.
57. Roco and Bainbridge (2002), *Converging Technologies for Improving Human Performance: Nanotechnology, Biotechnology, Information Technology and Cognitive Science*, p. 4.
58. Biotechnology Industry Organization (BIO) (2008), *Guide to Biotechnology 2008* (Washington, D.C.: BIO), http://bio.org/speeches/pubs/er/BiotechGuide2008.pdf, date accessed 17 August 2010, p. 68.
59. Shelley (2006), *Nanotechnology: New Promises, New Dangers*, p. 91.
60. Roco and Bainbridge (2002), *Converging Technologies for Improving Human Performance: Nanotechnology, Biotechnology, Information Technology and Cognitive Science*, p. xi.
61. NASA (2002), 'Convergence of Biotechnology, Information Technology and Nanotechnology: A NASA Perspective', *Aerospace Technology Innovation*, Vol. 10, No. 4, July/August, 4.
62. Ibid.
63. S.L. Venneri (2001), 'Implications for Nanotechnology for Space Exploration' in Roco and Bainbridge (2001), *Societal Implications of Nanoscience and Nanotechnology*, p. 162.
64. 'Frequently Asked Questions – Molecular Manufacturing', Foresight Nanotech Institute Website, http://www.foresight.org/nano/whatismm.html, date accessed 12 August 2010.
65. Ibid.
66. Phoenix and Treder (2008), 'Nanotechnology as Global Catastrophic Risk', p. 489.
67. BIO (2008), *Guide to Biotechnology 2008*, p. 68.
68. Roco and Bainbridge (2002), *Converging Technologies for Improving Human Performance: Nanotechnology, Biotechnology, Information Technology and Cognitive Science*, p. xiii.
69. Ibid., p. x.
70. Ibid., p. xii.
71. Ibid.
72. Ibid.
73. Bainbridge (2007), *Nanoconvergence: The Unity of Nanoscience, Biotechnology, Information Technology, and Cognitive Science*, p. 11.
74. Budinger and Budinger (2006), *Ethics of Emerging Technologies: Scientific Facts and Moral Challenges*, p. 444.
75. Ibid.
76. Bainbridge (2007), *Nanoconvergence: The Unity of Nanoscience, Biotechnology, Information Technology, and Cognitive Science*, p. 92.
77. Budinger and Budinger (2006), *Ethics of Emerging Technologies: Scientific Facts and Moral Challenges*, p. 444.
78. 'Nanotechnology and Intellectual Property Issues' (2006), Nanowerk News Website, 26 December, http://www.nanowerk.com/news/newsid=1187.php, date accessed 12 August 2010.
79. Ibid.
80. Ibid.
81. B. Bastani and D. Fernandez (2002), *Intellectual Property Rights in Nanotechnology* (Menlo Park, CA: Fernandez & Associates), http://www.iploft.com/Nanotechnology.pdf, date accessed 12 August 2010, p. 5.

82. Ibid.
83. A.P. Alivisatos (2007), 'Less is More in Medicine', *Scientific American Reports*, Special Edition on Nanotechnology, 75.
84. Kurzweil (2006), *The Singularity Is Near*, p. 254.
85. Alivisatos (2007), 'Less is More in Medicine', p. 78.
86. Ibid.
87. Ibid.
88. Kurzweil (2006), *The Singularity Is Near*, p. 254.
89. De Micheli (2009), 'Material and Nano Technology'.
90. Bainbridge (2007), *Nanoconvergence: The Unity of Nanoscience, Biotechnology, Information Technology, and Cognitive Science*, p. 16.
91. F. Salamanca-Buentello, D.L. Persad, E.B. Court, D.K. Martin, A.S. Daar and P.A. Singer (2005), 'Nanotechnology and the Developing World', *PLoS Med*, Vol. 2, No. 5, http://www.plosmedicine.org/article/info%3Adoi%2F10.1371%2Fjournal.pmed.0020097, date accessed 18 August 2010.
92. C. Pellerin (2007), 'Nanotechnology Could Improve Health, Water in Developing Nations', America.gov Website, 5 March, http://www.america.gov/st/washfile-english/2007/March/20070305134101lcnirellep0.9842035.html, date accessed 12 August 2010.
93. Ibid.
94. De Micheli (2009), 'Material and Nano Technology'.
95. Kurzweil (2006), *The Singularity Is Near*, p. 246.
96. Deloitte Touche Tohmatsu, Technology, Media & Telecommunications (2008), *Technology Predictions: TMT Trends 2008* (United Kingdom: The Creative Studios), p. 9.
97. Cf. Salamanca-Buentello et al. (2005), 'Nanotechnology and the Developing World'.
98. Ibid.
99. De Micheli (2009), 'Material and Nano Technology'.
100. Ibid.
101. Ibid.
102. M. Scott (2008), 'Innovative Plans May be Key for Greener Future', *The Financial Times*, 18 April, http://www.ft.com/cms/s/0/afa073fa-0b6b-11dd-8ccf-0000779fd2ac.html, date accessed 12 August 2010.
103. R. Shapiro (2008), *Futurecast 2020: A Global Vision of Tomorrow* (New York: St. Martin's Press), p. 286.
104. Ibid., p. 305.
105. De Micheli (2009), 'Designing Nanosystems for a Safer Tomorrow', p. 146.
106. Ibid., p. 151.
107. Scott (2008), 'Business and the Environment: Innovative Plans May Be Key for Green Future'.
108. Shapiro (2008), 'Futurecast 2020: A Global Vision of Tomorrow', p. 306.
109. Ibid.
110. US Department of Defense (2007), *Defense Nanotechnology Research and Development Program*, 16 April, http://www.fas.org/irp/agency/dod/nano2007.pdf, date accessed 12 August 2010.
111. Shelley (2006), *Nanotechnology: New Promises, New Dangers*, p. 39.
112. Altmann (2008), 'Military Uses of Nanotechnology – Too Much Complexity for International Security', p. 64.

113. J. Altmann and M.A. Gubrud (2004), 'Military, Arms Control, and Security Aspects of Nanotechnology' in D. Baird, A. Nordmann and J. Schummer (eds) *Discovering the Nanoscale* (Amsterdam: IOS Press), p. 273.
114. Ibid., p. 272.
115. Ibid., p. 273.
116. Ibid., p. 274.
117. De Micheli (2009), 'Material and Nano Technology'.
118. Ibid.
119. Ibid.

# 8 Materials Science

1. Committee on Materials Science and Engineering, Solid State Sciences Committee, Board on Physics and Astronomy, Commission on Physical Sciences, Mathematics, and Resources, National Materials Advisory Board, Commission on Engineering and Technical Systems, National Research Council (1989), *Materials Science and Engineering for the 1990s: Maintaining Competitiveness in the Age of Materials* (Washington, D.C.: National Academies Press), p. 19.
2. Massachusetts Institute for Technology (MIT) Department of Materials Science and Engineering 'What Is Materials Science and Engineering?, http://dmse.mit.edu/about/whatis.html, date accessed 16 August 2010.
3. L.J. Pellack (2002), 'Introduction to Materials Science', Materials Science Resources on the Web, http://www.istl.org/02-spring/internet.html, date accessed 16 August 2010.
4. P. Patel (2008), 'Self-Assembled Organic Circuits', *MIT Technology Review*, 17 October, http://www.technologyreview.com/computing/21575/?a=f, date accessed 16 August 2010.
5. Ibid.
6. 'Organic Electronics as a Two-Way Street, Thanks to New Plastic Semi-Conductor' (2009), *Science Daily*, 7 September, http://www.sciencedaily.com/releases/2009/08/090817143606.htm, date accessed 16 August 2010.
7. K. Bullis (2008), 'Mass Production of Plastic Solar Cells' (2008), *MIT Technology Review*, 17 October, http://www.technologyreview.com/energy/21574/pagel, date accessed 16 August 2010.
8. D.R. Howell and R. Silberglitt (2006), 'Appendix C: Materials Science and Engineering Trends to 2020' in R. Silberglitt, P. Anton, D. Howell, A. Wong, N. Gassman, B.A. Jackson, E. Landree, S. L. Pfleeger, E.M. Newton and F. Wu (eds) *The Global Technology Revolution 2020, In-Depth Analyses: Bio/Nano/Materials/Information Trends, Drivers, Barriers, and Social Implications* (Santa Monica, CA: RAND), p. 177.
9. Biotechnology Industry Organization (BIO) (2008), *Guide to Biotechnology 2008* (Washington, DC: BIO), http://bio.org/speeches/pubs/er/BiotechGuide2008.pdf, date accessed 17 August 2010, p. 67.
10. 'How Green Are Green Plastics?' (2008), Mindfully Website, http://www.mindfully.org/Plastic/Biodegrade/Green-PlasticsAug00.htm, date accessed 16 August 2010.

11. T.A. Adams II (2000), 'Physical Properties of Carbon Nanotubes', Michigan State University Website, http://www.pa.msu.edu/cmp/csc/ntproperties/, date accessed 16 August 2010.
12. Ibid.
13. M. Dresselhaus and G. Dresselhaus (1998), 'Carbon Nanotubes', PhysicsWorld Website, 1 January, http://physicsworld.com/cws/article/print/1761, date accessed 16 August 2010.
14. J. Savage (2008), 'Silicon Nanowires Turn Heat to Electricity', IEEE Spectrum Website, January, http://www.spectrum.ieee.org/energy/renewables/silicon-nanowires-turn-heat-to-electricity, date accessed 16 August 2010.
15. J.R. Heath and M.A. Ratner (2003), 'Molecular Electronics', *Physics Today*, May, 43.
16. 'It's an Advanced Material World' (2006), Nova: Science in the News, Australian Academy of Science, http://www.science.org.au/nova/093/093glo.htm, date accessed 16 August 2010.
17. 'Spintronics' (2008), Nanotechnology Now Website, 29 March, http://www.nanotech-now.com/spintronics.htm, date accessed 16 August 2010.
18. Ibid.
19. Ibid.
20. 'Wide Bandgap Semiconductors' (2005), Information Society Technology (IST) World Website, http://www.ist-world.org/ProjectDetails.aspx?ProjectId=875e48af26924964ac9f294efa7dd914, date accessed 16 August 2010.
21. 'Wide Band Gap Semiconductors' (2005), European Space Components Information Exchange System Website, 1 November, https://escies.org/ReadArticle?docId=220, date accessed 16 August 2010.
22. Ibid.
23. 'It's an Advanced Material World' (2006), Nova: Science in the News, Australian Academy of Science.
24. 'Smart Materials', Discovery Channel Website, http://www.discoverychannel.co.uk/technology/basic_materials/smart/index.shtml, date accessed 16 August 2010.
25. 'It's an Advanced Material World' (2006), Nova: Science in the News, Australian Academy of Science.
26. Howell and Silberglitt (2006), 'Appendix C: Materials Science and Engineering Trends to 2020', p. 170.
27. Ibid.
28. Ibid.
29. R. Miller (2009), 'Scientists Develop Piezoelectric Motor for Medical Microbots', Endagadget Website, 20 January, http://www.engadget.com/2009/01/20/scientists-develop-piezoelectric-motor-for-medical-microbots, date accessed 16 August 2010.
30. C. Johnson (2004), ' "Smart" Textiles Emerge from Nanotech Labs', MSNBC.com Website, 15 December, http://www.msnbc.msn.com/id/6713188/, date accessed 16 August 2010.
31. Howell and Silberglitt (2006), 'Appendix C: Materials Science and Engineering Trends to 2020', p. 172.
32. 'On the Path to Sophisticated Fibers that Can Hear and Sing' (2010), *R&D Magazine*, 12 July, http://www.rdmag.com/News/2010/07/Materials-Material-Science-On-The-Path-To-Sophisticated-Fibers-That-Can-Hear-And-Sing/, date accessed 16 August 2010.

33. Howell and Silberglitt (2006), 'Appendix C: Materials Science and Engineering Trends to 2020', p. 172.
34. G.P. McKnight, 'Magnetostrictive Materials Background', University of California Website, http://aml.seas.ucla.edu/research/areas/magnetostrictive/overview.htm, date accessed 16 August 2010.
35. Howell and Silberglitt (2006), 'Appendix C: Materials Science and Engineering Trends to 2020', p. 171.
36. McKnight, 'Magnetostrictive Materials Background'.
37. 'Shape Changing Smart Material', Inventables Website, https://technology.inventables.com/technologies/terfenol-d-smart-material, date accessed 16 August 2010.
38. Ibid.
39. Howell and Silberglitt (2006), 'Appendix C: Materials Science and Engineering Trends to 2020', p. 171.
40. Ibid.
41. Ibid.
42. Ibid., pp. 171–172.
43. Ibid., p. 172.
44. J. Elliot and B. Hancock (2006), 'Pharmaceutical Materials Science: An Active New Frontier in Materials Research', *MRS Bulletin*, Vol. 31, 869.
45. Ibid., p. 873.
46. University of Cambridge 'The Pfizer Institute for Pharmaceutical Materials Science', http://www.msm.cam.ac.uk/pfizer/, date accessed 16 August 2010.
47. Elliot and Hancock (2006), 'Pharmaceutical Materials Science: An Active New Frontier in Materials Research', p. 873.
48. 'What Are Composites?' Society of Manufacturing Engineers Website, http://www.sme.org/cgi-bin/communities.pl?/communities/ema/what_is_pcc.htm&&&SME&, date accessed 16 August 2010.
49. Ibid.
50. US Department of Energy: Industrial Technologies Program (2006), 'Advanced Composite Coatings for Industries of the Future', http://www1.eere.energy.gov/industry/imf/pdfs/1791_advanced_compositecoatings.pdf, date accessed 16 August 2010, p. 1.
51. 'Metal, Heal Thyself' (2010), *The Economist*, 10 June, http://www.economist.com/node/16295654, date accessed 16 August 2010.
52. K.C. Jones (2008), 'Metamaterials Hold Promise for Invisibility Cloaks', *Information Week*, 11 August, http://www.informationweek.com/news/hardware/peripherals/showArticle.jhtml?articleID=210001982, date accessed 16 August 2010.
53. S. Markey (2006), 'First Invisibility Cloak Tested Successfully, Scientists Say', *National Geographic News*, 19 October, http://news.nationalgeographic.com/news/2006/10/061019-invisible-cloak.html, date accessed 16 August 2010.
54. 'Invisibility Cloak One Step Closer: New Metamaterials Bend Light Backwards' (2008), *Science Daily*, 11 August, http://www.sciencedaily.com/releases/2008/08/080811092450.htm, date accessed 16 August 2010.
55. Ibid.
56. Ibid.
57. Ibid.
58. Foundation for Fundamental Research on Matter (FOM Institute AMOLF) (2010), 'Optical Metamaterial with Negative Refractive Index for Visible

Light', 18 April, http://www.amolf.nl/news/detailpage/back_to/news/article/optical-metamaterial-with-negative-refractive-index-for-visible-light//chash/7df92703a2/, date accessed 16 August 2010.

59. 'Invisibility Cloak One Step Closer: New Metamaterials Bend Light Backwards' (2008), *Science Daily*.

60. K. Bourzac (2009), 'TR10: Nanopiezoelectronics', *MIT Technology Review*, March/April, http://www.technologyreview.com/printer_friendly_article.aspx?id=22118&channel=specialsections&section=tr10, date accessed 16 August 2010.

61. Ibid.

62. Ibid.

63. National Research Council (2005), *Globalization of Materials R&D: Time for a National Strategy*, (Washington, D.C.: National Academies Press), p. 120.

64. Ibid.

65. 'Army Selects MIT for $50 Million Institute to Use Nanomaterials to Clothe, Equip Soldiers' (2002), MIT News Website, 13 March, http://web.mit.edu/newsoffice/2002/isn.html, date accessed 16 August 2010.

66. 'Center for Computational Materials Website', Naval Research Laboratories Website, http://cst-www.nrl.navy.mil/, date accessed 16 August 2010.

67. Committee on Materials Research for Defense After Next, National Research Council (2003), *Materials Research to Meet 21st Century Defense Needs* (Washington, D.C.: National Academies Press), p. 13.

68. Ibid., p. 12.

69. 'Liquid Body Armor: Rheologists Apply Shear-Thickening Fluids to Protective Gear' (2006), *Science Daily*, 1 August, http://www.sciencedaily.com/videos/2006/0803-liquid_body_armor.htm, date accessed 16 August 2010.

70. M. Brant (2003), 'Sci-fi War Uniforms?', *Newsweek*, Vol. 141, No. 8, p. E2.

71. Committee on Materials Research for Defense After Next, National Research Council (2003), *Materials Research to Meet 21st Century Defense Needs*, p. 15.

72. 'Intelligent Structures: Structural Healing' (2005), *The Engineer*, 11 July, 24.

73. 'Span of Control' (2009), *The Economist*, 3 September, http://www.economist.com/sciencetechnology/tq/displaystory.cfm?story_id=E1_TQNJJGVQ, date accessed 16 August 2010.

74. Ibid.

75. Committee on Materials Research for Defense After Next, National Research Council (2003), *Materials Research to Meet 21st Century Defense Needs*, p. 22.

76. G. De Micheli (2009), 'Material and Nano Technology', Presentation at the 8th International Security Forum, Geneva, Switzerland, 19 May.

77. Ibid.

78. Ibid.

79. A. Kahära (2006), 'Smart Cold Protective Clothing for Military Use', Tampere University of Technology Website, http://www.tut.fi/index.cfm?MainSel=-1&Sel=6106&Show=5600&Siteid=54, date accessed 16 August 2010.

80. Committee on Materials Research for Defense After Next, National Research Council (2003), *Materials Research to Meet 21st Century Defense Needs*, p. 22.

81. Ibid.

82. European Space Agency (ESA) Materials Science in Space 'What is Gravity?', http://www.spaceflight.esa.int/users/materials/introduction/gravity/

gravity.html, date accessed 20 August 2010; Committee on Materials Research for Defense After Next, National Research Council (2003), *Materials Research to Meet 21st Century Defense Needs*, p. 22.

83. Madrid Institute of Material Science 'Nanostructured Materials for Space Applications: Sensors and Coatings', http://www.icmm.csic.es/solgel/sol-gel_space_applications.html, date accessed 16 August 2010.

84. R.W. Bruce, 'NASA 2004 SBIR Phase 1 Solicitation', National Aeronautics and Space Administration (NASA) Website, http://sbir.nasa.gov/SBIR/abstracts/04-1.html, date accessed 16 August 2010.

85. De Micheli (2009), 'Material and Nano Technology'.

86. Ibid.

87. G. De Micheli (2009), 'Designing Nano Systems for a Safer Tomorrow' in N.R.F. Al-Rodhan (ed.) *Potential Global Strategic Catastrophes: Balancing Transnational Responsibilities and Burden-sharing with Sovereignty and Human Dignity*, (Berlin: LIT), p. 152.

88. De Micheli (2009), 'Material and Nano Technology'.

89. Kahära (2006), 'Smart Cold Protective Clothing for Military Use'.

90. 'Trappings of Waste' (2009), *The Economist*, 3 September, http://www.econo-mist.com/sciencetechnology/tq/displaystory.cfm?story_id=E1_TQNJJGGQ, date accessed 16 August 2010.

91. Ibid.

92. Ibid.

# 9 Artificial Intelligence

1. R. Kurzweil (2006), *The Singularity Is Near* (London: Gerald Duckworth & Co., Ltd.), p. 7.

2. University of Michigan 'Intelligence and Artificial Intelligence', http://ai.eecs.umich.edu/cogarch0/common/theory/ai.html, date accessed 16 August 2010.

3. R. Kurzweil (2006), 'Reinventing Humanity: The Future of Machine-Human Intelligence', *The Futurist*, Vol. 40, No. 2, March/April, 39.

4. Ibid.

5. P. Domingos (2009), 'Solving AI', *Technology Review*, Vol. 112, No. 2, March/April, 10.

6. Ibid.

7. C. Goldberg (2007), 'Beyond AI: Creating the Conscience of the Machine', *Science News*, Vol. 172, No. 11, 15 September, 175.

8. P.W. Singer (2009), *Wired for War: The Robotics Revolution and Conflict in the 21st Century* (New York: Penguin Press), p. 75.

9. 'What Is the Singularity?' The Singularity Institute for Artificial Intelligence Website, http://singinst.org/overview/whatisthesingularity, date accessed 16 August 2010.

10. Ibid.

11. Ibid.

12. D.L. Waltz (1996), 'Artificial Intelligence: Realizing the Ultimate Promises of Computing', NEC Research Institute Website, http://www.cs.washington.edu/homes/lazowska/cra/ai.html, date accessed 16 August 2010.

13. 'Predicting AI's Future' (2001), *BBC News*, 21 September, http://news.bbc.co.uk/1/hi/in_depth/sci_tech/2001/artificial_intelligence/1555742.stm, date accessed 16 August 2010.
14. Singer (2009), *Wired for War: The Robotics Revolution and Conflict in the 21st Century*, p. 78.
15. Waltz (1996), 'Artificial Intelligence: Realizing the Ultimate Promises of Computing'.
16. G. Anthes (2009), 'AI Comes of Age', *Computerworld*, Vol. 43, No. 4, 16.
17. R.A. Palmquist (1996), 'AI and Expert Systems', The University of Texas at Austin Website, http://www.ischool.utexas.edu/~palmquis/courses/ai96.htm, date accessed 16 August 2010.
18. Anthes (2009), 'AI Comes of Age', p. 16.
19. Waltz (1996), 'Artificial Intelligence: Realizing the Ultimate Promises of Computing'.
20. 'Artificial Intelligence', British Computer Society Website, http://www.bcs.org, date accessed 16 August 2010.
21. Kurzweil (2006), *The Singularity Is Near*, p. 261.
22. Ibid.
23. 'AI', MSN Encarta Website, http://encarta.msn.com/encnet/features/dictionary/dictionaryhome.aspx, date accessed 16 August 2010.
24. University of Toronto 'Symbolic AI', http://www.psych.utoronto.ca/users/reingold/courses/ai/symbolic.html, date accessed 16 August 2010.
25. 'AI', MSN Encarta Website.
26. University of Toronto 'Symbolic AI'.
27. University of Toronto 'The Common Sense Knowledge Problem', http://www.psych.utoronto.ca/users/reingold/courses/ai/commonsense.html, date accessed 16 August 2010.
28. 'University of Toronto 'Symbolic AI'.
29. 'AI', MSN Encarta Website.
30. 'Artificial Intelligence (AI)', Encyclopedia Britannica Website, http://www.britannica.com/EBchecked/topic/37146/artificial-intelligence, date accessed 16 August 2010.
31. Kurzweil (2006), *The Singularity Is Near*, p. 269.
32. Ibid.
33. 'AI', MSN Encarta Website.
34. 'Artificial Intelligence (AI)', Encyclopedia Britannica Website.
35. B. Faltings (2009), Personal Email Communication with Author, 3 August.
36. 'Kernel Machines' (2007), Kernel Machines Website, 1 February, http://www.kernel-machines.org/, date accessed 20 August 2010.
37. Faltings (2009), Personal email communication, 3 August.
38. 'Evolutionary AI', Herself's Artificial Intelligence Website, http://www.herselfsai.com/2007/02/evolutionary-ai.html, date accessed 16 August 2010.
39. 'AI', MSN Encarta Website.
40. Ames Research Center (2004), 'NASA Evolutionary Software Automatically Designs Antenna', 15 June, http://www.nasa.gov/mission_pages/st-5/main/04-55AR.html, date accessed 16 August 2010.
41. Ibid.
42. Kurzweil (2006), *The Singularity Is Near*, p. 271.

43. B.J. Copeland (2000), 'What Is Artificial Intelligence?' AlanTuring Website, http://www.alanturing.net/turing_archive/pages/Reference%20Articles/ what_is_AI/What%20is%20AI14.html, date accessed 16 August 2010.
44. 'Strong and Weak AI', PhilosophyOnline Website, http://www.philosophyon-line.co.uk/pom/pom_functionalism_AI.htm, date accessed 16 August 2010.
45. University of Texas 'Philosophical Arguments Against "Strong" AI', http:// www.cs.utexas.edu/users/mooney/cs343/slide-handouts/philosophy.4.pdf, date accessed 20 August 2010.
46. Copeland (2000), 'What Is Artificial Intelligence?'.
47. E. Naone (2009), 'TR: 10: Intelligent Software Assistant', *MIT Technology Review*, March/April, http://www.technologyreview.com/printer_friendly_ article.aspx?id=22117&channel=specialsections&section=tr10, date accessed 16 August 2010.
48. Ibid.
49. Ibid.
50. C. Thompson (2010), 'What Is I.B.M.'s Watson?' *The New York Times*, 14 June, http://www.nytimes.com/2010/06/20/magazine/20Computer-t.html, date accessed 17 August 2010.
51. E. Harrell (2009), 'A Robot Performs Science', *Time Magazine*, 8 December, http://www.time.com/time/specials/packages/article/0,28804,1945379_194 4416_1944423,00.html, date accessed 17 August 2010.
52. Ibid.
53. B. Hibbard (2006), 'The Singularity Summit and Regulation of AI', Email communication, 10 May, http://www.ssec.wisc.edu/~billh/g/singularity_ summit.html, date accessed 16 August 2010.
54. Ibid.
55. J. Hughes (2007), 'Hughes on Regulating AI at the Singularity Summit', Singularity Institute for Artificial Intelligence Website, 21 September, http:// ieet.org/index.php/IEET/more/2021/, date accessed 17 August 2010.
56. B. Joy (2004), 'Why the Future Doesn't Need Us', *Wired*, August, http://www. wired.com/wired/archive/8.04/joy_pr.html, date accessed 17 August 2010.
57. Ibid.
58. Ibid.
59. Faltings (2009), Personal email communication, 3 August.
60. Ibid; P.W. Singer (2010), 'War of the Machines', *Scientific American*, Vol. 303, No. 1, 59.
61. Singer (2009), *Wired for War: The Robotics Revolution and Conflict in the 21st Century*, p. 398.
62. Ibid., p. 399.
63. 'Terminate the Terminators' (2010), *Scientific American*, July, http://www. scientificamerican.com/article.cfm?id=terminate-the-terminators&page=2, date accessed 17 August 2010.
64. Singer (2009), *Wired for War: The Robotics Revolution and Conflict in the 21st Century*, p. 403.
65. Ibid.
66. Singer (2010), 'War of the Machines', p. 56.
67. 'Military Applications of AI', Oracle ThinkQuest Library Website, http:// library.thinkquest.org/18242/app_military.shtml, date accessed 16 August 2010.

68. Kurzweil (2006), *The Singularity Is Near*, p. 280.
69. Singer (2009), *Wired for War: The Robotics Revolution and Conflict in the 21st Century*, pp. 36–37.
70. Singer (2010), 'War of the Machines', p. 56.
71. W.S. Weed (2002), 'This Year in Ideas: Robotic Warfare', *The New York Times*, 15 December, http://www.nytimes.com/2002/12/15/magazine/the-year-in-ideas-robotic-warfare.html, date accessed 17 August 2010.
72. S. Ackerman (2010), 'Air Force Wants Drones to Sense Other Planes's Intents', *Wired*, 23 July, http://www.wired.com/dangerroom/2010/07/air-force-wants-drones-to-sense-other-planes-intent/, date accessed 17 August 2010.
73. Weed (2002), 'This Year in Ideas: Robotic Warfare'.
74. 'AI Topics: Military' (2008), Association for the Advancement of Artificial Intelligence Website, http://www.aaai.org/aitopics/pmwiki/pmwiki.php/AITopics/Military, date accessed 17 August 2010.
75. Singer (2010), 'War of the Machines', p. 57.
76. F. Reed (2005), 'Robotic Warfare Drawing Nearer', GlobalSecurity Website, 10 February, http://www.globalsecurity.org/org/news/2005/050210-robotic-warfare.htm, date accessed 17 August 2010.
77. Ibid.
78. L. Greenemeier (2009), 'Researchers Turn to Artificial Intelligence and Real Data to Improve War Games', *Scientific American*, 26 November, http://www.scientificamerican.com/article.cfm?id=virtual-war-games, date accessed 17 August 2010.
79. Ibid.
80. W. Warner (1993), 'Send in the Machines: War in the Age of Intelligent Machine', *MIT Technology Review*, Vol. 96, No. 3, 75.
81. Kurzweil (2006), *The Singularity Is Near*, p. 330.
82. Ibid., p. 332.
83. Singer (2009), *Wired for War: The Robotics Revolution and Conflict in the 21st Century*, p. 398.
84. Reed (2005), 'Robotic Warfare Drawing Nearer'.
85. R. Finkelstein (2009), 'Robotics in Future Warfare', Presentation at the U.S. Army War College Strategic Studies Institute, 14–16 April.
86. Kurzweil (2006), *The Singularity Is Near*, p. 421.
87. Weed (2002), 'This Year in Ideas: Robotic Warfare'.
88. Singer (2009), *Wired for War: The Robotics Revolution and Conflict in the 21st Century*, p. 212.
89. Ibid., p. 213.
90. C.D. Walton (2007), *Geopolitics and the Great Powers in the 21st Century* (New York: Routledge), p. 94.
91. Kurzweil (2006), *The Singularity Is Near*, p. 280.
92. Ibid., p. 281.
93. NASA: Jet Propulsion Laboratory, California Institute of Technology 'Current Projects', http://ai.jpl.nasa.gov/public/projects/current.html, date accessed 17 August 2010.
94. Kurzweil (2006), *The Singularity Is Near*, p. 282.
95. 'AI in Medicine: An Introduction', Open Clinical Website, http://www.openclinical.org/aiinmedicine.html, date accessed 17 August 2010.

96. W.H.W. Ishak and F. Siraj (2002), 'Artificial Intelligence in Medical Applications: An Exploration', *Health Informatics Europe Journal*, 30 June, http://74.125.155.132/scholar?q=cache:O4B5JcShRjMJ:scholar.google.co m/+%E2%80%98Artificial+Intelligence+in+Medical+Applications:+An+E xploration%E2%80%99,+Health+Informatics+Europe+Journal,&hl=de&a s_sdt=2000&as_vis=1, date accessed 17 August 2010.

97. 'Doctors Team Up with Computers' (2000), *The Futurist*, Vol. 34, No. 5, September/October, 13.

98. T. Horowitz (2010), 'Cyber Care: Will Robots Help the Elderly Live at Home Longer?', *Scientific American*, 21 June, http://www.scientificamerican.com/ article.cfm?id=robot-elder-care, date accessed 17 August 2010.

99. A. Harmon (2010), 'A Soft Spot for Circuitry', *The New York Times*, 4 July, http://www.nytimes.com/2010/07/05/science/05robot.html, date accessed 17 August 2010.

100. 'Artificial Intelligence in Medicine', Cedars-Sinai Website, http://www. csmc.edu/5835.html, date accessed 17 August 2010.

101. C. Whelan (2010), 'The Doctor is Out, But New Patient Monitoring and Robotics Technology is In', *Scientific American*, 25 March, http://www.scien-tificamerican.com/article.cfm?id=patient-monitoring-tech, date accessed 17 August 2010.

102. Ishak and Siraj (2002)'Artificial Intelligence in Medical Applications: An Exploration'.

103. Faltings (2009), Personal email communication, 3 August.

104. Ibid.

105. O. Port, M. Arndt, and J. Carey (2003), 'Smart Tools: Companies in Health Care, Finance, and Retailing are Using Artificial Intelligence Systems to Filter Huge Amounts of Data and Identify Suspicious Transactions', *Business Week*, Issue 3826A, 25 March, 154.

106. A. Anshum and P. Siddharth (2009), 'Applications of AI in Finance', Indian Institute of Technology Bombay, http://www.cse.iitb.ac.in/~cs344/2009/ seminars/, date accessed 17 August 2010.

107. Ibid.

108. Ibid.

109. Ibid.

110. Faltings (2009), Personal email communication, 3 August.

111. Kurzweil (2006), *The Singularity Is Near*, p. 286.

112. Ibid.

113. R. Wray (2006), 'Google Users Promised Artificial Intelligence', *The Guardian*, 23 May, http://www.guardian.co.uk/technology/2006/may/23/ searchengines.news, date accessed 17 August 2010.

114. H. Green (2005), 'Building a Smarter Search Engine', *Business Week*, 11 January, http://www.businessweek.com/technology/content/jan2005/ tc2005014_2937.htm, date accessed 17 August 2010.

115. 'Search Engine Privacy Tips', World Privacy Forum Website, http://www. worldprivacyforum.org/searchengineprivacytips.html, date accessed 17 August 2010.

116. Ibid.

117. Kurzweil (2006), *The Singularity Is Near*, p. 420.

118. D.M. Ewalt (2001), 'Stephen Hawking Warns of 'Terminator'-Style Menace', *Information Week*, 5 September.
119. Kurzweil (2006), *The Singularity Is Near*, p. 420.
120. J. Markoff (2010), 'Scientists Worry Machines May Outsmart Man', *The New York Times*, 25 July, http://www.nytimes.com/2009/07/26/science/26robot. html, date accessed 17 August 2010.

## 11 Introduction: Definitions, Terms and Concepts

1. N.R.F. Al-Rodhan (2008), *'Emotional Amoral Egoism': A Neurophilosophical Theory of Human Nature and its Universal Security Implications* (Berlin: LIT), p. 65.
2. 'Transhumanist FAQ', Extropy Institute Website, http://www.extropy.org/faq.htm#1.1, date accessed 17 August 2010.
3. Ibid.
4. N. Bostrom (2010), 'Transhumanist FAQ', Humanity+ Website, http://humanityplus.org/learn/transhumanist-faq/, accessed 29 July 2010.
5. J.J. Hughes and N. Bostrom, 'What Is Transhumanism?' World Transhumanist Association Website, http://www.transhumanism.org/resources/PressIntro.ppt#3, date accessed 17 August 2010.
6. A. Sandberg, 'Introduction to Transhumanism', Transhumanist Resources Website, http://www.aleph.se/Trans/, date accessed 17 August 2010.
7. 'Transhumanist FAQ', Extropy Institute Website.
8. Bostrom (2010), 'Transhumanist FAQ'.
9. M. More (1994), 'On Becoming Posthuman', Max More Website, http://www.maxmore.com/becoming.htm, date accessed 17 August 2010.
10. 'Transhumanist FAQ', Extropy Institute Website.
11. Ibid.
12. D. Ust (2001), 'What Is Posthumanism?' Daniel Ust Website, http://mars.superlink.net/~neptune/Posthuman.html, date accessed 17 August 2010.
13. Bostrom (2010), 'Transhumanist FAQ'.
14. More (1994), 'On Becoming Posthuman'.
15. 'Definition of Eugenics' (2001), MedicineNet Website, 5 August, http://www.medterms.com/script/main/art.asp?articlekey=3335, date accessed 17 August 2010.
16. Bostrom (2010), 'Transhumanist FAQ'.
17. F. Galton (2009), quoted by L. Koch, 'Eugenics' in P. Atkinson, P. Glasner and M. Lock (eds) (2009), *Handbook of Genetics and Society: Mapping the New Genomic Era* (New York: Routledge), p. 437.
18. L. Koch (2009), 'Eugenics' in Atkinson et al. (2009), *Handbook of Genetics and Society: Mapping the New Genomic Era*, p. 437.
19. Ibid., p. 438.
20. 'Definition of Eugenics' (2001), MedicineNet Website.
21. D.B. Resnik (2006), 'The Moral Significance of the Therapy-Enhancement Distinction in Human Genetics' in H. Kuhse and P. Singer (eds) *Bioethics* (Malden, Oxford and Victoria: Blackwell Publishing Ltd), p. 216.
22. Bostrom (2010), 'Transhumanist FAQ'.
23. M. Foucault (1990), *The History of Sexuality: An Introduction* (New York: Random House), p. 138.

24. Ibid., p. 141.
25. Ibid., p. 138.
26. Ibid., p. 140.
27. Ibid., p. 137.
28. Ibid.
29. T. Ellis-Christensen, 'What Is Biopower?' Wisegeek Website, http://www. wisegeek.com/what-is-biopower.htm, date accessed 17 August 2010.
30. Foucault (1990), *The History of Sexuality: An Introduction*, p. 143.
31. Ibid., p. 146.
32. Ibid., p. 148.
33. N.R.F. Al-Rodhan (2008), *'Emotional Amoral Egoism': A Neurophilosophical Theory of Human Nature and its Universal Security Implications'* (Berlin: LIT), p. 67.
34. Ibid., pp. 85–92.
35. Ibid., p. 104.
36. Ibid.
37. Ibid., p. 65.
38. Ibid.
39. Ibid., p. 73.
40. N.R.F. Al-Rodhan (2009), *Sustainable History and the Dignity of Man: A Philosophy of History and Civilisational Triumph* (Berlin: LIT), p. 86.
41. Ibid., p. 97.
42. Ibid.
43. United National General Assembly (1948), *Universal Declaration of Human Rights*, 217 A (III), 10 December, http://daccess-dds-ny.un.org/doc/RESOLUTION/GEN/NR0/043/88/IMG/NR004388.pdf?OpenElement, date accessed 20 August 2010.
44. Ibid.
45. N.R.F. Al-Rodhan (2009), *Sustainable History and the Dignity of Man: A Philosophy of History and Civilisational Triumph*, p. 185.
46. Ibid.
47. Ibid., p. 186.
48. Ibid.
49. Ibid., pp. 186–187.
50. Ibid., p. 181.
51. Ibid., p. 183.
52. Ibid.
53. Al-Rodhan (2008), *'Emotional Amoral Egoism': A Neurophilosophical Theory of Human Nature and its Universal Security Implications*, p. 65.
54. Ibid., pp. 81–82.
55. Al-Rodhan (2009), *Sustainable History and the Dignity of Man: A Philosophy of History and Civilisational Triumph*, p. 437.
56. Ibid., p. 187.
57. Ibid., p. 188.
58. Ibid.
59. Ibid., p. 190.
60. Ibid., p. 191.
61. Ibid., p. 194.
62. Ibid., p. 13.

63. Ibid.
64. Ibid., p. 14.
65. Ibid., p. 27.
66. Ibid., pp. 39–40.
67. Ibid., p. 424.
68. Ibid., p. 419.
69. Ibid.
70. Ibid.
71. Ibid., p. 420.
72. Ibid., p. 39.
73. Ibid., p. 34.
74. Ibid., pp. 96–99.

## 12   Human Enhancement: The Nature of the Debate

1. Cf. The President's Council on Bioethics (2003), *Beyond Therapy: Biotechnology and the Pursuit of Happiness* (Washington, D.C.), http://bioethics.georgetown. edu/pcbe/reports/beyondtherapy/beyond_therapy_final_webcorrected.pdf, date accessed 26 August 2010.
2. M. Anissimov, 'What Is Bioconservatism?' Wise Geek Website, http://www. wisegeek.com/what-is-bioconservatism.htm, date accessed 17 August 2010.
3. Ibid.
4. Ibid.
5. 'Bioconservative', Institute for Ethics and & Emerging Technolologies (IEET) Technoprogressive Wiki Website, http://ieet.org/index.php/tpwiki/ Bioconservative/, date accessed 17 August 2010.
6. Ibid.
7. Ibid.
8. Ibid.
9. 'Bioconservatism', StateMaster Website, http://www.statemaster.com/ency-clopedia/Bioconservatism, date accessed 17 August 2010.
10. 'Bioconservative', IEET Technoprogressive Wiki.
11. F. Fukuyama (2002), *Our Posthuman Future: Consequences of the Biotechnology Revolution* (New York: Farrar, Straus and Giroux), p. 159.
12. Ibid.
13. Cf. The President's Council on Bioethics (2003), *Beyond Therapy: Biotechnology and the Pursuit of Happiness*.
14. Ibid.
15. F. Allhoff, P. Lin and J.Steinberg (2009), 'Ethics of Human Enhancement: An Executive Summary', December, *Science and Engineering Ethics*, Vol. 16, No. 2, 1–12.
16. M.J. Sandel (2009), 'The Case Against Perfection: What's Wrong with Designer Children, Bionic Athletes, and Genetic Engineering' in N. Bostrom and J. Savulescu (eds) *Human Enhancement* (Oxford: Oxford University Press), p. 78.
17. Ibid.
18. T. Assenheuer and J. Jessen (2002), 'Interview: Auf schiefer Ebene', *Zeit Online*, May, http://www.zeit.de/2002/05/200205_habermasint..xml, date accessed 17 August 2010.

19. P. Brey (2009), 'Human Enhancement and Personal Identity' in J.K.B Olsen, E. Selinger and S. Riis (eds) *New Waves in Philosophy of Technology* (New York: Palgrave Macmillan), p. 170.
20. Ibid., p. 181.
21. R. van Est, P. Klaassen, M. Schuijff and M. Smits (2008), *Future Man – No Future Man* (The Hague: The Rathenau Institute), p. 17.
22. C.A.J. Coady (2009), 'Playing God' in Bostrom and Savulescu (2009), *Human Enhancement*, p. 179.
23. Cf. The President's Council on Bioethics (2003), *Beyond Therapy: Biotechnology and the Pursuit of Happiness*.
24. Fukuyama (2002), *Our Posthuman Future: Consequences of the Biotechnology Revolution*, p. 149.
25. F. Fukuyama (2004), 'Transhumanism', *Foreign Policy*, September/October, http://www.foreignpolicy.com/articles/2004/09/01/transhumanism, date accessed 17 August 2010.
26. N. Bostrom (2005), 'In Defense of Posthuman Dignity', *Bioethics*, Vol. 19, No. 3, 202–214.
27. N. Bostrom (2008), 'Smart Policy: Cognitive Enhancement in the Public Interest' in L. Zonneveld, H. Dijstelbloem and D. Ringoir (eds) *Reshaping the Human Condition: Exploring Human Enhancement* (The Hague: Rathenau Institute,), p. 29.
28. Ibid., p. 30.
29. Allhoff et al. (2009), 'Ethics of Human Enhancement: An Executive Summary'.
30. S. Cave (2001), 'The Most Dangerous Idea on Earth', *The Financial Times*, 28 May; J. Harris (2007), *Enhancing Evolution: The Ethical Case for Making Better People* (Princeton, NJ: Princeton University Press).
31. 'Who's Afraid of Human Enhancement? (2006), Reason Website, January, http://www.reason.com/news/show/33064.html, date accessed 17 August 2010.
32. P. Lin and F. Allhoff (2006), 'Nanoethics and Human Enhancement: A Critical Evaluation of Recent Arguments', *Nanotechnology Perceptions*, Vol. 2, 47.
33. Bostrom (2005), 'In Defense of Posthuman Dignity'.
34. Lin and Allhoff (2006), 'Nanoethics and Human Enhancement: A Critical Evaluation of Recent Arguments', p. 47.
35. Ibid.
36. Cf. R. Persaud (2006), 'Does Smarter Mean Happier?' in J. Wilsdon and P. Miller (eds) *Better Humans? The Politics of Human Enhancement and Life Extension* (London: Demos).
37. F. Allhoff, J. Moor, P. Lin and J. Weckert (2009), 'Ethics of Human Enhancement: 25 Questions and Answers', submitted to *Studies in Ethics, Law and Technology*, Manuscript 1110 (Berkeley, CA: The Berkeley Electronic Press), http://files.allhoff.org/research/Ethics_of_Human_Enhancement_SELT.pdf, accessed 30 July 2010, 22.
38. Ibid.
39. N. Bostrom and R. Roache (2007), 'Ethical Issues in Human Enhancement', Personal website, http://www.nickbostrom.com/ethics/human-enhancement.pdf, date accessed 17 August 2010, p. 20.

40. N. Bostrom (2003), 'Human Genetic Enhancements: A Transhumanist Perspective', *Journal of Value Inquiry*, Vol. 37, No. 4, 493–506.
41. Bostrom (2005), 'In Defense of Posthuman Dignity'.
42. N. Daniels (2009), 'Can Anyone Really Be Talking About Ethically Modifying Human Nature?' in Bostrom and Savulescu (2009), *Human Enhancement*, p. 31.
43. Ibid.
44. Ibid.
45. A. Caplan (2009), 'Good, Better, or Best?' in Bostrom and Savulescu (2009), *Human Enhancement*, p. 204.
46. Ibid., p. 205.
47. P. Hagoort interviewed by L. Zonneveld and M. Slob (2008), 'Cognitive Perfection is not the Optimal Condition' in Zonneveld et al. (2008), *Reshaping the Human Condition: Exploring Human Enhancement* (The Hague: Rathenau Institute), p. 92.
48. R. Roache and S. Clarke (2009), 'Bioconservatism, Bioliberalism, and the Wisdom of Reflecting on Repugnance', *Monash Bioethics Review*, Vol. 28, No. 1, 1–21.
49. Ibid.
50. J. Hughes (2006), 'Human Enhancement and the Emergent Technopolitics of the 21st Century', IEET Website, 19 May, http://ieet.org/index.php/IEET/more/hughes20060519/, date accessed 17 August 2010.
51. Ibid.
52. D. Carrico (2006), 'Technoprogressivism: Beyond Technophilia and Technophobia', IEET Website, 12 August, http://ieet.org/index.php/IEET/more/carrico20060812/, date accessed 17 August 2010.
53. F. Baylis and J.S. Robert (2004), 'The Inevitability of Genetic Enhancement Technologies', *Bioethics*, Vol. 18, No. 1, 1.
54. Cf. N.R.F. Al-Rodhan (2008), *Emotional Amoral Egoism: A Neurophilosophical Theory of Human Nature and its Universal Security Implications* (Berlin: LIT).
55. 'Overview of Biopolitics', IEET Website, http://ieet.org/index.php/IEET/biopolitics, date accessed 17 August 2010.
56. Ibid.
57. Ibid.
58. Ibid.
59. Ibid.
60. Ibid.
61. Ibid.
62. Ibid.
63. Ibid.

# 13   The Science and Technology of Human Enhancement

1. E.A. Williams (2006), 'Good, Better, Best: The Human Quest for Enhancement', Summary Report of an Invitational Workshop Convened by the Scientific Freedom, Responsibility and Law Program, American Association for the Advancement of Science, 1–2 June, http://www.aaas.org/spp/sfrl/projects/human_enhancement/pdfs/HESummaryReport.pdf, date accessed 17 August 2010.

2. Ibid.
3. N. Bostrom and R. Roache (2007), 'Ethical Issues in Human Enhancement', Nick Bostrom Website, http://www.nickbostrom.com/ethics/human-enhancement.pdf, date accessed 17 August 2010, p. 7.
4. Mayo Clinic 'Performance-enhancing Drugs: Are They a Risk to Your Health?' http://www.mayoclinic.com/health/performance-enhancing-drugs/hq01105, date accessed 17 August 2010.
5. Ibid.
6. Ibid.
7. Ibid.; 'Human Enhancement: Making People Better or Making Better People?', The Irish Council for Bioethics Website, http://www.bioethics.ie/uploads/docs/Humanenh.pdf, date accessed 17 August 2010.
8. A. Bhargava (2010), 'All's Fair in Botox and Evolution', Institute for Ethics and Emerging Technologies (IEET) Website, 20 July, http://ieet.org/index.php/IEET/more/bhargava20100720/, date accessed 17 August 2010.
9. Bostrom and Roache (2007), 'Ethical Issues in Human Enhancement', p. 3.
10. R. Kurzweil and T. Grossman, 'What Is Longevity?' Ray Kurzweil and Terry Grossman's Longevity Products Website, http://www.rayandterry.com/wellness_information.asp?section=Resources&question=142, date accessed 17 August 2010.
11. Ibid.
12. Ibid.
13. US Department of Energy Office of Science 'Artificial Retina Project: Restoring Sight Through Science', Artificial Retina Project Website, http://artificialretina.energy.gov/about.shtml, date accessed 17 August 2010.
14. Ibid.
15. P. Belluck (2009), 'Burst of Technology Helps to See', *The New York Times*, 26 September, http://www.nytimes.com/2009/09/27/health/research/27eye.html?pagewanted=1, date accessed 17 August 2010.
16. F. Allhoff, P. Lin and J. Steinberg (2009), 'Ethics of Human Enhancement: An Executive Summary', December, *Science and Engineering Ethics*, Vol. 16, No. 2, 2.
17. A. Sandberg, 'Physical Improvements of Humans', Anders Sandberg Personal Website, http://www.aleph.se/Trans/Individual/Body/improvements.html#Eyes, date accessed 17 August 2010.
18. R. Radebaugh (2002), 'About Cryogenics', Cryogenics Technologies Group Website, http://cryogenics.nist.gov/, date accessed 17 August 2010.
19. 'Cryonics', Cryonics Institute Website, www.cryonics.org, date accessed 17 August 2010.
20. Ibid.
21. 'Crynoics; A Basic Introduction', Cryonics Institute Website, http://cryonics.org/prod.html, date accessed 27 July 2010.
22. Ibid.
23. J. Altmann (2008), 'Military Uses of Nanotechnology – Too Much Complexity for International Security', *Complexity and Security*, Vol. 14, No. 1, 64.
24. Centre for Disease Control and Prevention (CDC) (2007), '2007 Art Report Section 5- ART Trends 1998 – 2007, Figure 49' http://www.cdc.gov/ART/ART2007/section5.htm, date accessed 27 July 2010.
25. C. Stolba (2002/03) 'Overcoming Motherhood', *Policy Review*, Vol. 116, December/January, 31.

26. P. Singer (2009), 'Parental Choice and Human Improvement' in N. Bostrom and J. Savulescu (eds) *Human Enhancement* (Oxford: Oxford University Press), p. 279.
27. 'Trying to Conceive: Artificial Insemination', WebMD Website, http://www.webmd.com/infertility-and-reproduction/guide/artificial-insemination, date accessed 17 August 2010.
28. Ibid.
29. Ibid.
30. 'In Vitro Fertilization', eMedicineHealth Website, http://www.emedicine-health.com/in_vitro_fertilization/article_em.htm, date accessed 17 August 2010.
31. 'IVF', Shared Journey Website, http://www.sharedjourney.com/ivf.html, date accessed 17 August 2010.
32. University of California Los Angeles, Earth anf Space Sciences 'Germline Gene Therapy', http://www.ess.ucla.edu/, date accessed 17 August 2010.
33. N. Bostrom (2003), 'Human Genetic Enhancements: A Transhumanist Perspective', *Journal of Value Inquiry*, Vol. 37, No. 4.
34. University of California Los Angeles, Earth and Space Sciences 'Germline Gene Therapy'.
35. D.W. Brock (2009), 'Is Selection of Children Wrong?' in Bostrom and Savulescu (2009), *Human Enhancement*, p. 252.
36. M.D. Lemonick, D. Bjerklie and A. Park (1999), 'Designer Babies', *Time Magazine*, 11 January, http://www.time.com/time/magazine/article/0,9171,989987,00.html, date accessed 17 August 2010.
37. T. Assenheuer and J. Jessen (2002), 'Interview: Auf schiefer Ebene', *Zeit Online*, May, http://www.zeit.de/2002/05/200205_habermasint.xml, date accessed 17 August 2010.
38. M. Henderson (2010), 'Demand for 'Designer Babies To Grow Dramatically', *The Times of London*, 7 January, http://www.timesonline.co.uk/tol/news/science/genetics/article6978400.ece#cid=OTC-RSS&attr=1515793, date accessed 17 August 2010.
39. Ibid.
40. N. Bostrom and A. Sandberg (2009), 'Cognitive Enhancement: Methods, Ethics, Regulatory Challenges', *Science and Engineering Ethics*, Preprint, 2.
41. M.C. Roco and W.S.Bainbridge (2002), *Converging Technologies for Improving Human Performance: Nanotechnology, Biotechnology, Information Technology and Cognitive Science* (Arlington: National Science Foundation), p. 97.
42. W.J. Riedel (2008), 'Psychopharmaceutical Cognition Enhancement' in L. Zonneveld, H. Dijstelbloem and D. Ringoir (eds) *Reshaping the Human Condition: Exploring Human Enhancement* (The Hague: Rathenau Institute), p. 115.
43. Ibid., p. 116.
44. Ibid., p. 120.
45. Roco and Bainbridge (2002), 'Converging Technologies for Improving Human Performance', p. 154.
46. T. Sejnowski (2010), 'When Will We Be Able to Build Brains Like Ours?' *Scientific American*, 27 April, http://www.scientificamerican.com/article.cfm?id=when-build-brains-like-ours, date accessed 17 August 2010.

47. Roco and Bainbridge (2002)'Converging Technologies for Improving Human Performance', 54
48. Ibid.
49. Cf. U. Lee, H.J. Lee, S. Kim and H.C. Shin (2006), 'Development of Interacranial Brain-Computer Interface System Using Non-Motor Brain Area for Series of Motor Functions', *Electronics Letters*, Vol. 42, No. 4, 98–200.
50. Roco and Bainbridge (2002), 'Converging Technologies for Improving Human Performance', p. 99.
51. Ibid., p. 155.
52. Bostrom and Sandberg (2009), 'Cognitive Enhancement: Methods, Ethics, Regulatory Challenges', p. 6.
53. Columbia University 'History of Neuroscience', http://www.columbia.edu/cu/psychology/courses/1010/mangels/neuro/history/history.html, date accessed 17 August 2010.
54. E.H. Chudler, 'The Hows Whats and Whos of Neuroscience', University of Washington Website, http://faculty.washington.edu/chudler/what.html, date accessed 17 August 2010.
55. 'What Is Neuroscience?' Society for Neuroscience Website, http://www.sfn.org/index.cfm?pagename=whatIsNeuroscience, date accessed 17 August 2010.
56. Ibid.
57. Columbia University 'History of Neuroscience'.
58. Ibid.
59. 'What Is Neuroscience?' Society for Neuroscience Website.
60. C. Dackis and C. O'Brien (2005), 'Neurobiology of Addiction: Treatment and Public Policy Ramifications', *Nature*, November, 1436.
61. 'Principles of Drug Addiction Treatment: A Research Based Guide', National Institute on Drug Abuse Website, http://www.nida.nih.gov/podat/faqs.html#faq2, date accessed 17 August 2010.
62. 'Neural Engineering', Weldon School of Biomedical Engineering at Purdue University Website, https://engineering.purdue.edu/BME/Research/NE, date accessed 17 August 2010.
63. 'Neural Engineering', University of Florida College of Engineering Website, http://www.bme.ufl.edu, date accessed 17 August 2010.
64. K.W. Horch and G.S. Dhillon (eds) (2004), 'Neuroprosthetics: Theory and Practice', *Series on Bioengineering and Biomedical Engineering*, Vol. 2, February, http://www.worldscibooks.com/engineering/4987.html, date accessed 17 August 2010.
65. Siao (2007), 'Neuroprosthetics: Restoring Function in the Bionic Human', Harvard Science Review, Spring, http://www.hcs.harvard.edu/~hsr/pdf-sspring2007/SiaoGE.pdf, date accessed 5 September 2010, p. 46.
66. S.L. Nasr (2008), 'How Brain Mapping Works', HowStuffWorks Website, http://health.howstuffworks.com/brain-mapping.htm, date accessed 17 August 2010.
67. M. Brain (2007), 'How Uploading Works', *Journal of Geoethical Nanotechnology*, Vol. 1, No. 2, Spring, http://ieet.org/index.php/IEET/more/brain20070409/, date accessed 17 August 2010.
68. Ibid.
69. Ibid.

70. Ibid.
71. 'Mind Transfer', EconomicExpert Website, http://www.economicexpert. com/a/Mind:transfer.htm, date accessed 17 August 2010.

## 14 The Geopolitics of Human Enhancement: Applying the 'Multi-Sum Security Principle'

1. Cf. N.R.F. Al-Rodhan (2007), *The Five Dimensions of Global Security: Proposal for a Multi-sum Security Principle* (Berlin: LIT).
2. Ibid., p. 16.
3. 'New Dimensions of Human Security', Human Development Report Office Website, http://hdr.undp.org/en/reports/global/hdr1994/, date accessed 20 August 2010; United Nations Development Programme (UNDP) (1994), *Human Development Report 1994: New Dimensions of Human Security* (New York: Oxford University Press).
4. UNDP (2005), *The Human Security Report 2005: War and Peace in the 21st Century* (New York: Oxford University Press).
5. Ibid.
6. 'What Is Environmental Security?', Institute for Environmental Security Website, http://www.envirosecurity.org/activities/What_is_Environmental_ Security.pdf, date accessed 17 August 2010, p. 1.
7. P. Collier (2004), 'Development and Conflict', Centre for the Study of African Economies, Department of Economics, Oxford University, 1 October, http:// www.un.org/esa/documents/Development.and.Conflict2.pdf, date accessed 17 August 2010, p. 3.
8. Al-Rodhan (2007), *The Five Dimensions of Global Security: Proposal for a Multi-sum Security Principle*, p. 73.
9. Ibid.
10. N.R.F. Al-Rodhan (2009), 'Multi-sum Security: Five Distinct Dimensions', International Relations and Security Network (ISN) Website, http://www. isn.ethz.ch/isn/Current-Affairs/Special-Reports/Safeguarding-Security-in-Turbulent-Times/Overview, date accessed 17 August 2010.
11. Al-Rodhan (2007), *The Five Dimensions of Global Security: Proposal for a Multi-sum Security Principle*, p. 79.
12. Al-Rodhan (2009)'Multi-sum Security: Five Distinct Dimensions'.
13. N. Bostrom (2005), 'In Defense of Posthuman Dignity', *Bioethics,* Vol. 19, No. 3, 202–214.
14. J. Hughes (2006), 'Human Enhancement and the Emergent Technopolitics of the 21st Century', Institute for Ethics and Emerging Technologies (IEET) Website, 19 May, http://ieet.org/index.php/IEET/more/hughes20060519/, date accessed 17 August 2010.
15. Ibid.
16. Ibid.
17. Bostrom (2005), 'In Defense of Posthuman Dignity', pp. 202–214.
18. 'Human Cloning and Human Dignity: An Ethical Inquiry' (2002), *Atlantic Monthly*, Vol. 290, No. 3, 42.
19. N. Bostrom (2010), 'Transhumanist FAQ', Humanity+ Website, http:// humanityplus.org/learn/transhumanist-faq/, accessed 29 July 2010.

20. I.A. Colquhoun (2008), 'Life Expectancy in Prehistory: How Long Did Our Prehistoric Ancestors Live?', Suite101 Website, 24 October, http://archaeology.suite101.com/article.cfm/prehistoric_population, date accessed 17 August 2010.
21. World Health Organisation (WHO) 'Global Life Expectancy Reaches New Heights but 21 Million Face Premature Death This Year, Warns WHO', http://www.who.int/whr/1998/media_centre/press_release/en/index.html, date accessed 17 August 2010.
22. R.G. Lipsey, K.I. Carlaw, and C.T. Bekar (2005), *Economic Transformations: General Purpose Technologies and Long-Term Economic Growth* (Oxford: Oxford University Press), p. 213.
23. Bostrom, 'Transhumanist FAQ'.
24. Ibid.
25. M. More (1994), 'On Becoming Posthuman', Max More Website, http://www.maxmore.com/becoming.htm, date accessed 17 August 2010.
26. M. Shindikar, S. Jadhav, R. Karpe, M. Lale, P. Tetali and V.R. Gunale, 'Quantification Studies on the Accumulation of Non-Biodegradable Solid Waste Material in the Mangroves of Thane Creek', University of Pune Website, http://wgbis.ces.iisc.ernet.in/energy/water/proceed/section4/paper1/section4paper1.htm, date accessed 18 August 2010.
27. N. Shachtman (2007), 'Be More Than You Can Be', *Wired,* Issue 15, 3 March, http://www.wired.com/wired/archive/15.03/bemore.html, date accessed 17 August 2010.
28. Ibid.
29. N. Shachtman (2008), 'Top Pentagon Scientists Fear Brain-Modified Foes', *Wired*, 9 June, http://blog.wired.com/defense/2008/06/jason-warns-of.html, date accessed 17 August 2010.
30. Ibid.
31. W. Evans (2007), 'Singularity Warfare: A Bibliometric Survey of Militarized Transhumanism', *Journal of Evolution and Technology*, Vol. 16, No. 1, 2.
32. Ibid.
33. The President's Council on Bioethics (2003), *Beyond Therapy: Biotechnology and the Pursuit of Happiness* (Washington, D.C.).
34. Ibid.
35. P.W. Singer (2009), *Wired for War: The Robotics Revolution and Conflict in the 21st Century* (New York: Penguin Books), pp. 416–417.
36. T. Kirkwood (2008), 'Changing Expectations of Life' in L. Zonneveld, H. Dijstelbloem and D. Ringoir (eds) *Reshaping the Human Condition: Exploring Human Enhancement* (The Hague: Rathenau Institute), p. 98.
37. Ibid., p. 97.

## 15 Criteria for a Regulatory Framework of Human Enhancement

1. United National General Assembly (1948), *Universal Declaration of Human Rights*, 217 A (III), 10 December, http://daccess-dds-ny.un.org/doc/RESOLUTION/GEN/NR0/043/88/IMG/NR004388.pdf?OpenElement, date accessed 20 August 2010.

2. N.R.F. Al-Rodhan (2009), *Sustainable History and the Dignity of Man: A Philosophy of History and Civilisational Triumph* (Berlin: LIT), p. 180.
3. R. Bailey (2006), 'Human Rights and Human Enhancement: Is Genetic Modification of People Moral?', *Reason Online*, 29 May, http://www.reason.com/news/show/117339.html, date accessed 18 August 2010.
4. Ibid.
5. M. Brockman (2009), 'A Limited View of the Future', *Nature*, 459.7246, 28 May, 511.
6. The White House (2009), 'Executive Order 13521: Establishing the Presidential Commission for the Study of Bioethical Issues, November 24, Bioethics.gov Website, http://www.bioethics.gov/documents/Executive-Order-Establishing-the-Bioethics-Commission-11.24.09.pdf , date accessed 17 August 2010.
7. P. Shanks (2010), 'President Obama's Bioethics Commission', Biopolitical Times, Center for Genetics and Society Website, 13 April, http://www.bio-politicaltimes.org/article.php?id=5154, date accessed 17 August 2010.
8. Bioethics.gov (2010), 'The Presidential Commission for the Study of Bioethical Issues', http://www.bioethics.gov/, date accessed 30 July 2010.
9. M. Smits (2009), *STOA Workshop in the European Parliament: A European Approach to Human Enhancement* (Den Haag: Rathenau Institute).
10. Ibid.
11. Ibid.
12. The President's Council on Bioethics (2003), *Beyond Therapy: Biotechnology and the Pursuit of Happiness* (Washington, D.C.).
13. The European Group on Ethics in Science and New Technologies (EGE) (2007), 'The European Group on Ethics in Science and New Technologies Adopted on 17 January 2007 Opinion No. 21 on the Ethical Aspects of Nanomedicine and Presented It Today to President Barroso', 24 January, http://ec.europa.eu/european_group_ethics/activities/docs/press_release_op_nano_en.pdf, date accessed 17 August 2010.
14. F. Allhoff, P. Lin, and J.Steinberg (2009), 'Ethics of Human Enhancement: An Executive Summary', December, *Science and Engineering Ethics*, Vol. 16, No. 2, 3.
15. M. A. Rorty (2003), 'The Future of Human Nature', Notre Dame Philosophical Reviews, 2 December, http://ndpr.nd.edu/review.cfm?id=1291, date accessed 30 July 2010.
16. Cf. Allhoff et al. (2009), 'Ethics of Human Enhancement: An Executive Summary'; P. Lin and F. Allhoff (2008)'Against Unrestricted Human Enhancement', *Journal of Evolution &Technology*, Vol. 18, No. 1, 35–41.
17. Ibid., p. 185.
18. Ibid., p. 210.
19. Ibid., pp. 187–191.
20. G. Dvorsky and J. Hughes (2008), 'Postgenderism: Beyond the Gender Binary', Institute for Ethics and Emerging Strategic Technologies (Hartford, CT: IEET), http://ieet.org/archive/IEET-03-PostGender.pdf, date accessed 18 August 2010, p. 2.
21. Ibid.
22. Cf. Smits (2009), *STOA Workshop in the European Parliament: A European Approach to Human Enhancement*.

# Index